The Hawaiian Spinner Dolphin

Figure 1. A school of resting spinners over the sand patch at Kealake'akua Bay.

The Hawaiian Spinner Dolphin

Kenneth S. Norris
Bernd Würsig
Randall S. Wells
Melany Würsig

WITH

Shannon M. Brownlee · Christine M. Johnson
Jody Solow

Illustrations by Jenny Wardrip

UNIVERSITY OF CALIFORNIA PRESS
Berkeley   Los Angeles   London

University of California Press
Berkeley and Los Angeles, California

University of California Press
London, England

Copyright © 1994 by The Regents of the University of California

Library of Congress Cataloging-in-Publication Data

The Hawaiian spinner dolphin / Kenneth S. Norris . . . [et al.] ;
illlustrations by Jenny Wardrip.
      p.   cm.
  Includes bibliographical references and index.
  ISBN 0-520-08208-7
  1. Stenella longirostris—Hawaii. 2. Stenella longirostris—
Behavior—Hawaii. 3. Stenella longirostris—Research—Hawaii
I. Norris, Kenneth S. (Kenneth Stafford)
QL737.C432H39 1994   93-38911
599.5'3—dc20   CIP

Printed in the United States of America

1 2 3 4 5 6 7 8 9

The paper used in this publication meets the minimum requirements of American
National Standard for Information Sciences—Permanence of Paper for Printed Library
Materials, ANSI Z39.48-1984 ∞

DEDICATION

*This volume is dedicated to our colleague Dr. William F. Perrin,
whose more than two decades of work and leadership of studies on
tropical dolphins has brought them, including the spinner dolphin, from
being among the least known to among the best known of cetaceans.*

CONTENTS

FIGURES

Introduction

Kenneth S. Norris

This book is the result of a 25-year-long attempt to learn about the lives of wild dolphins. From the late 1940s to the 1960s, a scattering of behavioral observations of captive bottlenose dolphins accumulated (McBride and Hebb 1948, Tavolga and Essapian 1957, Tavolga 1966, Bel'kovich and Yablokov 1963, 1969). The conclusion from these studies was that dolphins are social mammals exhibiting a variety of patterns known from terrestrial species. They do not behave like fish but instead have set up reasonably typical mammalian societies in captivity.

Although captive studies allowed consecutive and detailed observations over long periods of time, they still left the behaviorist to wonder what it was truly like to live in a wild dolphin school. The degree to which the captive environment had warped natural patterns remained unassessed. What, we wondered, was it like for a dolphin to live in a society of many dozens of animals, to swim many kilometers a day, and to dive a dozen times an hour to depths of 100 m (330 ft) or more? Who were the dolphin's enemies and how were they dealt with? How did wild dolphins catch food, sleep, communicate, give birth, mature, and die out in the ocean? How did they use both their vision and their echolocation at sea?

The captive studies could not include observations in the deep water where most dolphins live, nor the external dangers that must have been central in shaping their societies.

In the late 1960s, I became convinced that while we could learn much about dolphins from captive studies, somehow observers had to go to sea with the animals if we were ever to learn how they structured their societies. At the time, it was unclear if this could be done to any useful degree.

My home base at the time was Hawaii, so I sought a place around those islands where I could hope to find the same animals over and over again in a workable stretch of sea. I began by attempting to follow radiotagged schools from a shore-based triangulation station located on Oahu's Waianae Mountains. At the same time, I constructed my first dolphin-viewing vessel, which was officially named the *Mobile Observation Chamber (MOC)* but came to be called by one and all the "semisubmersible seasick machine" (or "SSSM") (see chap. 3).

We did spend much time with the dolphins along that coast and we made some preliminary findings, but the animals proved to be wary of people and boats and the area of calm water was small. Although we could often see the dolphins' backs and fins above the water as they rose to breathe and we did learn a little of the structure of their schools, nearly all that they did beneath the sea surface remained hidden to us. In the end, the vagaries of weather and the need for extensive vessel time made the effort prohibitively costly.

But like most such observational puzzles—whether the scientist is attempting to learn about orangutans or dolphins—once the uncertainties are brushed aside and one begins to search for opportunities to observe, one finds them. In time, we found ways to identify many individuals in a school and we could sometimes sex them. Our best opportunity for observation, to our surprise, proved to be the Hawaiian spinner dolphin *(Stenella longirostris),* a member of a mostly open ocean genus. Some of its populations were found to frequent the clear water coves of the various Hawaiian Islands on a daily basis. This discovery allowed us to do the first phase of this work (Norris and Dohl 1980a). We found a population of these dolphin along the extensive lee shore of the largest of the Hawaiian Islands, the island of Hawaii (see Appendix B).

These spinner dolphins did something else for us. They allowed small craft and even swimmers in their midst, and they usually swam in clear waters (clearer than we usually found along the shore of Oahu) where they could be identified, photographed, and observed underwater. Other studies of wild dolphin behavior had lacked, or nearly lacked, this underwater observational dimension. This chance allowed us to observe them in the context of a wild school, complete with predators, food sources, and the physical world of the sea. We could see the dolphins' reactions to the topography of deep island slopes and to sandy-bottomed coves where we learned their schools came to rest.

This penchant for coming into shallow bays gave us our only truly bias-free chance to watch them. From the nearly vertical lava bluffs backing Kealake'akua Bay, we could see and track wholly undisturbed schools as they swam over a white coral sand patch located at the deepest part of the bay (see fig. 1, frontispiece).

We soon learned how extremely sensitive spinner dolphins are to intruders. We learned that our underwater work with swimmers and viewing vehicles did indeed influence the schools we watched. At times when the dolphins had not yet settled into deep rest, the intrusion of a boat or swimmer caused them to inch away. If pursued tenaciously, they sometimes moved out of the rest cove altogether. At first, this effect was not clearly evident. On the water, we did not easily perceive the reactions of the entire school relative to us. But from the clifftop, their edginess toward one of our teams on the water was clearly observable against the dimension of the entire bay. In time, even from close up these reactions became obvious.

Nonetheless, especially when the spinners were active, they swirled around our observation vessel and sometimes peered through the windows at us. We finally came to understand that the spinners' descent into rest is a touchy time when they seem to be making the decision of whether or not to remain in a given rest area. But once the die had been cast, our intrusion seemed to be of much less concern to them.

Awakening proved to be a very interesting time. Repeatedly, we saw the dolphins' activity state change in a few minutes from deep rest to high activity. By classifying the aerial patterns of the spinners, we were soon able to judge the activity state of such dolphin schools. A hierarchy of different aerial patterns proved to be arrayed along a gradient of increasing activity.

Mother–young pairs cruised by our viewing capsule on a frequent basis. Sometimes we could hear their vocalizations in the context of their lives at sea. Jody Solow of our team was able to attract some active dolphins to her and exchange play objects with them as she swam along with their schools.

It is probably fair to say that more than in any study to date, we came to know the detailed cycling of the daily activity patterns of wild dolphins with some confidence. Much intimate school behavior began to fall into a broad contextual framework. For example, we could eventually match the frequency and kind of sound emissions against our knowledge of the daily patterns of spinner dolphin life. It is quite another matter to assign specific meanings to these sounds, although a speculative framework for the context of some of these signals is erected here. It is presented as a template for other workers to consider when they probe beyond the limits of our work.

What we did is what I call *natural history*. That is, we attempted to look at a single species from as many viewpoints as we could contrive to understand what the totality of the dolphin's life is really like. To me, this approach provides the ultimate in intellectual excitement that one can derive from field biology. Everything one sees is worth thinking about.

Then one can extend beyond these observations to consider a larger view of the animal's life and evolution, as I attempt to do here.

Animal behaviorists working with terrestrial animals can sometimes sit quietly watching their animals for long periods, but almost always, we dolphin watchers had to be content with brief encounters. We were seldom granted the chance to observe even modestly long sequences of behavior except from the cliffside (and there we could not see any details of underwater activity). Instead, we had to piece together behavior from film clips that typically lasted from just a few seconds to about a minute in length. The dolphins, always on the move even during rest, simply faded into the blue murk beyond the limits of our vision. This happened no matter how unobtrusive we tried to be in following them. They had dolphin business to do and we were not part of it, and furthermore, we could not easily follow them.

We began our work by determining the daily progression of behavior, which ultimately put our short behavioral sequences into context. But before events could be related contextually, they had to be seen over and over again. Surely, we have just scratched the surface because new insights came as fast when we were packing up our camp as they did when we first began.

We came at our spinners from every observational and conceptual angle we could devise. This book, as a result, outlines the broad aspects of a school-dwelling mammalian society living in three-dimensional open space. It ties spinner dolphin behavior to the ecology of a marine environment, and it begins to sketch how sounds, respiration, locomotion, and vision are used by these dolphins.

But it is also, in part, an "old man's book." I began my work with marine mammals four decades ago and I will not attempt any effort such as this again. At about my age, people begin to fall off of boats. As I wrote, I found I wanted very much to spell out how I had come to view dolphin life. This seemed especially urgent in view of the popular myths that had built up around dolphins. As the reader will find, dolphins are regarded here as a perfectly reasonable twig on the mammalian tree, albeit one that extends the group's range of possibilities in certain directions.

By the time I wrote my part of this book, years of rumination were behind some of the constructs we present here. I wanted to spell out the most important of these constructs while I had the chance (for example, I have been thinking about how dolphin schools work since I was a graduate student). The bulk of the long term piling of fact upon fact was work done by my younger colleagues who operated our field camp and could stay there all the time. Most of the inevitable routine work and most of the statistical tests are theirs.

Members of my team and I sometimes had quite different ways of extracting information from nature, and the work is much better for this diversity. My usual method is to spend time (often months or years) observing, hypothesizing, and checking my hypotheses against nature before I begin real quantification (beyond the normal run of daily field notes, which are full of little quantifications). I built up an overview of the dolphin's patterns and then began to ask pointed questions. I depended a great deal on this knowledge of my animal and upon this slow hypothesis building, testing, and retesting. It can be an almost magical experience to watch the structure of nature unfold as one comes to see more and more deeply into how things work. From chaos at the start, one can sometimes come to see meaning in most events. My method lets me refine my personal biases as I shape and reshape them. It works because nature is not reluctant to point out my mistakes, and because I have trained myself to discard fallen hypotheses. But one must reconcile oneself to creeping up on truth through a thicket of these fallen hypotheses. Most of the time the process is mental and observational and, only at the last, statistical. Statistics becomes a method to validate your conclusions for others.

Others in my team took a radically different route. They sought to quantify what was in front of them at the outset, to build up anonymous data sets free of the inevitable personal biases implicit in my method. Then, with these "true" numbers before them, they looked for patterns in the data sets. What emerged from these two methods performed side-by-side was, I thought, fascinating. We tended to see different things. The statistical search could uncover things I could not see with just eyes.

My method allowed us to set the events of breathing, spinning, schooling and school structure, and the progress of acoustic behavior throughout the 24-hr cycle in a contextual framework subject to further testing and refinement by other observers at other times. The method of quantifying at the start showed unexpected things: the relationship between departure from rest coves and day length, the fluid nature of dolphin society, and the necessary erection of an offshore superschool as a larger unit of organization at the island of Hawaii. In short, this book benefitted because we used both paradigms in its construction.

Our study, like any piece of fieldwork, raises more questions than it answers. Intriguing questions about the balance between kinship and cooperation beyond family boundaries in an extreme of mammalian social order are implicit in what we found. It seems true that no other mammalian species (including birds and treetop monkeys) exists all its life in an open three-dimensional world to the extent that these and other oceanic cetaceans do. For this reason, how schools work as protective systems is a major question in dolphin behavior and an important focus here.

Not surprisingly, we feel that this work is preliminary. It has been primarily an attempt to outline the basic behavior patterns of a dolphin species so that later, more sophisticated questions can be asked. Our efforts to relate spinner dolphin life to the important theories of sociobiology and behavioral ecology are preliminary. Instead of plunging directly into such questions, we first had to learn how to sex dolphins, to understand their mating system, to define the memberships and boundaries of the groups we saw, to make some broad sense of the babble of sounds they emit, and to see, first hand, how their schools were arranged.

My earlier and even more preliminary study with Thomas Dohl (Norris and Dohl 1980a) showed only the outlines of what we could expect to learn about spinner dolphins. We did the work during our spare time while we both worked at the Oceanic Institute, on Oahu, Hawaii. Between that work and this present study, I had returned to my mainland post at the University of California. Thus, any new work became, perforce, a real expeditionary effort and not something that could be pieced together on weekends spent at the island of Hawaii. So my first effort for this work was to build a team (fig. 2).

I had learned of a pair of promising young workers from cetologist Roger Payne. Bernd and Melany Würsig had worked for three years at Payne's right whale camp at Golfo San José, Argentina, living in a generator house because the main camp was too crowded. The two of them had taken Roger's innovation of tracking marine mammals from shore with a surveyor's theodolite and had made it into an art. Preeminent among a small handful of workers, they had advanced the art of photo-identification of individual cetaceans into a major tool for the analysis of cetacean populations and movements. Their work and that of Dr.'s Michael Bigg and John Ford, and to a minor extent my colleague Thomas Dohl and myself, had begun to demonstrate during the early 1970s that if one were persistent, it was possible to learn a considerable amount about any dolphin school by scars and marks on backs and dorsal fins. Then, over time, the social structure of a dolphin society could be unraveled bit by bit.

Bernd and Melany Würsig did another thing that attracted me to them. Among behavioral cetologists, more than anyone preceding them, they put numbers on everything they saw. Their papers were peppered with probabilities that let the reader assess the validity of their assertions. Because of them and a few others, cetacean science leapt from the anecdotal to the verifiable. I wanted that hallmark on this new work.

Figure 2. The Hana Nai'a research team. Above: Kenneth Norris, Bernd and Melany Würsig; page 8: Randall Wells, Shannon Brownlee; page 9: Christine Johnson, Jody Solow.

I had initially corresponded with Bernd when he wrote from storm-swept Patagonia, Argentina, asking for advice on how best to begin his behavior studies on dusky and bottlenose dolphins. Four years later, as he was writing his Ph.D. dissertation, we laid plans for them to join the spinner dolphin team.

Bernd became my team leader. He was involved in nearly everything we did. He and Melany took as special tasks the establishment and operation of a theodolite tracking station one-third of the way up the 150-m Kealake'akua cliff. They led the enormous task of assembling,

Figure 3. A spinner in a spin at Kealake'akua Bay.

cataloguing, and interpreting the nearly 20,000 individual identification slides on which this work is based. Bernd spearheaded the hourly watch over Kealake'akua Bay from the dock at the field station that went on continuously for the first 12 months of the study. This chronicled the comings and goings of the dolphins into and out of the bay, their numbers, and their behavioral states. We learned to gauge behavioral state by observing the kinds and intensities of aerial behavior. An adult spinner is shown demonstrating its trademark spin in figure 3.

Bernd also made everything work, from pick-up trucks to electronic equipment. I had not bargained for that skill, but I found that I had brought into our camp one of the most skillful adjudicators with the mechanical world that I have ever met. This was fortunate because I am one of those in whose hands instruments and automobiles quickly disassemble following the second law of thermodynamics.

I met the other part of my field leadership team on board the old square-rigged whale research vessel *Regina Maris* as my wife Phyllis and I sailed along the outer coast of Newfoundland. I was on board to teach and help with observations of humpback whales along that rugged shore. Also on board was knowledgeable, organized Randy Wells. He had come aboard as part of a team attempting to radiotrack a humpback whale that had been released from a nearby pound net. As experienced as Bernd, from a decade of studies of bottlenose dolphins in Florida, he had yet to complete his degree. So he joined both my doctoral program and the Kealake'akua effort at the same time, and the spinner team leadership was complete. I could not have asked for better.

Randy had worked extensively with a population of bottlenose dolphins living along the Florida coast, among other projects. He knew a great deal about programs and boats and all it takes to run them safely, and so he had much to do with making our sea effort go. He also emerged as an excellent photographer. He and Bernd became a well-oiled team when photography of wild dolphins was needed. They balanced on opposite rails of our vessel's bow, and their motor-driven cameras soon caught every animal that surfaced. He designed and led the flight program that allowed us to locate, count, and resight schools around Hawaii. He traveled every two weeks to Oahu Island to take blood samples and observe behavior from a captive spinner population so that we could pin down the annual hormonal cycle of spinners. He and Bernd captured and radiotracked dolphins and obtained underwater films of their schools. Together they kept the entire complex operation going in every aspect of its accounting, logistics, and science.

There is nothing that pleases me more than seeing the talents of some young person unfold when opportunity is provided. So I had much to do with populating the Kealake'akua camp with promising students who I thought could help with key parts of our program or who wanted to try new things. Here I mention the four who share the masthead with the Würsigs, Randy Wells, and myself. Many others who had lesser roles, some of great importance to our effort, are acknowledged later.

I had known Shannon Brownlee for a long time, in fact, since she was a young child. Years later, she became one of my students at the University of California, Santa Cruz. By then, she was an artist of skill and style much like her father, but I did not fully understand that she had also developed powerful analytical, mathematical, and writing skills. The task was placed in remarkably capable hands when I suggested that she attempt the difficult task of unraveling the daily cycle of sounds of spinner dolphins.

The second student, Jody Solow, joined us to see if a swimmer could make friends with a dolphin school, just as Jane Goodall had done with her chimpanzees. Jody is as fearless as any fieldworker I have every known, which was necessary because swimming with a warm water dolphin school also means swimming with their attendant sharks. Jody applied her inventive mind and great empathy for animals into an effort that came close to succeeding. Her work contributed important insights about how dolphins live. Spinner dolphin schools proved to be vital to their members to a far greater degree than to dolphins of shallower or more enclosed waters. They would not lag behind their traveling schoolmates for long to play with a plodding human. But they swam with her repeatedly, if only briefly, even allowing her in the midst of their schools at times, but they always continued on their way.

I met Christine Johnson as one of the many young students interested in dolphins who have passed through my office. She showed special mettle in a long observational study of two Pacific white-sided dolphins (*Lagenorhynchus obliquidens*) at Steinhart Aquarium in San Francisco. I found her to be a student of exceptional theoretical capability and one wholly devoted to learning about the minds of animals and people, an ability we now lump under the term *cognitive science*. When I met her, I had begun to edge my own interest over in that direction. I, too, wanted to learn in cognitive terms how dolphin societies work. So Chris became a explorer with me of the intricacies of dolphin behavior as seen from underwater. More than anyone else, she spent her days down in the viewing compartment of the *Maka Ala,* our underwater observation vehicle, watching the dolphins of Kealake'akua Bay. The bulk of what we say in this book about that effort evolved between us. It is still an incomplete effort, a reconnaissance, but it clearly shows the value of underwater observation for learning about the societies of wild dolphins. In fact, I venture to say that the new frontier in wild dolphin behavioral studies lies in such underwater efforts.

Nearly everything we saw from the *Maka Ala* was new. What Tom Dohl and I had done earlier, and what others such as Bill Evans with his viewing vehicle *SeeSea* had produced, were really just snapshots. The *Maka Ala* study reported in this book is a little better, even though the vessel was uncomfortable and unable to negotiate even modest seas. It was also very difficult to keep her ports clean, and only occasionally could we sex a dolphin through her downward-sloping windows. Nonetheless, she showed us what a better vessel could teach us.

After the Hana Nai'a camp was closed down and we had returned to Santa Cruz, Randy and I began to design a new and better viewing vessel with a hoistable chamber, bearing in mind what we had learned from the *Maka Ala.* My doctoral student Jan Östman joined us and took a major burden in the construction and early operation of this new craft, which is described in chapter 3. Jan has the new vessel at sea off the Kona Coast as I write. We have found that we can easily hoist the observation cylinder up into the hull, go 12 knots into a reasonable sea until we find dolphins, lower the cylinder down into its coaming through the hull, descend a ladder into it for observations, and when done, hoist it again for the trip home. Sexing dolphins is no longer the difficult problem it was in the present work because one can look upward through the vertical windows of the cylinder and see the genitalia of the animals swimming diagonally above.

Much unraveling of the details of spinner dolphin life lies ahead with this new craft, which Jan, a Swedish national, has given the almost unpronounceable name of *Smyg Tittar'n,* which he says means "tiptoeing looker" but which we Americans translate into "tiptoeing peeping Tom."

My part has been to shape all this, to share in the field experience, and to watch my young professional colleagues go about their work, doing much of it better than I could. I have also thought hard about the trove of information we have assembled and have tried to make a coherent story of it.

As it turns out, I chose my colleagues well. Dr. Bernd Würsig is now a major worker on wild dolphins and whales, directing a large marine mammal program at Texas A&M University, Galveston, Texas. Melany raises their two children and helps, as she always has, with Bernd's various projects. Dr. Randall Wells is now the world authority on the structure of wild bottlenose dolphin schools and a staff behavioral ecologist at the Brookfield Zoo, Chicago. Shannon Brownlee has won two national awards as a science writer and is now Senior Editor for Science at the magazine *U.S. News and World Report* in Washington, D.C. Jody Solow is a Ph.D. candidate in cultural ecology at Cambridge University, England, and in the field in the Solomon Islands. Christine Johnson has completed a doctoral program in cognitive psychology at Cornell University in New York and teaches that subject at the University of California, San Diego.

We live in an exciting time. This century, this decade, this year, legions of scientists around the world probe into "the way things are," seeing aspects of how our world works for the first time in history. It is breathtaking to have been a tiny part of this great uncovering. This book includes the first attempts by scientists to learn about oceanic dolphin lives underwater, at sea, where this lineage of animals has lived for several times as long as the human race has existed.

ONE

The Spinner Dolphin

Kenneth S. Norris, Bernd Würsig, and Randall S. Wells

Most of this chapter describes spinner dolphins throughout the world. The last section introduces our studies of the Hawaiian subspecies, the main topic of this book.

The spinner dolphin takes its common name from its unusual aerial behavior. Wherever the species *Stenella longirostris* (Gray 1828) occurs around the tropical and subtropical world, school members are reported to burst from the water and to rotate rapidly about their longitudinal axis for as many as about four revolutions before falling back into the water (Hester et al. 1963). On reentry after these leaps, some part of the body is slapped against the water—the dorsal fin, the flukes, or the back. This produces a sharp, smacking sound that we speculate marks the location of the spinning animal to schoolmates swimming nearby (Norris and Dohl 1980a) (see chap. 4 and figs. 3 and 53).

Wherever populations of these dolphins associate with land, at about dusk after a daytime rest period is over, these predictable animals move a short distance offshore to feed. In Hawaii, we followed them by radio-telemetry as they swam back and forth along the steep island slope during the night. Such dolphins seem to be taking advantage of the "island effect" in which the bulk of an island rising from the seafloor blocks oceanic circulation and concentrates a halo of nutrients and pelagic food organisms at specific points along its shores.

WORLDWIDE DISTRIBUTION

Spinners are found in oceanic populations around the tropical and subtropical world and also frequent the shores of continents, oceanic islands, and reefs. The spinner dolphin appears to be an abundant though

14

rather sketchily known form (Perrin and Gilpatrick in press) throughout this range. It is reported to be abundant in the Caribbean (Caldwell et al. 1971, Erdman et al. 1973). Mead et al. (1980) summarize strandings for the Gulf of Mexico, and Leatherwood et al. (1983, p. 244) describe its total range as follows:

> Spinner dolphins are found in the Atlantic, Indian, and Pacific oceans, where they are restricted to tropical, subtropical, and, less often, warm temperate regions. . . . Specimens have been collected from waters near various South Pacific Islands, Australia, Solomon Islands, New Guinea, Indonesia, Japan, Ceylon [presently Sri Lanka], Madagascar, eastern and western Africa, the Caribbean Sea, the coast of eastern United States, and the Gulf of Mexico.

Robineau and Rose (1983, 1984) reported on spinners from Aden at the mouth of the Red Sea and at nearby Djibouti on the African coast. These dolphins proved to be of a patterned form not very different in appearance from Hawaiian animals, although apparently slightly smaller.

Two other geographic variants of the spinner dolphin are worthy of mention here. Perrin et al. (1987) have recognized, but not named, a dwarf form of spinner from the very warm water of the Gulf of Thailand that apparently differs from other spinners in body and skull size ($n = 10$; total length $= 129$–137 cm, including physically mature individuals). Four adult males weighed only 21.5–26.5 kg (47–58 lb), representing perhaps the smallest of all spinners and among the smallest of all dolphins. The distinct short-beaked spinner-like species *Stenella clymene* occurs in the tropical Atlantic Ocean (Perrin et al. 1981), but it apparently does not spin.

HISTORY OF SPINNER DOLPHIN STUDY

Until the early 1970s, the genus *Stenella* was considered to be a taxonomic wastebasket from which it was not possible to separate the various species in any logical way (Fraser 1966). The problem was that, although collections of considerable size existed in the museums of the world, most specimens were simply skulls without any accompanying data on the external appearance of the animal. The majority lacked precise locality data, and only a few coherent series existed. Even today we know rather little about most of the world's spinner dolphin populations.

A sharp change in this situation for the eastern tropical Pacific occurred when William Perrin (1969*a*) first outlined the extent of incidental mortality of these dolphins in the eastern Pacific yellowfin tuna seine fishery. He estimated that about 250,000 of these dolphins were being killed annually during seining operations. The vast and almost unknown eastern tropical Pacific very quickly became a battleground between the conservationists and the fishermen. At stake was the crown jewel of American fishing efforts, the tuna seine fleet, with its fast, highly mechanized ships that deployed very large nets capable of encircling thousands of dolphins and tons of tuna. Because of a severe kill of dolphins, this fishing activity resulted in the collection of a large series of dolphins from known localities, from which many features of morphology, anatomy, and population dynamics were determined (Perrin 1972*b*).

The news of the kill reached legislators in Washington, D.C., just as a major piece of legislation, the U.S. Marine Mammal Protection Act of 1972, was being drafted to help stem the kill of the great whales. Quickly, the act was redrafted and new sections designed to control events in the tuna fishery. Soon, a concerted effort by the U.S. government to reduce the dolphin kill began. An extensive research effort followed under the direction of Southwest Fisheries Center of the National Marine Fisheries Service, initially led by Dr. William Perrin. The effort has since been assisted by scientists of the Interamerican Tropical Tuna Commission.

This total effort was designed to define the populations, life history, parameters, and trends in dolphin abundance in the vast tract of ocean where tuna seining took place. Remarkably, Perrin and his colleagues were quickly able to turn this giant *aqua incognita* into a known territory of shifting currents, thermoclines, and sea bottom over which a whole fauna of cetaceans moved. They were able to extract reliable data from that huge remote territory and do so with sufficient precision that, in spite of constant challenges by the fishery, their figures generally "stood up in court." Repeatedly, the fishery was required to institute changes. By 1980, data from the fishery and from extensive collections and special cruises to the fishery area had resulted in 457 papers and analyses on aspects of the problem and the animals involved in it (see Holbrook 1980).

Their findings showed that eastern tropical Pacific spinner dolphin populations congregate in an arrow-shaped wedge of ocean, widest off the Mexican, Central American, and northern South American coasts (fig. 4) and narrowing to the west. Within this area, the thermocline is shallow at the coast and deepens to the west. The thermocline concentrates pelagic food organisms in and above it, upon which the dolphins and the associated tuna feed. The importance of the thermocline depth to the fishery is that wherever it is reasonably shallow, fishermen can set their nets to hang down through it, corraling both fish and dolphins. Both dolphins and fish can be seined because neither will move much below the sharp temperature interface. Dolphins die because they will

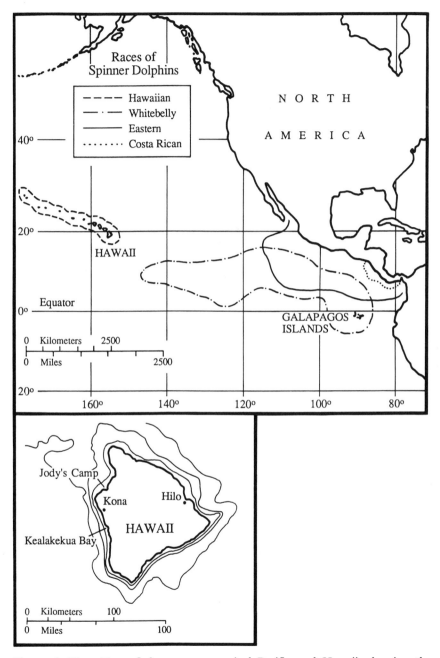

Figure 4. Top: Map of the eastern tropical Pacific and Hawaii, showing the ranges of spinner dolphin races. Bottom: Study area.

not leave the net, even through rather large openings (Perrin and Hunter 1972). This unwillingness to leave one by one is apparently related to the structure of spinner dolphin schools, which require several animals for their full expression (see chap. 13).

ENVIRONMENT OF THE SPINNER DOLPHIN IN THE EASTERN TROPICAL PACIFIC

Within this large fishery area, biological oceanographers have been able to discern local environments within which certain dolphin species tend to congregate. For example, divergence zones at current margins and "current ridges" both concentrate food organisms and are heavily frequented by dolphins (Au et al. 1979, Reilly 1990). Two dolphin species pairs have been defined, occurring modally in different water masses of this tropical sea. Although there is broad overlap at times, one such pair, the spinner and spotted dolphins (*Stenella longirostris* and *S. attenuata*) is primarily associated with tropical surface water, centered at about 10°N latitude off the coast of southern Mexico, and with seasonal tropical waters south of the Galapagos Islands on the other side of the Equator. The other less well defined species pair, the common and striped dolphins (*Delphinus delphis* and *S. coeruleoalba*) occurs most abundantly in transition water to the north and south of the primary spinner–spotted dolphin habitat (Au et al. 1979). Striped dolphin distribution can be considered as intermediate between the distribution of common and spinner–spotted dolphins. This pattern of separation occurs throughout the year and appears to be related to thermocline depth and surface water density (Reilly 1990, Reilly and Fiedler 1991), and perhaps to other factors.

These faunal differences appear to have a trophic basis in that 70.7% of spinner–spotted dolphin pairs were found to be associated with bird flocks, while only 30.6% of common and 1.6% of striped dolphin schools were associated with birds (Au and Pitman 1986). The implication is that spotted dolphins, which feed during the day near the surface, drive food organisms into the range of plunging birds, while the other species feed deeper beneath the surface, or at night, and do not contribute food to the birds.

OCEANIC FAUNAL RADIATION

Spinner dolphin populations are part of an oceanic odontocete adaptive radiation of worldwide scale within which the scientist can recognize different food-related niches. For example, members of the warm water oceanic dolphin fauna, such as the false killer whale (*Pseudorca crassidens*) and the pygmy killer whale (*Feresa attenuata*), are predators upon medium-sized to large prey, a proclivity reflected by their very large teeth. Typical prey items range from large fish to other dolphins. Others, such as the sperm whale (*Physeter catodon*) and the beaked whales (family Ziphiidae), are deep-diving teuthovores. The diet of the sperm whale can be remarkably varied but generally emphasizes medium-sized squid. The spinner dolphin and Fraser's dolphin (*Lagenodelphis hosei*) feed largely upon smaller mesopelagic food items.

ADAPTATION TO LOCAL RESOURCES

Within the restraints of tooth size, cetaceans generally seem capable of some adaptation to local food resources, so that much variation in food composition can be expected at different places in a species' range. For instance, in the open eastern tropical Pacific, spinner and spotted dolphins are frequently found swimming together during the day in large aggregated schools, but they feed at different times during the day–night cycle while consuming different assemblages of food organisms (Fitch and Brownell 1968, Perrin et al. 1973, Scott and Wussow 1983). The spinners are nocturnal feeders, usually upon vertical migrant prey, and they may dive 200 m or more, while spotted dolphins, at least on the eastern Pacific tuna grounds, are daytime or crepuscular feeders within about 30 m of the surface (Fitch and Brownell 1968, Scott 1991*b*).

In Hawaii, where spinners feed over an island slope at a depth that can usually be reached during their dives, some of the diet consists of organisms that frequent the bottom or near-bottom waters (Norris and Dohl 1980*a*) (see also chap. 12). The shallow water dwarf Gulf of Thailand spinner feeds on reef animals (Perrin et al. 1987).

Spinners caught late in the afternoon in the eastern tropical Pacific typically have empty stomachs, while spotted dolphins caught at the same time contain freshly caught food (Perrin et al. 1973, Norris and Dohl 1980*a*), highlighting a difference in feeding periods (see chap. 10).

Norris and Dohl (1980*b*) have speculated that the open water association between spinner and spotted dolphins might be an arrangement allowing protection of a resting species from predators, while the other member of the pair is wholly alert and foraging. Although spinners associated with spotted dolphins during the day in the eastern tropical Pacific, it is not clear whether a reciprocal relationship exists, that is, whether spotted dolphins rest with spinners during the night.

The mixing of spinner and spotted dolphins in aggregated schools is not complete. When the two species have been caught together in the same tuna seine, their schools retain their individual integrity even though associated closely together. The usually more numerous spotted dolphins appear to be nuclear in netted schools, while spinner schools move mostly around the periphery of such groups (Norris et al. 1978, Pryor and Kang 1978, Pryor and Shallenberger 1991).

VARIATION

Perrin and his colleagues discovered that, although some populations of spinner dolphins live wholly in the open ocean long distances from land, they exhibit regional differences in morphology (fig. 5). The eastern spinner dolphin is a striking differentiate (Perrin 1975*a,b*, Perrin et al. 1979, Perrin 1990), which occurs in a triangle of sea from near the tip of Baja California, Mexico, to 10°S latitude off the coast of Peru, and offshore to about 145°W, 10°N. Perrin (1990) has given it the name *Stenella longirostris orientalis*. It is a largely gray animal, with the basic spinner pigment pattern only faintly expressed, while nearly all other races of spinners are boldly patterned. In both sexes of eastern spinners, the white of

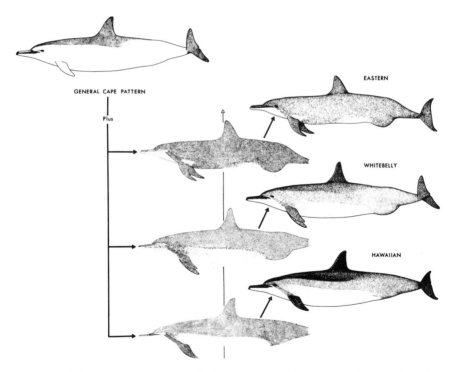

Figure 5. Three races of eastern Pacific spinners: (a) eastern spinner, (b) whitebelly spinner, and (c) Hawaiian spinner. (Courtesy William Perrin)

the ventral field is restricted to patches in the axillary and genital areas (Perrin et al. 1991).

Furthermore, both the postanal hump (a connective tissue protuberance located just posterior to the anus; see fig. 6) and the dorsal fin are sexually dimorphic. In about 70% of adult male eastern spinners, the dorsal fin is canted forward, as if it had been "put on backwards," and the postanal hump is typically very large (figs. 5 and 6). The canted fin occurs in much reduced frequency in adjacent populations of spinners and has been found in a single adult female whitebelly spinner at the western edge of the tuna grounds (Perrin et al. 1991).

The geographic range of eastern spinners, while perhaps stable, is not a sharply bounded exclusive range. On its western portion, both boldly patterned and gray forms can be caught together in a single seine haul. Perrin and Gilpatrick (in press) suggest that this emphasized difference in pattern and morphology may be an example of character displacement. The forms may have originated in isolation at some time in the geologic past and then come into contact in more recent times.

© 1991 Jenny Wardrip

Figure 6. Characteristics of male and female Hawaiian spinner dolphins; top, male; bottom, female.

Mitochondrial DNA analysis fails to show genetic separation among the various stocks of eastern tropical Pacific spinners (Dizon and Perrin 1987, Dizon et al. 1990), as if introgression had occurred or as if the differences are too slight for this analysis to reveal. We suggest that in an area heavily fished by the yellowfin tuna fishery, schools may be more cosmopolitan than in areas of less fishing pressure. This effect may result from either heavy mortality in these areas and in-migration of peripheral dolphins into the heavily fished area or from the mixing of schools by the chase, capture, and release operations. This view of the spinner dolphin school is one supported by our work even in the absence of fishing pressure. In other words, while kin-related association patterns may exist in spinner schools, the larger school is mostly formed of neighboring dolphins beyond effective kin boundaries. The most interesting feature of this arrangement is that the dolphins seem able to assemble viable schools from a large pool of individuals. They do not seem to require genetically close lineages in forming such schools.

ECOTYPES

We suggest that both the coastal Central America spinner *Stenella centroamericana* and the eastern spinner *S. l. orientalis*, which is described later, may be ecotypes whose color differences from other spinners are direct responses to the habitats in which they live. The most likely adaptation would seem to be related to the use of body pattern in the signaling among school members in murky or dark water, a feature discussed later in chaps. 7 and 14. Both of the eastern Pacific forms are largely unicolored, relatively light-colored spinners, while the other forms of the spinner dolphin, including the Hawaiian spinner, have well-differentiated patterns featuring a dark cape, a lateral gray panel, and a white belly, as well as other smaller refinements of pattern (dark beak tips and detailed pattern around pectoral fins and genital area, for example) (Perrin 1969*b*).

Spinner dolphins dive to feed, while the spotted dolphin with which it tends to associate in the eastern tropical Pacific is primarily a near-surface feeder (see chap. 12). The two unicolored spinners live in waters where the thermocline is shallow or at times reaches the surface. Because water below the thermocline is rich in nutrients (see chap. 2), productivity may typically be high in the lower reaches of the photic zone, a zone that these dolphins frequent. The result may be that these spinners tend to swim in waters of reduced visibility during the beginning and end of the day. Other taxa of spinners outside the ranges of the two gray forms, in contrast, may spend their days in clearer, less productive seas, with the thermocline deep enough not to affect productivity markedly.

If these postulates are true, the patterns of the two pale spinner taxa may be muted in favor of a total body flash pattern, while their patterned shallower feeding associates, the spotted dolphins, may be able to use their more complex pattern in near surface waters where visibility remains good (see Fitch and Brownell 1968). This hypothesis suggests that other clearer water spinner populations, by contrast, can make use of more refined signaling systems. Also, they may require protection of a dark dorsal cape from the remarkably intense sun of clear tropical surface seas, which includes an important ultraviolet component (McFarland 1991), or perhaps they use it for protective countershading (see chap. 7). Only at night may pattern no longer matter in the adaptive equation.

There are strong parallels to this situation found in the patterns of other odontocetes. Wherever odonocete species dive deeply or tend to live in murky water, large areas of their bodies, or even their entire bodies, tend to be pale. This holds true for such murky water taxa as the belukha (*Delphinapterus leucas*), which lives in arctic–subarctic nearshore and tidal environments; the species of *Sousa* that live in a similar environment in warm waters; and the Irrawaddy River dolphin (*Orcaella brevirostris*) found in rivers, mangrove swamps, and muddy shallow nearshore seas. Even the bottlenose dolphin (*Tursiops truncatus*) is paler nearshore than in clearer open waters.

The killer whale (*Orcinus orca*), while cosmopolitan, is most abundant in nearshore productive waters. Both it and the Dall's porpoise (*Phocoenoides dalli*), which occupies a similar visual habitat, have large white patches on otherwise dark bodies, suitable for signaling through relatively murky waters, but which are also capable of disrupting the animal's body outline. Conversely, species from clearer surface waters, especially in the clear, open tropical sea, are uniformly countershaded with more complex body patterns.

In Hawaiian waters, spinner and spotted dolphin schools swim separately, at least at the surface, although both species may occur close together in the same area. On the Kona coast of the island of Hawaii and off the lee shore of Oahu, we have noted schools of the two species swimming within less than 1 km (0.6 mi) of one another. The distribution of sightings of Hawaiian spinner schools is shown in figure 7.

POPULATION PATTERNS

Perrin et al. (1991) have studied the variation in color pattern, dorsal fin shape, and body length in eastern Pacific spinners. They have concluded that outside the core area of the eastern spinner (*S. l. orientalis*) there are a variety of morphologic features that show varying gradients of change across the eastern Pacific range of the species. The predominant range

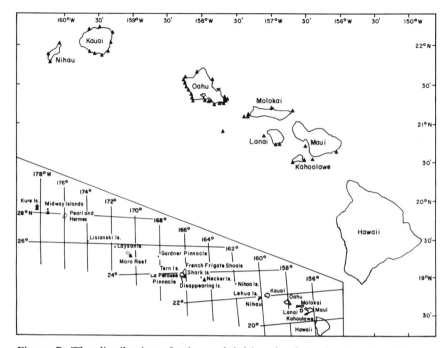

Figure 7. The distribution of spinner dolphin schools in the Hawaiian Islands.

of the eastern spinner seems to be a relatively stable portion of this total eastern tropical Pacific spinner range: north of 10°N latitude and east of 120°W longitude (see fig. 4). Within this core, members of the subspecies are common and can be identified as gray animals with disjunct axillary and genital white patches on the belly. The dorsal fin shape of eastern spinners changes with approaching maturity, until in old males, about 70% show canted fins, a feature also shown by the Central American spinner. Canted fins are evident in a small percentage of spinners outside the ranges of the Central American and eastern spinners.

Young spinners across the entire eastern Pacific area are patterned with a distinct cape and a pale belly. This pattern changes in both sexes with the approach of maturity, with the Central American and eastern spinners becoming gray. Falcation of the dorsal fin and the percentage of animals showing erect fins show other geographic trends, as does body length.

Perrin and Gilpatrick (in press) note intergradation between the gray and more sharply patterned forms known as whitebelly spinners on the west edges of the eastern spinners' range in the ocean south and east of Hawaii. They say of this form:

The dorsal overlay is less dense and less extensive, yielding a faint cape and a more extensive ventral field. The margin of the ventral field may be speckled. These highly variable animals, called "whitebelly spinner dolphins," may represent intergradation and/or hybridization between the distinctive spinner dolphins to the east and those of the Central Pacific.

Perrin (1990) has also recognized as a differentiate of the eastern spinner the gray dolphin living close along the Central American shore, which he calls the Central American spinner dolphin,[1] or *Stenella longirostris centroamericana*. This animal, while generally gray like the eastern spinner, is longer and more attenuate and has a proportionately longer and narrower skull than other eastern Pacific spinners.

Akin (1988), who studied variation in spinner tooth morphology and dentinal patterns, found that these features supported the general variation pattern of spinner dolphin stocks found by Perrin and his associates and especially reinforced the north–south gradient of change across the equator. More recently, Perrin (1990), while recognizing that much variation exists, lumps nearly all of the patterned spinners, including the Hawaiian spinner, into a single subspecies, *S. l. longirostris*, pending collection of adequate samples that might allow recognition of further subspecific entities.

Spinners outside the eastern tropical Pacific, although rather poorly known, all seem to be boldly patterned (Cadenat and Doutre 1959). We have casually observed recognizable geographic differences outside the eastern Pacific. For example, Hawaiian animals and those from the South Pacific, although similar, appear separable to casual inspection. In contrast, spinners recorded from Aden and Djibouti in the Indian Ocean are also boldly patterned and seem much like Hawaiian animals (Robineau and Rose 1983, 1984).

LOCAL OCEANOGRAPHY

Spinner dolphin abundance and distribution is related to certain local oceanographic phenomena. For example, divergence zones at current margins and current ridges both concentrate food organisms and are heavily frequented by dolphins of various species, including spinners (Au et al. 1979, Reilly 1990).

The bottom topography of the abyssal ocean seems to have an effect on dolphin distribution even though it may be thousands of meters below the surface. Dolphins tend to be few in number over abyssal plains

1. The terms *porpoise* and *dolphin* have often been used interchangeably in the scientific literature, while U.S. fishermen usually refer to all small delphinids as porpoise. In 1976, the U.S. Marine Mammal Commission standardized all marine mammal names and applied the term *porpoise* only to members of the family Phocoenidae.

and more abundant over rough bottom topography (Evans 1974, Hui 1979). Evans (1974) found common dolphin schools moving quickly over abyssal plains, while they lingered and dove over rougher bottom topography. However, Perrin et al. (1991) suggest that the patterns of distribution of the eastern spinner (*S. l. orientalis*) and the whitebelly intergrades or hybrids (*S. l. longirostris*) to the west and south remain relatively geographically stable from year to year, but that the northern boundary of the whitebelly population may possibly fluctuate from year to year by as much as 5° of latitude. These authors also present data suggesting that this apparent shift might be due to geographic changes in fishing effort. Reilly (1990), however, found differences in species abundance to be a year-round pattern, with density differences moving from summer to winter in relation to oceanographic variables.

The question is an interesting one because other workers (Au et al. 1979) have previously suggested considerable movements of dolphin populations correlated with water mass changes in the area occupied by the whitebelly spinner. Clearly, the water mass changes associated with El Niño events affect tuna behavior since fishing patterns change greatly, shifting away from fishing on dolphins to seining surface feeding tuna schools directly (Cosgrove 1991). One view suggests that populations of oceanic eastern spinner dolphins somehow remain geographically stable over the bottom, while the other view suggests that at least some populations may move widely without reference to the bottom.

Where a warm current swings away from the tropics along an ocean margin—for example, where the Kuroshio Current moves northward along the eastern shore of Japan—ocean dolphin populations, including spinners, migrate in such water masses and move considerable distances away from the true tropics (Kasuya et al. 1974).

PATTERN

The typical spinner pattern has been described by Perrin and Gilpatrick (in press) as follows:

> . . . tripartite, consisting of a dark gray dorsal field or cape, lighter-gray lateral field and white or very light-gray ventral field. This field is yielded by interaction of the dorsal cape with a "dorsal overlay". . . . The ventral margin of the cape dips over the eye, is lowest below the dorsal fin, passes dorsally about halfway between the dorsal fin and the flukes and is parallel to the ventral margin of the lateral field on the anterior half of the animal. A medium-gray flipper stripe, part of the dorsal overlay system, passes anteriorly to the eye. Flippers and flukes are medium to dark gray on both surfaces. A dark-gray eye stripe passes forward from the eye patch to or near the apex of the melon. The blowhole stripe extends anteriorly onto

the rostrum as a mesial dark stripe that expands at the rostrum tip to yield a black-tipped beak. This system of dark-gray stripes is continuous with a dark-gray lip mark bordering the gape. The white ventral field may be flecked with medium-gray, especially in the gular region and posterior to the flipper stripe laterally. There is a more-or-less faint medium-gray stripe adjacent to the genital region, parallel with the ventral margin of the lateral field.

PHYSICAL VARIATION

Perrin (1975*b*) notes a slight sexual difference in length at sexual maturity for the oceanic spinners of the eastern tropical Pacific (females, 1600 mm total length, versus males, 1625 mm or 63–64 in.). Adult Hawaiian spinners are large, reaching about 1700–2000 mm (67–79 in.) in length. With experience, differences between male and female Hawaiian spinners have become increasingly obvious to observers of living animals (fig. 6).

Weights of spinner dolphins have seldom been measured. Seven sexually mature males from Florida, of about the same length as the Hawaiian form (1920–2080 mm or 75–82 in. total length), ranged from 66 to 75 kg (145–165 lb) in weight, while females of 1870–2040 mm (74–80 in.) length weighed 55–65 kg (121–143 lb) (Perrin and Gilpatrick in press).

Perrin (1975*b*) did exhaustive analyses of spinner dolphin osteology and its geographic variation. Skull measurements in particular reinforce the patterns of racial differentiation we present here (see also Schnell et al. 1982).

In general, wherever the range of spinner dolphin populations abuts a shore, their schools frequent the coast, often coming into very shallow water (a few meters deep) for part of the day (Tomich 1986). As we document in this book, the places where spinners come close to shore seem related to closely available deep water nearshore and locally abundant dependable food resources. The result seems to be that, on a given atoll or island, certain passes and coves are noted as places where spinners are often seen while others may not be frequented at all.

BEHAVIORAL STUDIES

While knowledge of the morphology of the spinner dolphin has been accumulating since the 1800s, knowledge of its behavior is less than three decades old. In the late 1960s and early 1970s, Norris and Dohl (1980*a*) carried out a reconnaissance of the behavior of Hawaiian spinner dolphins. They established a camp and observation post atop the lava cliff backing Kealake'akua Bay on the island of Hawaii. They also worked from shipboard, radiotracked dolphins offshore, made a reconnaissance by air of the entire Hawaiian chain for spinner schools, and began the first detailed underwater analysis of life in a wild dolphin school.

Spinner schools were found at a number of locations throughout the Hawaiian chain from Kure Atoll (Ocean Island) at the northwestern limit of the Hawaiian chain to the southern tip of the island of Hawaii. Association with shore was found to be related to the occurrence of sand-bottomed coves and banks located near deep water. Where atolls were involved, as at Midway Island and Kure Atoll, the dolphins were found to enter passes through the reef and to remain during part of the day in certain sites within the interior lagoon before leaving near dusk for offshore waters.

Fifty recognizable animals (known by scars and marks on their backs and dorsal fins) were catalogued and followed through the course of the reconnaissance work, mostly from sightings within rest coves. Seventy-six resightings allowed some conclusions to be reached about association patterns within schools and about daily movements. The data suggested that school composition was remarkably fluid. Only occasional repeated associations were noted, and even then, such animals were seen at other times with other partners.

The daily behavioral sequence of these spinner dolphins is quite predictable. Nighttime feeding over the island slope was followed toward dawn by a back-and-forth movement along the coast, edging toward shore and thence into shallow water at the nearest available rest cove. There, active schools swimming into the bay in broad dispersed ranks soon subsided into tight discoidal and very quiescent schools. Both phonation and aerial behavior nearly ceased in such schools. Typically, after 4–5 hr, the schools abruptly awoke in a flurry of aerial activity and then entered a pattern called *zig-zag swimming* in which the school moved back and forth along the shore with a notable oscillation of activity level.

Aerial behavior has been classified by Norris and Dohl (1980a) into seven general patterns, six of which produced slapping sounds upon re-entry into the water. These patterns could be arranged in order of activity level and could then be used to predict the state of arousal of the school as a whole. The sequence of such arousal states was quite regular throughout the dolphins' day, ranging from a low during rest to full activity as they left for the feeding grounds.

Underwater observations were made in the preliminary study from a small vessel named the *Mobile Observation Chamber* (or *MOC*) in which an observer watched through a band of windows from a chamber that formed the fixed keel of the vessel. The reconnaissance work set the stage for this more concerted effort (see Appendix B). First, we wanted to check and quantify the earlier conclusions. We hoped to extend underwater observations, do further radiotracking offshore, and check the tantalizing questions that surround school fluidity. Within these apparently fluid schools are kin relations maintained? If fluidity involves the whole Kona spinner population, it would occur among as many as 1000 animals and far transcend the boundaries of normal kin-related groups. This in turn would imply that some aspects of social order must be significantly maintained by learned relationships among nonkin. If so, this spinner dolphin society would resemble the "fission–fusion" societies reported for primates (Smuts et al. 1987).

Finally, the important dimension of acoustic communication had hardly been touched in the preliminary work. It was obvious to all observers that, in a dolphin school—most of whose members travel out of sight of one another—such signals are surely crucial to school organization and interindividual relationships, yet no one knew how they might be used.

These, then, were the considerations that led to this study. We have been able to achieve some of our goals, including the reaffirmation, refinement, and extension of most of what the reconnaissance had concluded. We took the first steps in understanding the uses of sound in school organization, defining the annual reproductive cycle, and documenting by theodolite the nearshore movements of spinner schools. Our database of recognizable animals and their movements and association patterns was greatly extended. The underwater observation work reported here extended considerably beyond the first efforts but remains as a promising reconnaissance.

Since the field portion of this work was completed, other studies of spinner populations have commenced or been reported. Focal animal analysis of netted dolphins schools has been reported by Pryor and Shallenberger (1991). Observation of a spinner population living around the islands of Moorea and Tahiti in French Polynesia has begun by Michael Poole of the University of California, Santa Cruz (see Pryor and Norris 1991). At a very preliminary stage, the work shows some of the same general movement patterns around the 12-km (7.4-mi) wide southern hemisphere island of Moorea as we have found around the 120-km (74-mi) wide island of Hawaii. For instance, the dolphin populations shift around both islands to the calm side during stormy periods.

An Atlantic spinner population at the Archipelago of Fernando de Noronha off the Brazilian coast has recently attracted attention because of its approachability and the clear water in which the dolphins swim (Pryor et al. 1990). Records of its behavior are reportedly being kept. Using our newest viewing vehicle, the *Smyg Tittar'n* (described in chap. 3), the male role in the population of spinners at the island of Hawaii is being examined by Jan Östman of the Long Marine Laboratory, University of California, Santa Cruz. Also, the details of spinner dolphin zig-zag swimming, using the same population of dolphins, are presently being investigated by Ania Driscoll, also of Long Marine Laboratory. Studies related to the eastern Pacific tuna fishery continue to be pursued by scientists and on-board observers of the Interamerican Tropical Tuna Commission from their laboratory in La Jolla, California.

TWO

The Island Habitat

Randall S. Wells and Kenneth S. Norris

One tends to think of spinner dolphins as mammals of the open sea. Yet wherever land punctuates the water masses within which spinner dolphins live, the land seems to attract them. Populations are now known to occur near many oceanic atolls, reefs, and high islands (see chap. 1).

The size of the emergent land mass seems not to matter to dolphins so long as its shores offer adequate daytime shelter. Probably more important is a sufficiently large island pediment to influence the regular accumulation of foods (figs. 8 and 9). Schools of spinners come regularly to tiny Kure Atoll (Ocean Island) and to the slightly larger Midway Island, which is only about 15 km (9.3 mi) in diameter (Norris and Dohl 1980a). They also visit the high island of Moorea, of about the same dimensions, where the dolphins enter passes through the reef and form resting schools in the calm water inside. They also come daily to the open roadsteads and coves of most of the Hawaiian Islands, including the island of Hawaii, where schools can sometimes be found at many different locations.

Although the emergent portions of these islands may be small, their pediments are of much more impressive dimensions. At Kure and Midway, the local pediment stretches west–northwest from Ladd Seamount to Kure, a distance of about 400 km (250 mi). The pediment is even more impressive in the lower Hawaiian Islands, which includes our study site. There, the dissected pediment is more than 700 km (435 mi.) in length and is oriented roughly west–northwest across the general path of enriched cold, deep water movement from the Antarctic Ocean (fig. 9).

This chapter first discusses the general physiographic and oceanographic features of the Hawaiian spinner dolphin habitat. Then we de-

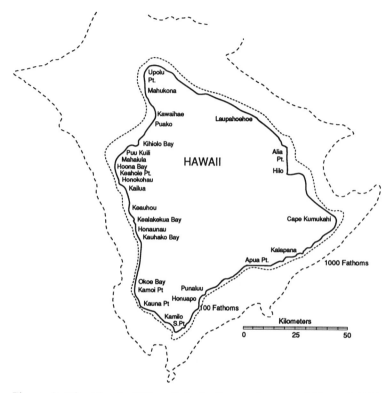

Figure 8. The island of Hawaii, including underwater 100- and 1000-fathom contours and place names.

scribe our attempts to understand how spinner dolphins use the shores of the island of Hawaii.

THE ISLAND OF HAWAII

Macdonald et al. (1983, pp. 7–8) describe the island of Hawaii as follows:

> Mauna Kea and Mauna Loa, two of the five mountains that make up the southernmost island, Hawaii, stand more than 9,000 meters above the adjacent ocean floor, and rise higher above their bases than any other mountain on earth. . . . The island of Hawaii is among the world's largest volcanic islands. Its area of 10,478 square kilometers (4,030 square miles) is greater than all the other Hawaiian Islands put together. The bulk of Mauna Loa [against whose southwestern flank this study is carried out] is estimated to be 42,000 cubic kilometers (10,000 cubic miles). . . . The truly enormous size of this mountain (probably the largest single mountain on earth) can be appreciated when it is compared with the great volcanic

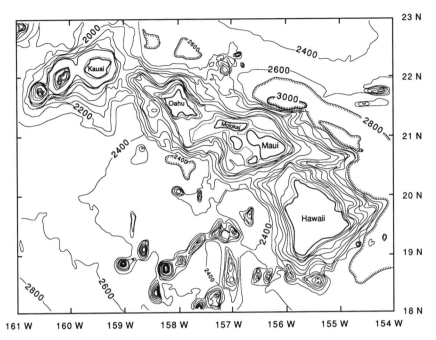

Figure 9. Submarine topography of the Hawaiian Islands area.

mountains Shasta and Fuji, each of which has a volume of about 420 cubic kilometers. Mauna Loa is one hundred times larger than either of them.

The bulk of the island of Hawaii is so great that it seems to have depressed the adjacent abyssal plain on which it rests to greater than average depth, simply because of its weight (Stearns 1985).

Hyperbolic or not, these statements about the island of Hawaii also describe an island that produces by far the largest lee shore in the Hawaiian Islands. It is a place where we could seek out spinner dolphins in often glassy calm coves or travel in relatively calm water far out to sea. (Place names along the lee or Kona coast of Hawaii are given in fig. 10.) Only after traveling a dozen or more kilometers offshore of Kealake'akua did we typically begin to meet the increasingly large swells and winds that had refracted around the island from the rough windward environment.

For spinner dolphins, this huge island size seems to translate into two important things. First, it seems to represent a varied shoreline where dolphin schools can find suitable places to rest between nighttime feeding bouts. Second, it seems to some degree to anchor the dolphins to a specific region of the ocean because of the food resources the island provides. Because the island bulk blocks the deep flow of water masses,

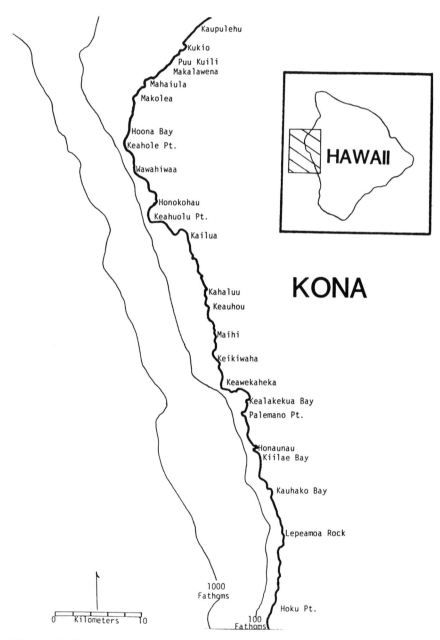

Figure 10. Place names on Kona Coast from Upolu Point (North Point) and south to Ka Lae (South Point).

island-induced and wind-driven upwelling can deflect water from below the nutrient-impoverished photic layer toward the surface sea, where it causes local enrichment (Simpson et al. 1983, Roden 1987).

The slow movement of Antarctic-derived water masses may also concentrate the mesopelagic faunas they carry against the submerged flanks of a mountain such as the island of Hawaii in what is usually called the *island effect* (Kennett 1982 refers to the source of this effect as *obstruction upwelling*). The spinner's main food supply comes from the mesopelagic fish, squid, and shrimp fauna that are restricted during the day to such intermediate depth water moving below the thermocline (see chap. 12, section on Food and Feeding).

OCEANOGRAPHIC CORRELATES

Kennett (1982) has outlined the major water mass characteristics and sources of water masses found in the northern Pacific. He notes that no important supply of deep water reaches the northeastern and central Pacific from the Pacific arctic. Surface salinities in the Bering Sea are too low to create a major dense water mass, even after freezing. The only current (the Oyashio Current) that flows out of the Bering Sea moves in deep water around the western end of the Aleutian chain, near Kamchatka and northern Japan and not near Hawaii.

As a result, the deep water that flows around the Hawaiian Islands and most of the north Pacific comes from the Antarctic. Mean northward current velocities of 9.3 cm/sec have been measured, requiring about 1000 years for this water to move from its source to the north Pacific (Kennett 1982). In this long traverse, it flows through a series of basins that constrain flow below about 4000 m (13,120 ft). One deep passageway into the northeast Pacific exists in the Line Islands, south of Hawaii. The water in this deep layer originates in the Antarctic because, due to freezing and other factors, it becomes very saline and very heavy and it sinks as it begins its long journey northward. The water, starting out with a high oxygen content from its origins at the surface, is gradually depleted as it moves northward, until a thick oxygen-minimum layer is produced at Hawaiian latitudes. At the limits of the north Pacific, a layer of intermediate depth water reflects southward above about 2500 m (8200 ft) (Longhurst and Pauly 1987).

These two water masses, initially enriched from continental erosion of Antarctica, are much richer that Hawaiian surface waters, which have been severely depleted by photosynthetic organisms to the point that they can be considered oligotropic. Slightly deeper, the water has become all but devoid of oxygen, especially because of the daily passage of vertical migrant organisms. Wherever the intermediate depth and mixed layer waters reach the surface in this region, they create localized plankton blooms.

It has been more difficult for oceanographers to sort out the contribution of the water masses below the thermocline to the surface productivity of islands in the mid-ocean. The problem seems to be that wherever an island penetrates the surface, a series of near-surface processes come into play of such magnitude that they obscure the slower, deeper water processes. In particular, these include the dynamic effects of the wind and tide as modified and ordered by the Coriolis force.

Nonetheless, other important effects acting upon the deeper water flowing around oceanic islands is likely to occur and to be significant, as shown by studies of productivity over seamounts. In such circumstances, the top of the land mass may terminate well below the thermocline and yet still produce localized enrichment in the euphotic zone. Boehlert and Genin (1987, p. 319) summarize these effects as follows:

> Seamount effects, which include internal wave generation, eddy formation, local upwelling, and closed circulation patterns called Taylor columns, have important effects upon pelagic and benthic ecosystems over seamounts. The biological effects of these current–topography interactions are poorly understood. Flow acceleration on upper flanks of seamounts may lead to low sedimentation but areas of high standing stocks of benthic fauna, particularly filter feeders. Other effects extend into the water column; nutrient enrichment and enhanced primary productivity occur over some seamounts.

They go on to say (p. 320) that

> Such upwelling at seamounts, like coastal upwelling, will transport nutrients into the euphotic zone where primary productivity is nutrient-limited; an analogous situation exists around islands in stratified seas [such as Hawaii] where increased tidal mixing may stimulate primary productivity (Simpson et al. 1983). Observations on a variety of seamounts supports this contention; Bezrukov and Natarov (1976) suggested that vertical velocities on the order of 0.00003 to 0.0008 cm/sec over various seamounts and that differences in the magnitude of upwelling may explain variability in productivity.

Thus, although it is difficult to sort out by standard observational means, the deep waters around islands such as Hawaii can be expected to contribute to the general productivity of the nearby sea. Also, because the northward drift of Antarctic water never stops, the effect is continuous. This clearly happens at Hawaii and elsewhere where the deep currents impinge upon the pediment of islands and are deflected upward in obstruction upwelling. This slow vertical movement of nutrients is doubtless also affected near the surface at islands such as Hawaii where trade winds drive normal wind-driven upwelling. One such place is located at Keahole Point, where an escarpment blocks the northward movement of intermediate depth water and plays a part in producing a locally enriched area well known to local fishermen. It is also the site of the largest congregation of spinner dolphins around the island of Hawaii.

The shallow surface layer at Keahole Point has been measured at about 80 m (260 ft) thick with a temperature of about 26°C (fig. 9) (Campbell and Erlandson 1979). Below the warm surface layer in the thermocline, temperature drops off fairly rapidly until at 600-m (1970 ft), it has declined to about 6°C. Then, in the intermediate water mass below the mixed zone, temperature declines much more slowly until at 1100-m (3600 ft) depth, it averages about 4°C.

The life on which spinner dolphins feed is thought to be concentrated during the day in or below the thermocline and to move toward the surface in a series of bands, some of which actually reach the surface and some of which remain at various depths beneath the surface. These then become accessible to diving or surface-feeding spinner dolphins.

The zone below the shallow and warm surface water mixed layer is the site of a strong oxygen-minimum layer in Hawaii. Throughout the ocean, the oxygen of surface waters is continually renewed by storms and by photosynthesis of planktonic plants. However, the water just below the reach of these effects has been continuously depleted during its long journey northward from Antarctica by the zooplankton and nekton that gather there in close proximity to surface-dwelling food sources. In Hawaii, this oxygen-minimum layer is especially thick and depleted of oxygen.

Internal waves are an important feature of the Kona Coast of the island of Hawaii. Their occurrence can be noted by the wavering lines of smooth water on the surface that generally parallel the coast and move slowly toward shore. These are the signs of internal turbulence caused by water masses moving over one another in different directions within the water column. This turbulence creates what are called *Langmuir cells*, or the oscillations of water beneath the surface and the surface water held between the subsurface wave crests. The slicks produced by internal waves are very important features for the cetaceans of the Hawaiian Islands and are prominent along the Kona Coast. The slicks tend to concentrate surface nekton and flotsam between their crests, creating long, rich "trash lines" that some cetacean species patrol for food.

Predatory fish, such as the mahi mahi (*Coryphaena hippurus*), are among those species that gather in this way. They form an important food source for delphinids such as the false killer whale (*Pseudorca crassidens*) and the rough-toothed dolphin (*Steno bredanensis*). The false killer whale has been suspected of being a predator on spinner dolphins (see chap. 15).

STRUCTURAL HYDROGRAPHY OF KEALAKE'AKUA BAY

Extensive intersecting subsurface benches form the underwater pediments of most of the Hawaiian islands. On the southwest quadrant of the island of Hawaii, however, the slopes of the Mauna Loa volcano run from the mountain crest to the seafloor in a single descent representing the general angle of repose of the mountain. On a finer scale, this line is broken in many places by the intersections of the lava flows and by gravitational slumping. This slumping has produced a number of minor fault systems typically running circumferentially around the mountain slopes and defining minor underwater benches over which water depths change gradually. The form of Kealake'akua Bay is defined by such a fault system (Nordmark et al. 1978).

The cliff at the back of the bay is the scarp of the Kealake'akua fault. It runs along the island contour to the north and south, disappearing under a lava mantel for much of its length. On the north side of the bay, a lava flow has poured over the scarp and fanned out to form the Kaawaloa Peninsula. The southern limb of the bay is also formed by a lava fan that spreads out toward the north at the base of the scarp (fig. 11). The deepest water in the bay is found in a broad channel between the margins of these two flows, running in close to the Kaawaloa shore to a deep basin (57 m or 187 ft) just off the Cook Monument (figs. 11 and 12).

Doty (1968, p. 9) describes the bottom of the bay as follows:

> In general the shore drops sharply beyond the ten-fathom line. At times this drop is quite precipitous. There seems to be very little coral or other organisms growing on the steep outer slopes, and SCUBA divers noted a marked sterility at greater depths associated with the substrate, which is generally sandy.
>
> The general biotic picture then is one in which the vast bulk of the living organisms is concentrated along a shallow rim of the bay. . . .
>
> Tows were made for zooplankton, and the area of the bay where they were found most abundantly was Kaawaloa Cove, near Cook Monument. This further indicates its sheltered nature, that this water mass does not flush as rapidly as do other parts of the bay. . . . Seventeen species of coral of 13 genera were collected, one of which was a new record for Hawaii. . . .
>
> There is a marked zonation of the coral with depth due to the steepness of the bay and its protection. Four distinct coral zones were distinguished in Kaawaloa Cove, for example, and the extent of their vertical distribution was thought limited by light penetration, temperature and possibly food.

Doty's team also recorded a number of brackish groundwater plumes entering the bay, the largest near Kaawaloa on the north and near Kahauloa Cove on the south.

Figure 11. Kealake'akua Bay from the air.

The bottom at the mouth of the bay is relatively flat and is between 60 and 90 m (197 and 295 ft) deep. The sandy patch over which spinner dolphins rest is located in water of 20 m (65 ft) depth or less, off Napoopoo Beach Park, and extends as a narrow nearshore band below the southern portion of the fault scarp cliff that defines the back of the bay. Dolphins were not seen resting in any other location within the confines of the bay, except temporarily while in transit.

BOTTOM TOPOGRAPHY AND DOLPHIN ABUNDANCE

To the north of Kealake'akua Bay, the 36-m (120-ft) contour defines a fairly broad (4.6 km or 2.5 nautical miles[2] wide), shallow shelf that extends all the way north to Kailua-Kona. This shallow water frequently harbors spinner schools during the day. To the south, the water deepens close to shore. The 36-m (118-ft) contour hugs the coast except in very small bays such as Hookena, all the way to Ka Lae, or South Point, on the island of Hawaii.

The 1000-fathom (1828-m) contour as depicted on the National Ocean Survey (NOAA) Nautical Chart (Catalogue No. 2, Panel C) of the

2. One nautical mile equals 1852 m. Although it is an outdated unit, it is still used on most nautical charts and is thus useful to report here.

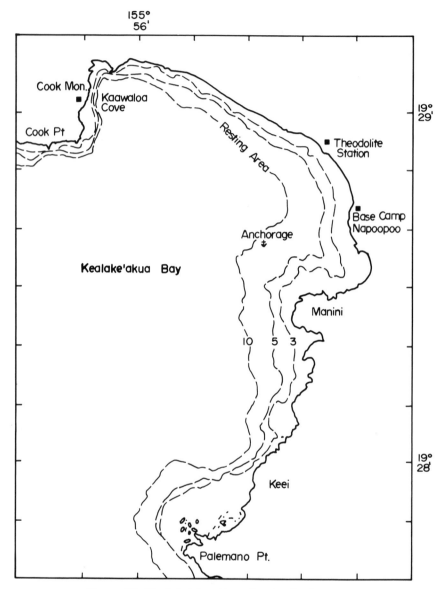

Figure 12. Kealake'akua Bay, showing bathymetry.

island of Hawaii (our fig. 8) gives a good idea of the subsurface shape of the island pediment. It also corresponds well to the waters that spinner dolphins require for feeding. During our radiotracks of feeding dolphins, they often approached or traveled over the 1000-fathom contour during the night. It is clearly a general feature that correlates best with the occurrence or nonoccurrence of spinner dolphin schools as recorded in our aerial survey. The closer to shore this contour runs, the more likely we were to sight spinner dolphin schools there.

This 1000-fathom contour comes closest to shore along the Kona coast and off Cape Kumukahi south of Hilo. Along these stretches of coast, it varies between about 5.5 and 11.1 km (3 and 6 nautical miles) offshore. To the north of Keahole Point on the Kona coast, the pediment broadens abruptly, with the 1000-fathom contour skirting nearly 68 km (35 nautical miles) offshore at a maximum from Kiholo Bay and then coming closer to shore (24 km or 13 nautical miles) at Upolu Point at the north end of the island. This relatively flat undersea bench seldom harbored spinner dolphins, except at its very northern end, and even there sightings were few.

The bottom generally slopes gradually away from the island along the northeast or Hamakua coast of the island. At the north end of this coastline, along the Kohala Mountains, the 1000-fathom contour is 12–24 km (6.5–13 nautical miles) offshore, but it narrows to only 6 km (3.2 nautical miles) north of Hilo at Kaloli, a place where spinners have been seen. Except at its southern end, this coastline seldom harbored spinner dolphins.

Along the southeast shore of the island, the 1,000-fathom contour is rather close to shore (9.3 km or 5 nautical miles) at Ka Lae, or South Point. Then it defines a broader shelf off the shores below Kilauea Crater (out to 26 km or 14 nautical miles) and narrows again along the Puna coast, to about 7.4–9.3 km (4–5 nautical miles) in width. Once again, the correlation between deep water close to shore and dolphin sightings was good.

AERIAL SURVEYS: LOCAL AND SEASONAL PATTERNS

We defined the local occurrence and seasonal abundance of spinner dolphins around the island of Hawaii by taking 23 biweekly dolphin spotting flights around the entire perimeter of the island (Appendix B). Our hope was that this could teach us several facts of importance to our study. First, we hoped to estimate the abundance of dolphins around the island of Hawaii and to learn if the number of dolphins remained the same or varied in some way throughout the year. Were we dealing with a closed population, or did animals come and go from some unknown

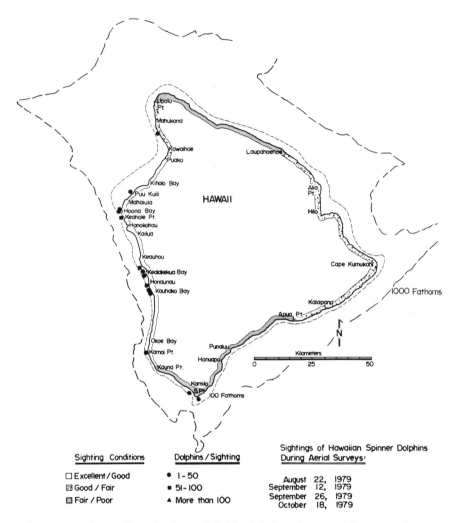

Figure 13. Distribution of spinner dolphin sightings from aerial surveys, August 22 to October 18, 1979.

adjacent population? Second, since we already knew that dolphins disappeared from Kealake'akua Bay for as long as two weeks at a time, we hoped to outline any population movements the dolphins might make around the perimeter of the island. Third, we hoped to learn what environmental features typified the localities where dolphins were found. Why did dolphins choose one cove and ignore others, a fact we knew to be true? Could we learn why some coves seemed to host many more dolphins than others?

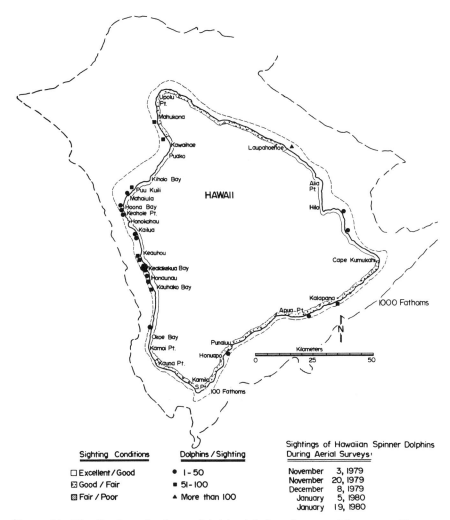

Figure 14. Distribution of spinner dolphin sightings from aerial surveys, November 3, 1979, to January 19, 1980.

As already mentioned, we found a positive correlation between the nearness to shore of deep water and the probability of sighting spinner dolphin schools in adjacent coves (figs. 12 and 13–17). During the year, we located 98 schools of spinner dolphins, 75% of which were seen on the usually calm Kona coast. Almost all of them were seen from Ka Lae to a few kilometers north of Keahole Point (to Puu Kuili), the section of coast where our other observations were being made. Six sightings were made near the north end of the island near Mahukona south of Upolu Point.

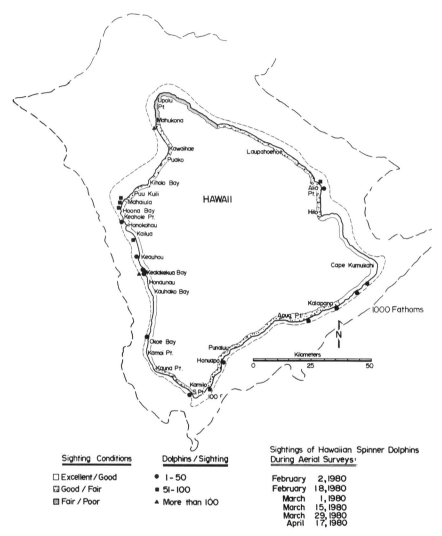

Figure 15. Distribution of spinner dolphin sightings from aerial surveys, February 2 to April 17, 1980.

Most of the rest (19 schools) were sighted along the southern coast of the island, around to the eastern shore at the town of Hilo and a little to the north at Alia Point. It was rare for us to see any spinner schools along the windswept northeastern coast (the Hamakua coast) north of Alia Point to the northern tip of the island at Upolu Point. In the year of flights, we sighted only two schools in this area, the northernmost at Laupahoehoe Bay (see fig. 13). Some part of this difference could have been due to difficult sighting conditions.

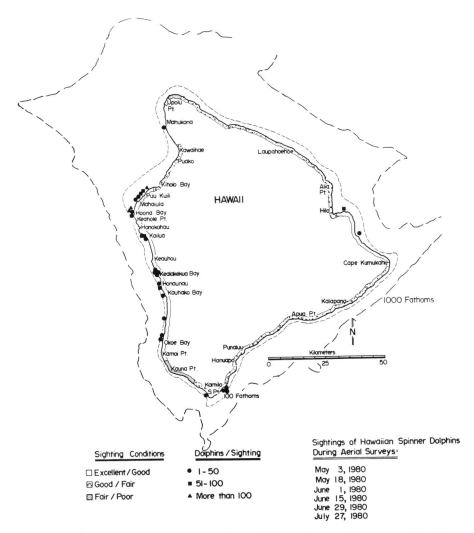

Figure 16. Distribution of spinner dolphin sightings from aerial surveys, May 3 to July 27, 1980.

One curiosity in the records was that we saw no spinner dolphins along what seemed to us an optimum stretch of shallow, sand-bottomed coves with calm water offshore from Kiholo Bay to the little harbor of Kawaihae. Only a few kilometers to the south of Kiholo Bay was the locality consistently frequented by the largest schools of spinners on the entire island, the Keahole Point—Hoona Bay area.

Another pattern emerged from flights taken on November 20, 1979, through March 1, 1980. Our records from these flights showed that an

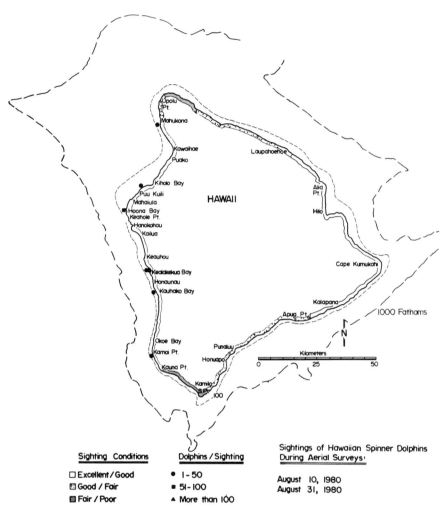

Figure 17. Distribution of spinner dolphin sightings from aerial surveys, August 10 to August 31, 1980.

obvious population shift had taken place in which the percentage of dolphins seen along the Kona coast declined from 100% sighted on the November flight to less than 5% sighted on February 18. At the same time, dolphin schools were appearing on the windward eastern shore south of the Hamakua coast, from Alia Point to Hilo Bay, at Cape Kumukahi, and along the southern shore. It appeared that a large part of the spinner population had switched sides of the island. It did so by moving away from the rough weather on the Kona coast. The effect was most pronounced during a period of southwesterly storms in February 1980 that swept into the Kona coast causing very turbulent seas throughout the Kona area (figs. 18, 19, and 20). At this time, spinner dolphins were difficult to find at Kona.

SIGHTING EFFICIENCY

The question of abundance of dolphins around the island and whether or not a stable population exists nearshore is obscured by our uncertain ability to sight a high percentage of schools on a reliable basis from flight to flight and to determine the composition of such schools. The flights were made in different weather conditions from month to month and in variable sightings conditions around the island. Sea state and sun angle were two major variables that we could not control for an entire flight. Such unavoidable, nonuniform sighting conditions around the island doubtless introduced bias into our counts, favoring the calm parts of the island. Nonetheless, especially because we alternated the flight directions of each succeeding flight, we feel that our counts provide a rough measure of the total momentary population, whether or not it is a fluid assemblage. But they are clearly minimum estimates because it was often difficult to sight schools.

Variables that clearly affected our total count were sighting efficiency, time of day (since dolphin schools rest nearshore for only a part of the day), normal cove occupancy rate, and percentage of dolphins that do not enter coves. To derive a reliable total momentary population figure for spinner dolphins associated with the island of Hawaii, these variables must be factored in.

We certainly missed some schools in our flights, although each search was as thorough as we could make it and was done by experienced aerial spotters. We expect that the effect was small, and that it was probably greatest for schools swimming some distance offshore. Our flight path followed the shoreline and was most effective within about 2 km (1.2 mi) of the beach.

The average number of hours a dolphin school occupied Kealake'akua Bay varied with the season from 6.9 hr in summer (May to September) to 5.1 hr in winter (October to March). On average, this means that we flew part of the time outside this general occupancy period and therefore could have been expected to miss some percentage of the dolphin schools that would have entered or left coves before or after we arrived. This should have produced an underestimate of schools coming into coves around the island. This assumes that the period of occupancy of Kealake'akua Bay is representative of the island, which may

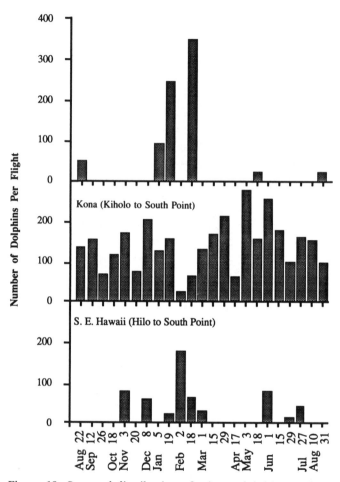

Figure 18. Seasonal distribution of spinner dolphin numbers seen during aerial surveys by coastal section.

not be perfectly correct because the travel distance to offshore feeding grounds varies from cove to cove.

IS THERE AN ISLAND OF HAWAII SPINNER POPULATION?

The flight data and that derived from marked animals produces an incomplete picture of the movements and composition of spinner schools

Distribution of Hawaiian Spinner Dolphin Sightings
(Aug. 22, 1979 – Aug. 31, 1980)

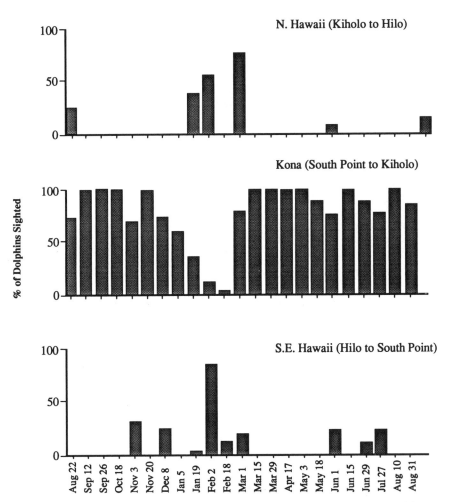

Figure 19. Seasonal distribution of spinner dolphin sightings during aerial surveys as a percentage of total sightings.

seen around the island of Hawaii. Left unresolved is the possibility that there may be a flux of dolphins between islands or with schools of dolphins living farther offshore. We cannot assess how many dolphins stay offshore of the island of Hawaii and only occasionally enter rest coves. This number may be considerable. From our few observations of the offshore school, especially from our radiotracks, all dolphins clearly did not

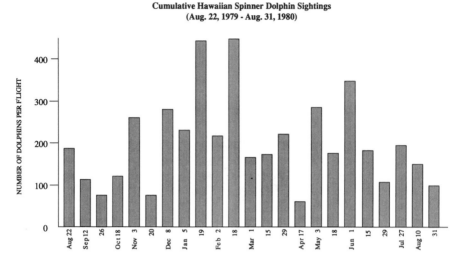

Figure 20. Cumulative sightings of Hawaiian spinner dolphins around the island of Hawaii.

enter rest coves every day. There did seem to be an offshore contingent of unknown but probably large size, and we have little idea of its possible change in composition.

The first question we asked about the island of Hawaii spinner population is did the shore population change in number throughout the year? The minimum and maximum number of dolphins we recorded for a given flight are depicted in figures 13 through 17. Our counts varied several-fold from month to month. Can we attribute this fluctuation to sighting error, or does some or all of it involve real changes in the number of dolphins occupying the shores of Hawaii (Appendix B)? Our speculation is that the population of spinners frequenting the shores of Hawaii may undergo fluctuations beyond those attributable to sighting error alone. Some counts, although taken in good sighting conditions with an experienced sighting crew, were much lower than others. It is difficult to ascribe all of this difference to the difficulties of performing our aerial survey. As we will document later (in chap. 6), certain members of the spinner population were consistently seen nearshore, suggesting a segregation of social elements within the entire population.

Circumisland Movements

Our second question about the island of Hawaii spinner population concerns the possibility of population movements around the island. The circumisland movement of dolphins indicated by the records taken in the winter portion of our survey is instructive because not all of these flights were taken in rough weather and yet the shift in sightings continued to exist in our records (figs. 14 and 15). Also, a distinctively marked individual whom we named Twin Peaks (see fig. 64), recorded south of Keahole Point area from November 1979 to April 1980, was resighted at Cape Kumukahi south of Hilo on June 16, 1981, having made a circuit from one side of the island to the other. To be sure, this is only a single sighting of one animal seen on both sides of the island and cannot, by itself, do more than suggest a circumisland movement.

Since our work was completed, Michael Poole (a doctoral candidate at the University of California, Santa Cruz), who is studying spinner school movements at Moorea in French Polynesia, has found a circumisland movement pattern around this much smaller island. As those on Hawaii appear to do, the dolphins circle the island, moving away from rough weather to enter calm coves (Michael Poole, pers. comm. 1992).

These bits of data taken together allow us to support the idea that spinner dolphins living along the shores of Hawaii do sometimes shift a significant portion of their population from one side of the island to another. For reasons outlined in chapter 4, these movements seem typified as being away from turbulent, dirty water in their rest coves.

Rest Coves

The third set of questions that we asked about the Hawaiian spinner population was this: Why should spinner dolphins choose one cove for rest and not others? Is it true that some coves typically house more dolphins than others? The figures from our flights did indeed support the idea that each cove had an approximate upper limit of dolphin numbers that could occupy it. For example, Kealake'akua Bay was observed to contain dolphin schools that ranged to about a maximum of 80 dolphins, with a mean number of 33.5 ± 26.9 ($n = 15$), while the Keahole Point schools frequently numbered as many as 180–200 animals.

The most obvious feature that could determine these "carrying capacities" of rest sites is the area of white sand bottom available to the dolphins for rest. Although precise measures of the sand areas were not made at these various coves (except at Kealake'akua Bay), casual inspection of the coves during our surveys suggests that the area of clear sand bottom is correlated with the number of dolphins that will rest in a given area. Our theodolite tracks and underwater observations of resting dolphins also show that such schools assiduously avoid dark coral bottom in favor of light sand generally free of coral heads.

At Kealake'akua, more than half of the available sandy area at the back of the cove was occupied by yacht moorings. During our observation period, the dolphins seldom entered among the boats and never seemed to rest there. Therefore, it is clearly possible that any future encroachment of such moorings at sandy-bottom sites could displace the dolphins from their habitual resting areas. We are pleased to note that since our work, the number of moorings has been restricted in the newly established Kealake'akua Bay Marine Reserve. The dolphins now have an even wider expanse of sandy bottom to rest over than during our work.

This use of shallow sand-bottomed areas along the Hawaiian shoreline by dolphins is discussed further in chapter 4. It is thought to relate to the cessation of echolocation by dolphins during the rest period and hence to their need to localize dangers visually, especially shark predators. When they enter the visual mode, they seem unwilling to occupy coves when underwater visibility in rest areas is poor. Perhaps this is also why the dolphins appear to circumnavigate the island when rough weather shifts from one side of the island to the other, in the process destroying both the sheltered lee in which they have been resting and good underwater visibility.

One final question is related to the location of coves and roadsteads that are used or left vacant by spinner dolphins. Why should one cove be used frequently and a nearby one, even with a large area of sandy bottom (such as at Kiholo Bay), remain virtually unused?

The answer to this question now seems clear: the most heavily frequented coves are all close to deep water. It seems clear that it is energetically expensive to swim to and from a cove a long distance from the feeding grounds. Also, the traverse from open water feeding grounds to a rest cove may take dolphins over variable distances of dangerous shallow water from which a shark could attack with more chance of success than in deep water (see chap. 13).

A dolphin school readying itself for sea typically has to traverse such dangerous water at dusk before reaching the relative safety of deep nighttime water where its echolocation can give early warning of predators in all directions and where it has a sensory advantage over antagonists that do not possess echolocation. Dusk, the time of the photopic–scotopic shift in vision (the shift from cone to rod vision), is a time utilized preferentially by some sharks for predation, and it is also the time that spinners are making their way out of rest coves and out to sea (see chaps. 7 and 13). At this point in our understanding, we cannot separate these possible explanations, and perhaps other factors are involved as well.

The Spinner Population: Open or Closed?

Our curve for the acquisition of newly identified dolphins in our scars and marks catalogue (see chap. 6) never leveled off during the course of this work in spite of concerted sampling. This would have been expected if our population were a closed one composed of a set number of animals frequenting the island shores. Instead, a steady increment of new, unidentified animals continued to appear in our records. But our aerial surveys showed that during the spring–summer and fall months, nearly all animals clustering around the island of Hawaii were on the Kona coast where our work was most intense. Given the modest estimated total numbers of dolphins frequenting shore areas and our strong effort at photoidentification (chap. 6), the most likely explanation is that the inshore dolphin population at the island of Hawaii draws members from a larger, almost unknown offshore reservoir of dolphins whose extent we do not know. It could even extend to other islands, but this seems unlikely. Probably the population is relatively stable in terms of numbers and members, but large enough that we had not neared completion of our task of individual identification. Probably the society is divided into portions that frequent the shore much more than others do. A suspicion (and it is not much better developed than that) is that nurture and instruction of young are better represented in cove-frequenting dolphins than are other parts of the social structure of the population.

THREE

Observing Dolphins Underwater

Kenneth S. Norris and Randall S. Wells

A major challenge of the study of dolphin natural history is to place an effective observer under the water in the ocean where dolphins live out their life patterns. Nearly all studies of wild dolphin societies have relied upon observations made above the water or those made by listening under the surface (Pryor and Norris 1991, Tyack 1991). We describe some of these approaches elsewhere in this book. We made attempts to design balloons or kites that carry cameras or video apparatus, and we once built a gyrocopter to cruise over dolphin schools. However, for our aerial observations, we always came back to standard aircraft that could take us aloft where we could see or photograph out a window or an open door. But such aerial approaches tell us only a fraction of what wild dolphins do. Usually, such observations are without the contexts that can explain them.

For the two decades of 1950–1970, the studies of captive dolphins by McBride (1940), McBride and Hebb (1948), Tavolga and Essapian (1957), Caldwell and Caldwell (1965, 1968, 1972a), Tavolga (1966) and Brown and Norris (1956) told us much of what we knew about the social arrangements of these animals. Yet this work always carried with it the question: "How are the patterns that are seen in captivity modified by the confinement? It was clear that wild dolphins swam long distances daily, some going from very shallow to very deep water each day. How did these excursions affect social patterns? How did depth affect them? How did living in three-dimensional, open water change their life patterns?

METHODS OF OBSERVING DOLPHINS

The options for new observational methods were rather few. One could attempt to swim or dive with dolphins, as we describe in this book, or a photographer could be placed within a school by being towed. One could instrument the dolphins, as we once attempted to do, by capturing them and placing a retrievable camera on them to film from within their schools. One could drive a remotely operated craft into their midst for filming or videotaping sessions. Or one could build a craft that let an observer see underwater while the boat driver sought to stay with the school.

The remote operated vehicle (ROV) operating from a fiberoptic tether has potential, but has not yet been exploited for cetacean observation to our knowledge. The ROV has three problems. Such vehicles have tended to be slow moving, but presumably could be made to move fast enough to keep up with a typical cetacean school (3–5 knots).[3] Most require an expensive mother ship capable of hoisting the vehicle on deck. Finally, only recently have they been able to carry video equipment that provides detailed images that rival the eye as an observation tool.

UNDERWATER VIEWING VEHICLES

Weighing these various alternatives, it became obvious to us that only one approach allowed the observer a direct chance to be with wild dolphins underwater for any length of time for frequent, even routine observation sessions, while also allowing use of instruments such as cameras, notebooks, and listening gear. And only this approach seemed to be within our budgetary limits. That was the underwater viewing vehicle. Such considerations led us to design and use three viewing vehicles specially built for the observation of wild dolphin behavior. As a result of our early attempts, the U.S. Navy built a fourth craft, the *SeeSea,* an approximately 12-m (36 ft) catamaran that featured a plastic observation sphere that was lowered between the hulls (Evans and Bastian 1969).

Nonetheless, these vehicles were also flawed in various ways. In particular there was the nexus of size, safety, and expense. A small vessel may be relatively cheap to build and operate, but it might also be dangerous in the open ocean where most oceanic dolphin life patterns are played out. A larger vessel would cost more to build, operate, store, and repair. So, the evolution of our various vessels has followed support budgets and the desire for information from previously unobserved portions of a dolphin's daily life.

It was not until the late 1960s that we began to go to sea in underwater vehicles to make direct observations of dolphin society in nature (Norris 1974, Norris and Dohl 1980*a*). The richness of underwater dolphin social arrangements was at once apparent. Behavioral events were

3. One knot equals 6076 ft, and 1 knot/hr equals 1.13 mi/hr.

immediately seen to be related to the geography of the dolphins' home areas and to their daily and seasonal cycles. It soon became obvious that the dolphin school, however fluid, was a social unit that enclosed and protected the life patterns of wild dolphins in three-dimensional space. The dolphins in a school obviously knew the details of their world in as much subtlety as mammals ashore do, although they often seemed to be responding to cues unfamiliar to us.

These underwater societies, as rich as we found them to be, nonetheless have continued to remain difficult to observe simply because the craft we used were also imperfect. Only now, for example, are we approaching a vessel capable of going very far to sea where half a spinner dolphin's life is spent. If the dolphin species one wishes to observe goes into open water offshore, questions of seamanship and safety at once begin to loom very large. This set of problems immediately tends to produce designs (such as larger size) that reduce the willingness of dolphin schools to tolerate an observer in their midst (see Appendix B).

APPROACHING DOLPHINS

Even with vessels as small as ours, insinuation into a school almost assumes the dimensions of courtship at times, as our observer teams sidle up to schools, watching intently for signs that the animals are reacting negatively to their presence. We began to learn that our hopes of habituating schools to our presence might work to a degree but that a very long period of association would be required, simply because the schools were so fluid that the animals we sought to befriend changed radically from day to day.

Early in our work, it became obvious that spinner dolphins were private animals at certain times of day, tending to move away from our approach, while at other times they would cluster around us avidly. At these times, an observer in the viewing chamber would be accompanied by dolphins slicing in close to peer in. This oscillation of behavior remains dimly understood.

Underwater observation also requires reasonably clear water, tending to restrict observations to the clear seas of the tropics or subtropics. Many events in dolphin lives we found to relate to the larger school because, above all, open water dolphins are group animals that depend heavily upon each other for many things, including survival, sociality, and food finding. So being able to see across a school allows the observer to put these matters into proper context.

LISTENING

Many events in dolphin schools, especially in dim light, are obviously mediated by sound signals. A viewing vessel not equipped for listening has cut off a major avenue of understanding for the observer. None-theless, to achieve three-dimensional listening, so that individual dol-phins making sound in a dense school can be identified, remains a formidable problem (Tyack 1991, Watkins and Schevill 1974). There are two general problems associated with this difficulty—water noise and triangulation.

Water noise, which can obliterate most of the incoming sound in a sea of masking noise, comes mostly from near-field water movements pro-duced directly adjacent to the hydrophone. This kind of disturbance op-erates only very close to the receiving instrument (mostly within the first centimeter) and can be greatly reduced by shielding the receiving hydro-phone inside a thin, oil-filled pipe. The mechanical disturbance of pass-ing water then occurs a distance away from the hydrophone and atten-tuates before reaching the sensitive element of the transducer. We have used a listening system in which the hydrophone was placed inside a thin plastic pipe filled with castor oil mounted under the vessel. Sensitivity is reduced, but noise is at much reduced levels, allowing a listener to hear events in a nearby school. Another helpful method is to use a high pass filter to cut off low frequency noise.

Our newest vessel is equipped with two through-hull pipes that allow mounting of hydrophones below the lateral sponsons on either side of a vessel. Thus, by using stereo headphones, we hope to achieve some sem-blance of directional listening. Such stereo listening requires a period of training during which the observer achieves a better and better sense of direction as he or she becomes adjusted to the time delays provided by a given system (Batteau 1968). But even without stereo listening, hear-ing sound emissions remains a crucial element in observing wild dol-phin schools.

THE *MOBILE OBSERVATION CHAMBER*

The first of our vessels, the *Mobile Observation Chamber,* or *MOC,* was con-structed for Norris from surplus aircraft parts by welder Jimmie Oku-dara at Kewalo Basin, Hawaii, in the mid-1960s (fig. 21). The hull was a 4-m (13-ft) long jettisonable jet aircraft fuel tank that was pierced by an observation cylinder in the bottom of which an observer sat in a swivel chair surrounded by a 360° ring of heavy, shatterproof glass windows. The vessel was designed to operate either by means of a 70-hp outboard motor set in a motor well at the rear of the vessel or by being towed by a larger vessel. Because there was some danger that the *MOC* might spin when under a high speed tow (greater than about 7 knots), the observa-tion cylinder was equipped with a waterproof hatch and a blower system to provide fresh air. It was also ballasted at the bottom of the observation chamber with 1000 kg (2200 lb) of lead pigs. The observer could cant the

Figure 21. The *Mobile Observation Chamber,* or *MOC* (otherwise known as the "semisubmersible seasick machine" or "SSSM").

vessel to one side or another by use of foot pedals that operated diving planes set on the sides of the chamber near the bottom. This allowed an observer, when under way, to tilt the vessel enough to see nearly straight down, or to swing to the side enough to look upward at a steep angle toward the surface.

This refinement, however, proved to be of limited value since the vessel was usually operated using its own engine, which restricted the speed to about 4 knots, not enough to tilt the vessel significantly. The *MOC* (which we all called the "semisubmersible seasick machine," or "SSSM") had a peculiar vessel motion due to placing the center of mass so far below the main hull. She tended to have a "snap roll" for operators on deck, as does the newest vessel. She was also very noisy, especially from engine noise and the swash of water in the ballasted flotation tanks.

The craft could be towed far offshore and proved able to operate in moderate sea states, albeit, at some cost to the equanimity of the observer. The major problems, aside from motion, were two. First, the surge of the heavy craft in a swell required heavy and secure towing gear and a relatively large towing vessel. Second, there was always the concern that the *MOC* would spin on the cable. Another feature of the higher speeds was the appearance of cavitation vortex trails that streamed from any projections on the skin of the craft, often flooding across the viewing windows with sufficient density to obscure vision.

The awkward craft required a crane to lift her from the water, and it proved a constant problem to clean the windows of the viewing chamber. Settling organisms, especially in the warm seas where the vessel was operated, quickly accumulated and cemented themselves firmly to the windows. Scrubbing them away quickly abraded the vessel windows and reduced visibility. Leaving them produced an even worse problem. Consequently, we designed a secondary window box that could be clasped over the viewing windows so that a strong cleaning solution, such as bleach, could be used regularly to prevent settling and that might help dissolve the calcareous skeletons of settling organisms. This refinement, however, was never given an adequate test before the *MOC*'s observation series were ended. This viewing problem was never adequately solved and remains a serious consideration for anyone planning to build a fixed-chamber viewing vehicle. The problem was solved for our last vessel in which the chamber is hoistable, bringing the windows out of the water at the end of a day's work, where they can be washed.

THE *MAKA ALA*

Our second vessel, the *Maka Ala* (Hawaiian for "watchful one"), was quickly built at the request of Norris by personnel at the Southwest Fisheries Center of the National Marine Fisheries Service, La Jolla, California. It was built for one of the Dedicated Vessel Cruises on the yellowfin tuna grounds, in which solutions to the massive kill of dolphins in the seine fishery were being sought (fig. 22).

The *Maka Ala* was built from the discarded fiberglass hull of a Boston Whaler cathedral hull sport craft. In essence, construction consisted of replacing the bottom of the vessel with an elongate viewing box in which an observer lay prone. The bow was replaced by a sloping transparent plastic box that conformed roughly to the old contour of the craft. Originally the craft was ballasted with 1000 kg (2200 lb) of lead bird shot in sacks, arranged under the observer's mattress. These were later melted down and arranged in racks along each side of the viewing chamber.

Two other refinements made the *Maka Ala* into a reasonably serviceable viewing vessel. First, we built a cover over the viewing vault that accomplished two major aims: it shielded the observer from the intense low latitude sun and it provided a dark space from which the observer could look into the better lit water. This allowed the observer to look out without interference from reflections.

The second improvement was to build an instrument deck partially over the observer's chamber. The observer slipped in on the mattress on

Figure 22. The viewing vehicle *Maka Ala*, our second viewing vessel.

the deck, with listening gear, tape recorders, and batteries above and only wires leading down to the observer. This arrangement worked well after adequate stowage containments were installed to keep gear from cascading onto the observer below.

Detractions of the vessel were numerous but not totally debilitating. We did achieve a good many hours of observation of dolphin schools with her, once again pointing to the value of the method, even though the craft was less useful than it could have been. For example, one problem was that the sloping plastic bow, although allowing visibility forward and downward, did not allow one to look up toward dolphins at the surface. Since sexual dimorphism in spinner dolphins is all but cryptic, this removed our best chance to sex animals, a key detriment in any behavioral study. Only through the small side windows could we hope to see genitalia.

The *Maka Ala* could achieve speeds of about 6 knots, making her much faster than the *MOC* under power. But she was a dangerous craft in even moderate seas, so she was restricted to observations in the confines of Kealake'akua Bay. Our fears were fully vindicated when she sank in a rainstorm, simply filling with rainwater until her weight made her prey to swells that splashed water over the rails. The added flotation we had installed in the bow, stern, and along the inner rails proved insufficient to keep her floating.

The old problem that we had faced with the *MOC* remained unsolved with the *Maka Ala*, that of keeping viewing windows clean. So every morning before observations could begin, a diver had to swim under the craft and wipe away any new accumulations. This was never entirely satisfactory since the act of cleaning always produced a haze of scratches on the plastic window after a time. Once such irregularities were created on the windows, these quickly filled with settling organisms, exacerbating the problem greatly.

THE *SMYG TITTAR'N*

The third vehicle was given the almost unpronounceable name *Smyg Tittar'n* (Swedish for "tiptoeing looker") by Jan Östman, who now uses her off the Hawaiian coast. The vessel solved many of the detractions of the earlier viewing vessels. She is larger, as large as can be towed legally on a trailer in California, 2.4 m (8 ft) wide and 7 m (23 ft) long (fig. 23).

We installed sponsons along both rails from bow to stern to give her lateral stability, and consequently, the vessel has proved remarkably stable. Only 3° of list is produced by standing people weighing a total of 180 kg (400 lb) on a rail, and even under full power and swung in a tight circle, the vessel heels over a maximum of only a few degrees.

Figure 23. The newest viewing vehicle, the *Smyg Tittar'n*. Above, vessel with an observer in the chamber; opposite, chamber hoisting design.

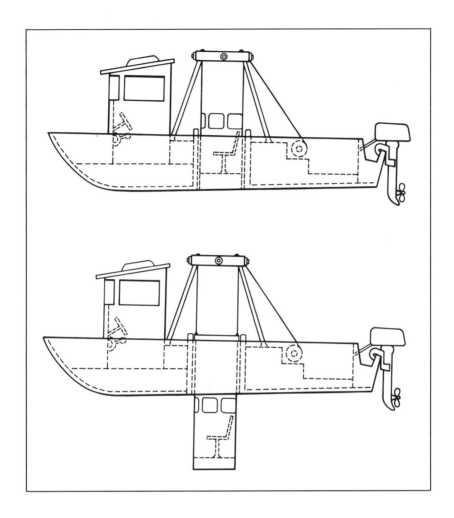

Unfortunately, one reason the vessel is so stable is because it is so heavy for its size. It weighs almost 4540 kg (10,000 lb), causing the boat to react rather slowly to oncoming waves. This has the effect of it tending to go through, rather than over, oncoming waves. *Smyg Tittar'n* is presently limited to operation within 1–2 km (about 1 mi) of the coast, that is, within swimming or rafting distance of shore.

Most important, the *Smyg* features a vertical 0.9-m (30-ft) diameter viewing cylinder that can be hoisted and lowered by means of a small auxiliary battery powered winch. The cylinder is inside a combing amidships and can be locked in either the up or down position (fig 23). In the up position, the bottom of the chamber is pulled up into the hull so that the chamber bottom, which has the contours of the old hull, produces no reduction of the vessel's speed. Powered by two 100-hp outboard engines, she can go 15 knots in a calm sea.

Fuel storage is in two built-in 25-gal (95-L) tanks positioned laterally to the chamber, with two 6-gal (23-L) external tanks as a backup. A battery bank and bilge pump are located just forward of the self-bailing engine well. These batteries are powered by two solar panels in addition to the engines.

The cylindrical chamber is pulled up inside the collar of an A-frame and bolted in place for travel. In the down position, the chamber lowers to the top of the fiberglass combing and once again is bolted in place. This allows an observer to step in at deck level and climb down to the viewing station and swivel chair, which is inside a 360° ring of heavy plastic windows at eye level.

The window-cleaning problem has been solved in the *Smyg Tittar'n* because when the chamber is in the up position, the windows that ring the chamber are in air just above the top of the combing where they can be cleaned easily and gently. Furthermore, the heavy Lexan windows are easy to replace by unbolting the inner frame that releases them from their polysulfide gasketed seats.

Communication between the boat captain and the observer is by headphones and a microphone so that the observer below can call for course changes in relation to the behavior of the dolphins. A speaker on the bow allows the driver and the observer in the chamber to coordinate activities above and below water.

The two hydrophone tubes that pierce the sponsons on each rail have yet to be tested for binaural listening, although we have hopes that, with training, our observers may be able to sense the direction of sounds emitted in the schools they observe.

Instrumentation is arranged on roll-protected shelving along one side of the main cabin where the driver can operate controls. Navigation gear such as a depth sounder, compass, and ship-to-shore radio are mounted overhead and forward of the driver's position. A life raft can be affixed to the top of the deck house or towed astern.

This evolution of underwater viewing capacity seems to be one central theme in the future of dolphin behavior study. Without observers watching the lives of wild dolphins over long periods of time, producing a coherent view of their lives can be expected to remain difficult. The naturalist's "feeling" about how their lives might work is what allows the observer to first ask sharp questions and then gather statistical data sets to test such ideas.

FOUR

A Spinner Dolphin's Day

Bernd Würsig, Randall S. Wells, Kenneth S. Norris,
and Melany Würsig

The normal daily routines of an animal's life provide the primary frame of reference within which most other behavior can be understood. For this reason, we begin our description of spinner dolphin behavior by outlining their movement patterns during an entire day–night cycle within the confines of Kealake'akua Bay.

METHODS OF OBSERVATION

To obtain the data, we first established two observation posts. One was near sea level in front of our base camp (a rented waterfront house in the village of Napoopoo on the south limb of Kealake'akua Bay); this post was monitored twice hourly during daylight hours for the first year of work. The other was a theodolite station 72 m (236 ft) above sea level on the cliff that backs the bay. The first let us document arrivals and departures and provided estimates of animal numbers. The second let us track dolphin schools with considerable precision by standard surveying techniques.

We usually worked at sea using one of two vessels: the 8-m (26-ft) *Nai'a*, which let us travel into relatively rough waters (fig. 24), or the viewing vessel *Maka Ala*, which allowed us to observe schools underwater in quiet bays. (A fuller description of these methods and equipment is given in Appendix B.)

AN OUTLINE OF THE DAILY CYCLE

At our Kealake'akua Bay observation posts, the day's first view of spinner schools often came not long after dawn (fig. 25). We could some

Figure 24. The vessels *Nai'a* and the *Maka Ala* in Kealake'akua Bay.

times see the alert rapidly moving schools swimming in deep water at the bay mouth, moving directly toward shore. Norris and Dohl (1980*a*), working from the much higher 150-m (490-ft) clifftop seem to have first sighted dolphins somewhat farther out at sea, but once the animals moved in under the cliff, our 72-m (236-ft) post allowed more detailed observation. Typically, entering schools were spread over a considerable swath of sea and were made up of obvious subgroups[4] whose members were more closely synchronized with one another than they were with the school as a whole. As the school moved into shallower water, such distinct subgroups became less and less evident. By the time the school was well within a bay, it moved synchronously as a tight unit, but with the subgroups still evident upon close inspection.

This increase in synchrony was accompanied by both tightening and slowing of the school. By the time the shallow waters deep within the bay were reached, what had been an open and rapidly moving assemblage, often broadly spread, had changed into a tight, slow-moving discoidal

4. Throughout this work, we refer to discrete aggregations of dolphins that move together within a recognizable boundary or envelope as a *school,* and to recognizable clusters of animals that move within such an envelope as *subgroups.*

TABLE 1 Hawaiian Spinner Dolphin Tagging Summary

Date	Tags	Location	Group Size	Sex	Measurements (cm)*			Comments
					1	2	3	
9/28/79	Freezebrand #1 Rototag, Yellow #5 V-notch, midfin	Keahole Point	150	Male	170	28.3	71.5	Resighted twice: (1) reported from Mahukona, 11/01/79, and (2) seen in Kealake'akua Bay, 11/14/79. Marks visible each time.
10/25/79	Radiotag #1 (RT-1)	Wawahiwaa Point	100	Female	183	34.0	85.0	Tracked from 10:05 until 15:40 on 10/26, when the signal was lost. The package was loose on the fin in photos near the end of tracking.
11/01/79	Radiotag #2 (RT-2)	Hoona Bay	30–40	Female	177	34.5	79.5	Tracked until morning of 11/06, when signal became weak and was lost. Seen in Kealake'akua Bay on 11/14; one of two mounting bolts was missing, but rototag was intact.

TABLE 1 (continued)

Date	Tags	Location	Group Size	Sex	Measurements (cm)*			Comments
					1	2	3	
4/26/80	Radiotag #3 (RT-3)	Kahaluu	30	Male	172.5	N/A	N/A	Tracked from 10:44 until 18:30 on 4/27, when the signal was lost after becoming weak and irregular. Resighted in Kealake'akua Bay: (1) 4/28 (antenna missing), (2) 4/29, (3) 4/30, and (4) 5/01. RT-3 was consistently in groups of 30–100. It appeared healthy, and the package seemed intact and well mounted, except for the broken antenna.

*1 = tip of rostrum to fluke notch; 2 = tip of rostrum to center of eye; 3 = tip of rostrum to leading edge of dorsal fin. All measurements are straight-line.

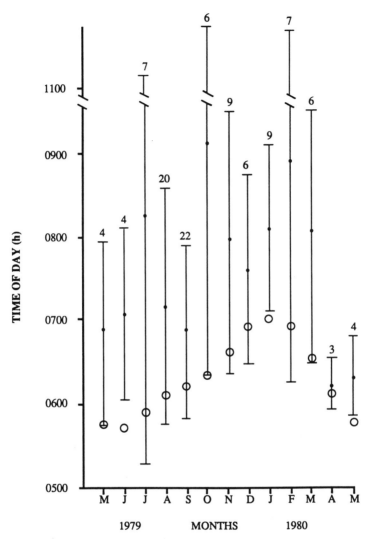

Figure 25. Average time of day of dolphin entry into Kealake'akua Bay, by month, relative to sunrise; circles represent sunrise.

school. The school then spent most of its time below water and surfaced almost surreptitiously before diving again. This overall pattern is what we and Norris and Dohl (1980a) have termed *descent into rest* (figs. 1, 26, and 27), which was accompanied by a change from predominantly acoustical cueing of school behavior to use of the visual mode.

Resting spinners move slowly back and forth, usually for about 4–5 hr. They typically swim deep in bays (fig. 28) or over more exposed sandy patches along open shorelines. As our radiotracks showed, they

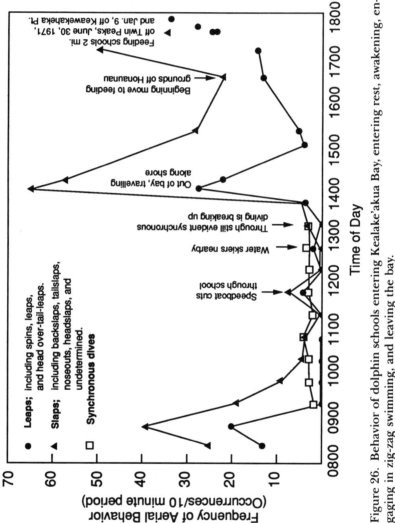

Figure 26. Behavior of dolphin schools entering Kealake'akua Bay, entering rest, awakening, engaging in zig-zag swimming, and leaving the bay.

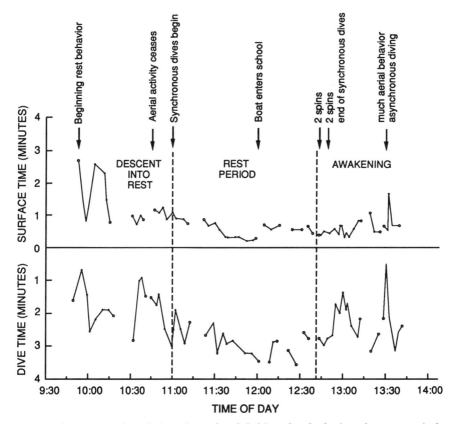

Figure 27. Dive and surfacing times for dolphin schools during the rest period.

also move along the coast as part of larger schools in open water near-shore (figs. 29–31). The average school size entering the bay showed no clear relation to month of entry (fig. 32).

Our Kealake'akua Bay base camp proved to be a fortuitous choice for these observations because the bay seemed to hold at most two coherent schools, each of which was independent of the other and had a maximum of about 40 animals. These schools were usually in a single behavioral state at any given time. In contrast, at locations such as Hoona Bay, where more than 250 animals regularly gathered, schools were typically in different behavioral states at any instant and the basic patterns were harder to sort out.

In the afternoon, the dolphins aroused abruptly and began a new pattern. It consisted of sorties of greatly fluctuating speed, back and forth, in a pattern we have termed *zig-zag swimming*. These movements were sometimes confined wholly to a bay, or at other locations where a deep rest cove was not present, they took the school back and forth close

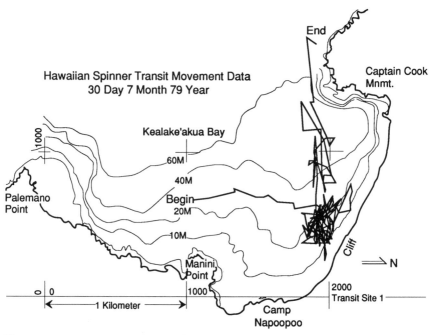

Figure 28. Movement pattern of a 40–60 spinner school, July 30, 1979, in Kealake'akua Bay, generated by theodolite tracking.

along an open coast. In this back-and-forth pattern, the school edged offshore. Both the average time dolphin schools spend in Kealake'akua Bay and the latest time of exit from the bay were related to sunset and daylength (figs. 33 and 34). Exit from the bay for the last occupying schools varied from about 2 hours before sunset to 1 hour after sunset, although many schools did not stay until late afternoon but rather moved out even before noon. Such short occupancy was usually correlated with a school of dolphins already occupying the bay (fig. 33). The speed of travel and dive duration of dolphin schools in the bay varied from slowest and shortest during the rest period, to an oscillation between slow and rapid swimming as dolphins left the bay, to rapid concerted swimming toward the offshore feeding grounds late in the day (figs. 35, 36, and 37). Dive durations were shortest during the rest period and tended to increase toward afternoon (fig. 38). As the dolphins left the shallow rest area and began their traverse to sea, the water deepened under them (fig. 39).

Once relatively deep water was reached, the school fragmented into subgroups that quite abruptly dispersed over a large area of water. These subgroups sometimes became so widely separated that most moved out of sight of the observer. Shortly thereafter, the school seemed

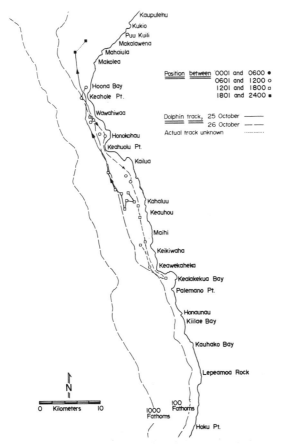

Figure 29. Movements of radiotagged spinner
RT-1 along the Kona coast of Hawaii, October 25–
26, 1979.

to reassemble again because a little later, as darkness fell, large feeding
schools were sometimes observed moving slowly off the coast. These
schools often seemed too large to be a single rest school and must have
represented a coalescence of schools. In the last light, one could sometimes see the black shapes of leaping subgroups silhouetted against the
fading western sky. These animals dove and surfaced subsynchronously
within the school envelope over the island slope about 200–2000 m
(100–1000 fathoms) below (Table 1). During the night, the now coalesced school moved back and forth along the coast, diving and feeding.
Toward dawn, it began to edge closer to shore for another day.

    With these overall events of a spinner dolphin's day in mind, let us
now look at the individual elements in more detail.

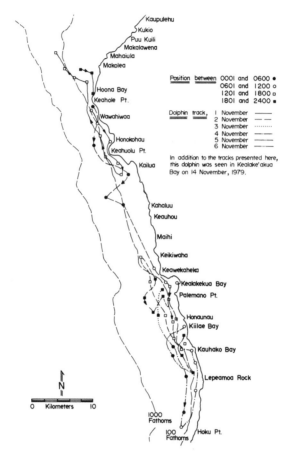

Figure 30. Movements of radiotagged spinner RT-2 along the Kona coast of Hawaii, November 1–6, 1979.

BAY RESIDENCE PATTERNS

Taking all our observations together, we found a significant relationship between sunrise and the time of entrance to Kealake'akua Bay, which averaged 1 hr and 12 min after sunrise (table 1 and fig. 25). During the fall and winter months (October to March), dolphins entered the bay significantly later in relation to sunrise than during the rest of the year.

There also proved to be a striking seasonal change in the amount of time spent per day in Kealake'akua Bay. Only 4–5 hr per day were spent in the bay in winter, and 7–9 hr per day in spring (fig. 34). In spring, the dolphins tended to enter early and leave late. In summer, on average

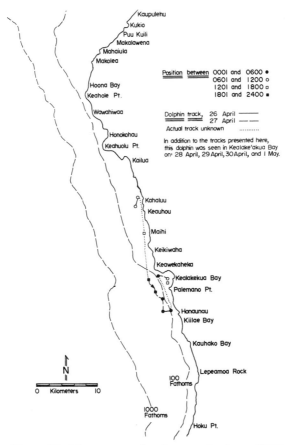

Figure 31. Movements of radiotagged spinner RT-3 along the Kona coast of Hawaii, April 26–May 1, 1980.

they entered not quite as early and they left quite early. May was the month of longest average daily residence in the bay.

The mean arrival and departure pattern is clearly seasonal and generally tracks the annual cycle of day length, although the longest residence was found in May, slightly before the longest days. But the relationship is not precise, as arrival and departure times for any given month are highly variable. At least two major classes of activity take place during bay residence: rest and active swimming–socializing. Rest may be the primary reason for residence in bays, and its duration of 4–5 hr coincides approximately with the minimum residence time (fig. 34).

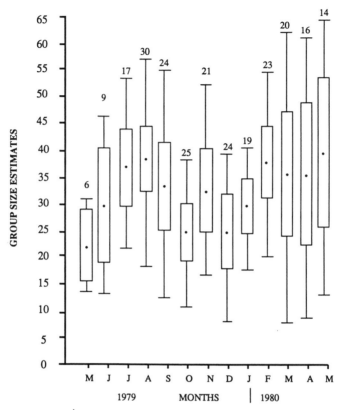

Figure 32. Average school size of spinner dolphins entering Kealake'akua Bay, by month.

During long days, several daylight hours intervene between the end of rest and the first possible time for feeding. This apparently happens because spinner dolphins feed on vertical migrant or scattering layer organisms that rise from deep water near dusk and on benthic organisms that may be hidden during the day and that are deep below the surface when the sun is high. Since movement toward the feeding grounds does not take place until near sunset, the ease of food capture may be causal.

The hours that intervene between the end of rest and departure for feeding grounds seem to be used in more than one way. The alert dolphins sometimes move away from the shallow sand bottom over which they have swum during rest and enter a period of socialization, still within the confines of the bay. Or these activities may take them beyond the confines of the bay and along the coast. This movement may continue until dusk, whereupon they angle offshore toward deep water.

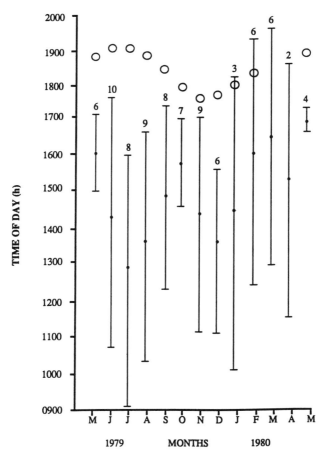

Figure 33. Average time of day of dolphin exit from Kealake'akua Bay, by month; circles represent sunset.

Another probably source of scatter in our data exists. Although the rest period (which is described next) is easy to see in moderate-sized schools and is of rather uniform length throughout the seasons, we sometimes could not determine its length with precision if the school was large. The various subgroups of the school were often in different stages of alertness. A final complication was that not all schools entering Kealake'akua Bay do so from directly offshore. Some make landfalls elsewhere along the coast and then travel an unknown distance along shore before arrival at the bay. Thus, events offshore seem to influence the time of entry and thereby residence time. From time to time, we sighted such schools rounding the far points of land of Kealake'akua as they entered the bay, rather than swimming in directly from the open sea.

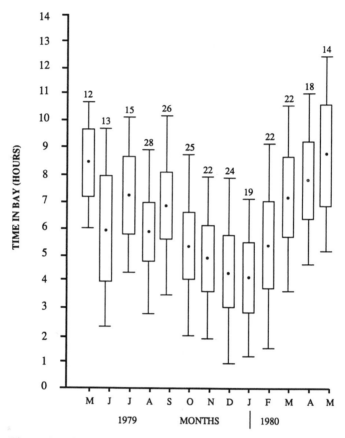

Figure 34. Average amount of time dolphins spent per day in
Kealake'akua Bay, by month.

    If food is more available close to the shoreline during one season than
another, less time may be required for hunting, leaving more time available for longer stays onshore. This may be true because the peak length
of time in the bay (occurring in May) seems to correspond to peak recruitment of reef fishes and the fishes and squids that feed on these lar-
val forms (Philip Lobel, Woods Hole Oceanographic Institution, pers.
comm.). This could mean that dolphins need to spend less time outside
the bay feeding than at other times. Unfortunately, we know nothing
about possible seasonal fluctuations in the diet of spinner dolphins.

DESCENT INTO REST

In the morning, once a spinner school passed in between the outlying
points of Kealake'akua Bay, it began to slow and contract (fig. 37), pro-

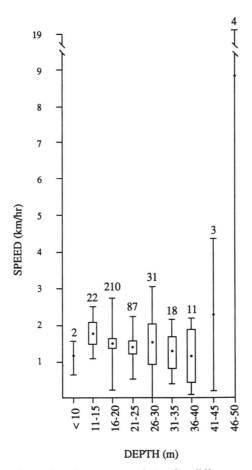

Figure 35. Average travel speed at different water depths, for aerially inactive schools, Kealake'akua Bay.

cesses that could consume 2 hr or more, before it entered the behavioral state we called *rest*. The process seems to include a suppression of echolocation, until in resting schools, movement appears to be largely oriented by vision. Norris and Dohl (1980a, pp. 839–840), describe the descent into rest this way:

Typically arriving schools . . . swam resolutely, with considerable aerial behavior. Little time was spent below the surface. Dives were brief [our fig. 26]. Once such a school arrived at the back of the bay, under the lava cliffs . . . the various classes of aerial behavior slowly disappeared in a rough graded series; first the most athletic patterns such as spins and tail-over-head leaps, then head- and backslaps, then tailslaps disappeared, and finally all but an occasional noseout was gone.

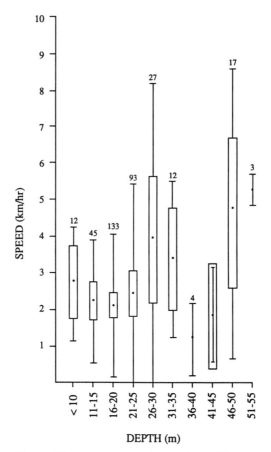

Figure 36. Average travel speed at different water depths of aerially active schools, Kealake'akua Bay.

School size and shape gradually changed at the same time. The ranked school shifted into a sub-discoidal shape and tightened markedly. For instance, a school of 30 animals that once [in deep water] formed a rank over 75 m of water might become concentrated in a 20-m diameter disc. Movements became leisurely; in fact surface excursions of the schools became almost surreptitious as the animals rose quietly from the depths, breathed once or twice, and descended again. It became very easy to overlook them, and two or three observers were needed to produce a complete record of their dive sequences.

Concurrent with these patterns, the school ceased producing its normal cascades of click trains and became all but silent (see chap. 8) as it

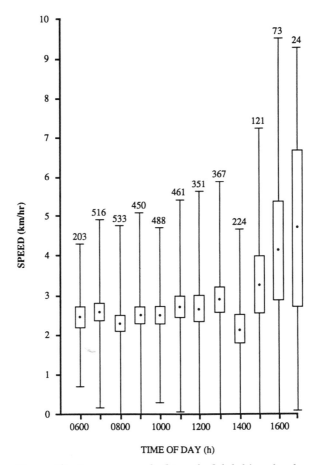

Figure 37. Average speed of travel of dolphin schools, by time of day.

came to depend upon the visual mediation of school movements. The school tightening we always observed seems to be a clear indication of this transition. This relationship was highlighted by the occasional observation that when the water in Kealake'akua Bay was unusually dirty, entering schools sometimes left instead of settling into rest, as if the locality did not provide conditions adequate for visual modulation of school activity.

REST

No pattern during the dolphin day is more distinct than rest. From a vantage point on shore, we were able to watch resting schools for hours

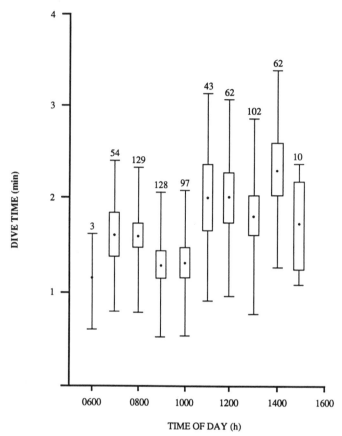

Figure 38. Average dive durations, by time of day, as determined by theodolite tracking in Kealake'akua Bay.

as they slowly made their way back and forth over a patch of white coral sand located in relatively shallow water deep in the curve of Kealake'akua Bay. Such schools, if composed of less than about 40 animals, moved as a single unit, with animals tightly arrayed but without any evidence of the exuberant socializing that marks other spinner school patterns. Instead, they swam close together but spaced just out of contact with one another. Resting schools proved to be quietly shy of intruders. If a swimmer or boater approached, typically the school edged slowly away. Their dives became longer and their surfacings brief (figs. 26, 38, and 39).

Underwater patterns were also quiet ones. Norris and Dohl (1980*a*, p. 841) describe resting spinner dolphins from their underwater observation vehicle as follows:

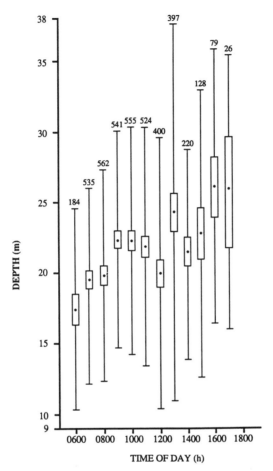

Figure 39. Average water depth over which dolphins traveled, by time of day.

Tight uniformly oriented groups of animals dove slowly, with measured tail beats, toward the sand bottom below; leveled out a few body lengths above the bottom and moved slowly along, schooled tightly; and swam largely without exploratory movements. Occasionally an animal descended to the bottom and beat boils of sand up into the water with its flukes. At the end of a dive the animals rose rather steeply to the surface, not as a single tightly integrated group, but more or less seriatim, as a column of sub-groups. Often after rising, the animals spread outward from this rising column a short distance before turning to define the compact confines of the surface school, like the petals of a flower opening. Once on the surface, group structure could be seen, but the animals seemed much more regu-

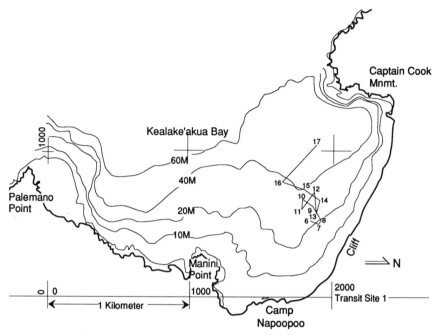

Figure 40. Cumulative average position of dolphin schools per hour, Kealake'akua Bay.

larly spaced than is the case in active schools. Diving, too, was steeper and slower than in traveling schools.

Resting spinner dolphins observed underwater often descended to swim just above the bottom, compressing the vertical aspect of the school into a sheet of animals, which we call a *carpet formation*.

Resting dolphins restrict their movements almost entirely to traverses over open sand bottom. Schools were watched underwater as they approached the dark coral bottom adjacent to the sand patch over which they swam, only to turn back over the sand at its margin. This correspondence was complete enough that we came to feel that spinner dolphins probably restrict rest periods along the shore solely to locations of patches of open sand bottom sufficiently large to support rest patterns. Wherever we have observed spinners, in the Hawaiian Islands and elsewhere, this relationship has proved to be correct (fig. 40). School size estimates made at all the rest coves along the Kona coast showed that the largest schools were consistently found at the coves with the largest area of sand bottom (fig. 41).

The close correlation of swimming during rest with a sandy bottom is shown in figures 28 and 40, in which a 40–60-member dolphin school

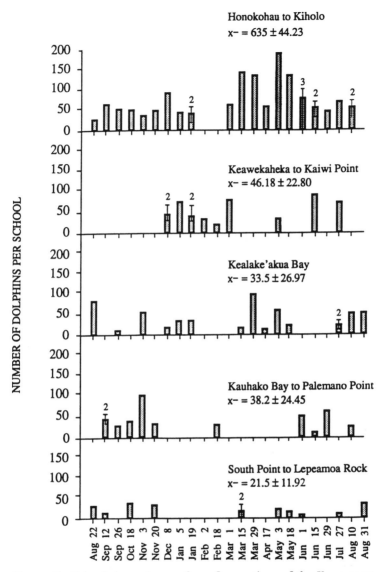

Figure 41. School size estimates from five sections of the Kona coast.

was tracked by theodolite from morning entry into the bay, through the rest period, and then to its exit from the bay in the afternoon. The entire rest period occurred over sandy bottom in 10–20 m (32–64 ft) depth. In this record, one can note the relatively rapid and direct entry of the dolphin school into the bay and its movement toward the deepest recesses of the bay, followed by its protracted period of slow swimming over a discrete area of the back bay. This location coincides with the occurrence of white sand bottom, the remainder of the shallower parts of the bay being cluttered with dark coral formations. As observed during radiotracking, not all spinners living along a given coast come daily into these rest areas. An unknown number stay offshore during the day.

The swimming patterns of resting schools are strikingly undolphin-like in that individuality is strongly suppressed in favor of group action. In fact, the way such resting schools swim and avoid obstacles is reminiscent of the behavior of schooling fish. For example, Norris and Dohl (1980a, p. 841) describe the movement of a resting school as follows:

> When a resting school cruised inshore of us near the cliffs [of Kealake'akua Bay], we waited in a quietly rocking skiff some 75 m offshore, and the school approached slowly as a discoidal group, thinned as it reached a point directly inshore of us, streamed between the skiff and the cliff as a long line of quietly moving animals, and reformed its discoidal group once past us. We found that our skiff or our anchored workboat could deform such discoidal groups from some distance, causing the side nearest the skiff to become dented or malformed as the entire school reacted to our presence. When a four-hydrophone array capable of sound triangulation was placed near the path of such resting schools, it was assiduously avoided and no animals were known to pass through it for 6 days after its placement (Watkins and Schevill 1974). A line stretched across the surface of the water was capable of deflecting such schools. In such cases, even though the animals moved slowly and other evidence of alertness, such as complex phonation or aerial behavior, was nearly absent, the school as a whole remained alert.

Our working hypothesis for the functioning of the spinner dolphin rest period is that it represents a time during which dolphins nearly suppress their acoustic orientation systems and come to rely on vision. This produces a series of predictable changes in their school structure and behavior. First, visual mediation requires that animals swim closer together than does the acoustic mode, so the schools tighten. Second, the change of modalities probably makes the dolphins more susceptible to shark attack. These predators can lurk in dark coral formations below and rush at dolphins passing overhead with some chance of success (see chap. 15). By restricting movement to areas of white coral sand and to times of clear water, we propose that the dolphins can use vision to pick out sharks against the light bottom, just as a human observer can.

Another feature of theodolite tracks of resting dolphins is the constantly changing heading of their schools (figs. 28 and 40). They not only move back and forth over the sand but are found to face in a different direction each time they surface. This, too, may be related to environmental surveillance.

In addition to the normal functions of sleep, it is possible that this daily period of relative silence is needed by dolphins for the maintenance of phonation structures that produce thousands upon thousands of intense yet precise sound impulses every day. Some of these are so intense that the question arises, how can living structures generate and propagate them without damage to themselves? (see Murchison 1980, Norris and Møhl 1983).

In the absence of an acoustic early warning system during the rest period, dolphins may have to rely heavily upon what Norris and Schilt (1988) have termed a *sensory integration system* (or SIS) as the major means of protection from predators. This system (described in chap. 13) is thought to be used in some form by perhaps all polarized animal groups. It is proposed to operate as a communications matrix across which information of various sorts may be passed, to be decayed or potentiated, depending upon message content. It is thought to allow the members of a school to gain an advantage in reaction time over a single predator.

AWAKENING

The termination of rest is typically abrupt. To a nearby observer, it is marked by a sharp increase in aerial behavior (figs. 26 and 27). Typically, within a few minutes, the quietly moving rest school becomes punctuated by one or more animals nosing out and then by one or two making desultory leaps. Soon more active patterns such as spinning occur across the dimensions of the school. The school then speeds up and at the same time begins to exhibit better defined subgroups. Residence times in the bay are longer for larger schools, possibly reflecting a heterogeneity of behavioral states in larger groups (fig. 42). Average school size varies little by month, and changes probably reflect more immediate fusion and separation of schools (fig. 32).

At this point, the behavior of the partially alert school now begins to merge into zig-zag swimming.

ZIG-ZAG SWIMMING

Zig-Zag Swimming, which follows the rest period, consists of a period of variable duration during which the school, now moving over considerable stretches of water, shows great changes in the speed and direction of movement. For example, in Kealake'akua Bay, a newly awakened school may begin to move out of the bay. Then, after several hundred meters of

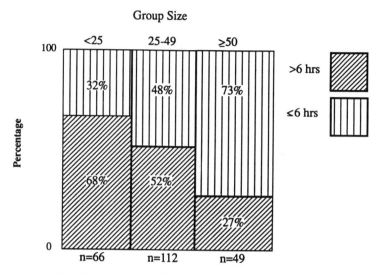

Figure 42. The relation of school size and residence time in Kealake'akua Bay.

TABLE 2   Average Speed, Water Depth, and Time of Day Dolphin Schools Entered Inner Kealake'akua Bay, Moved Within the Bay, and Exited

	Speed (km/hr)	Depth (m)	Time of Day
Enter bay	$\bar{x} = 4.9$	30.9	0730
	s.d. = 3.0	27.09	1 hr 44 min
	$n = 10$	10	106
Inside bay	$\bar{x} = 2.6$	21.3	—
	s.d. = 2.68	8.95	—
	$n = 3811$	4151	—
Exit bay	$\bar{x} = 5.9$	39.4	1442
	s.d. = 4.00	33.36	2 hr 58 min
	$n = 18$	18	85

travel, often with many animals engaged in high energy aerial behavior, school speed reduces, aerial behavior subsides, and the school may begin to mill. These features are reflected in the great afternoon change in the incidence of aerial behavior (see table on page 106).

As the dolphins slow, they may turn and make their way back into the bay again, sometimes completely back to the rest area or else into small coves along the bay margin where they mill quietly. Then, as suddenly as before, aerial behavior increases in frequency and the school begins another sortie toward the bay mouth. It is often striking that zig-zagging dolphins first face toward their destination at sea and then backward toward their resting grounds. In other words, they are oscillating both toward the new behavioral state and destination and back toward the old behavioral state of quiet swimming and its location deep in the bay.

This oscillation often continues, with school swimming speed gradually increasing on the outward legs. Near dusk (and often directly toward the setting sun on the Kealake'akua coast), swimming velocity (as measured by theodolite) approaches the crossover speed, when it is energetically cheaper for a dolphin to leap than swim beneath the surface. This speed is about 7 knots (Au and Weihs 1980, Hui 1992) (also see chap. 10). The dolphin's fins begin to trail pennants of water as they lunge forward, and then they leave the water altogether, with animals leaping free as the school races along (see chap. 5, section on Aerial Behavior). Their outward course is often a direct one out to sea, with all animals moving very swiftly in beautiful concerted leaping locomotion, arcing completing out of the water. Conversely, they may slow down again and turn back toward the bay.

If zig-zag swimming begins early in the day, the school often leaves the rest cove and travels along the shore for a time before edging out to sea. Where rest coves are shallow indentations close to deep water, as near Keahole Point, zig-zagging dolphins have been followed as they swam back and forth for a few kilometers, moving close along the open coast before they angled offshore toward the open sea. This daily cycle of events is reflected in dive times that occur from bay entry to departure (figs. 27 and 38). At entry, dives are short as the dolphins move rapidly at the surface. The dives then lengthen during rest, only to shorten again upon wakening and the beginning of zig-zag swimming.

We suggest that zig-zag swimming is a period of social facilitation that ensures behavioral synchrony of all school members before the school enters the deep sea. This is akin to the suggested meaning of the whistling choruses of African wild dogs (Estes and Goddard 1967), which are thought to ready packs for the hunt, and to the choruses of jackdaws (Lorenz 1952), which are thought to represent social decision making about roost selection for the night. We propose that zig-zag swimming is a group process that operates without unitary school leadership. In this way, the school as a whole seems to test its integrity as an alert unit capable of functioning in deep water. In fact, described later (in chap. 14) with regard to bout behavior, any decision making that affects the whole school may operate in such an oscillatory fashion.

Sounds, too, oscillate during zig-zag swimming. Resting animals are singularly quiet, emitting only sporadic low-level clicks and occasional whistles. But during zig-zagging, all of these sounds rise and fall from crescendos of sounds during rapid locomotion to periods of near silence when the school stops (chap. 8). The vocal periods include oscillatory clusters of whistles, barks, and showers of clicks, until a maximum is reached in the last sortie as the dolphins finally make their high activity dash out to sea.

An experimental study of whistle emission and chorusing by the closely related common dolphin (*Delphinus delphis*) (Caldwell and Caldwell 1968) may be related to the social facilitation function we propose for spinner zig-zag swimming. In common dolphins, the whistle choruses between two animals and their propagation to other animals within the small experimental group were found to be related to the timing of whistle initiation. If one animal whistled and the second responded within 0.7–0.8 sec, the first animal would continue whistling and the two would chorus. If the response of the second animal to the first was too rapid or too slow, the first animal fell silent.

If our spinners are responding like Caldwell's *Delphinus*, whistle choruses during zig-zag swimming could well be the means by which schoolmates in spinner schools test one another's alertness. The unanimity of a chorus may be an indication of readiness for offshore activity, possibly both with regard to cooperative feeding and protection from predators.

Another possibility is that such testing of one animal by another may relate to the sound generation and assessment systems themselves. The dolphins may be literally "tuning up" by testing the functioning of their echolocation systems against one another after a period of silence. It is highly likely that the extremely precise timing requirements placed on sound emission and its control by the task of feeding are central to coordinated feeding by the school. The location and assessment of small prey swimming in water of almost the same acoustic impedance, and the discrimination of predators from the bottom below, require very precise control, as does the operation of the school as a protective and social unit. Testing these ideas will require considerable ingenuity.

SPREAD FORMATION

The preliminary study of Norris and Dohl (1980*a*) did not recognize the spread formation, a distinctive part of the diurnal sequence. We were able to observe it only a few times in this work. The behavior occurs in a narrow time period just at dusk before the last chance for visual observation disappears with the onset of darkness.

As zig-zag swimming ends and the dolphins speed up and begin to leap free of the water, the school becomes oriented sharply offshore. This dash is typified by much aerial behavior, especially such directional patterns as head and back slaps and salmon leaps. The travel is so resolute and so rapid that an observer has to be very alert and ready to travel or the animals will be out of sight in a few moments. As the water deepens beneath the dolphins, which now travel in a swiftly moving column or rank, they may stop suddenly, and within moments, the subgroups of the school disperse in many directions. These subgroups, which sometimes include isolated adults with young, quickly spread over many square kilometers of sea. The observer, who had moments before been traveling with a coherent school may look up in bewilderment, wondering where the school went. Dolphins have gone in all directions, and they travel far enough that not all can be seen from a single location.

The function of spread formation is not wholly obvious, but we do know that it occurs near dusk when the dolphins are presumably locating food sources and just before they aggregate again into coherent feeding schools. By means of such spreading, a rapid assessment of local abundance of food could be made, with animals later coalescing upon that source and then reforming as a feeding school. Coalescence in the black water surely requires use of acoustic signals, and they are abundant. Sound emission level is high, especially whistles, which we later (in chap. 8) describe as *phatic* or *place signals* (Jakobsen 1960). We think these allow dolphins swimming out of sight of one another to assess the location and state of schoolmates and to synchronize the actions of a school.

NIGHTTIME PATTERNS

While following dolphin schools offshore until dusk provided much information about events nearshore, radiotracking was the only effective means of keeping track of individuals or schools after dark and continuously for several days. Occasionally, we were able to see feeding schools at dusk and thus obtain some idea of their deployment and diving patterns. However, what we know about them after dark stems mainly from following individuals by radio.

What we learned was that dolphin schools from the various rest areas move a rather short distance offshore and then traverse back and forth along the island slope all night, edging toward shore as dawn approaches. The changing size of schools containing our tagged animals indicated that offshore schools could be formed by a coalescence of schools into sometimes large nighttime aggregations. It was also clear from these radiotracks that some dolphins stayed offshore during the day and did not enter the rest areas at all. Spread formation seemed to be a fairly common feature of these schools, as did the more coherent subsynchronously diving feeding school. To demonstrate these points, we describe here the tracks of three marked or radiotagged animals and their asso-

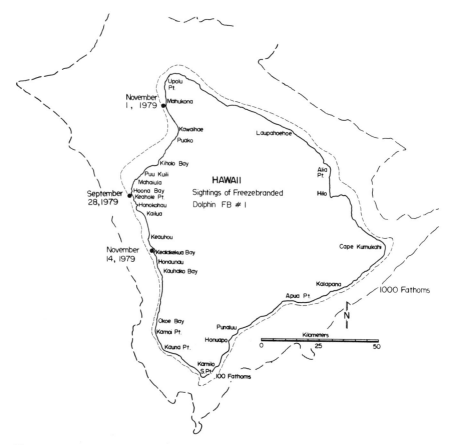

Figure 43. Movements of marked spinner dolphin FB-1 along the Kona coast of Hawaii.

ciates in some detail. The reader can follow their course by reference to the map in figure 9 and to the individual tracks shown in figures 29–31 and the plot of water depth, time of day, and distance from shore as shown in fig. 39.

Four dolphins were captured for such tracking. The first of these, which tested capture methods and release behavior, did not receive a radiotransmitter but was equipped with a visual tag and freeze brand markings on both sides of the dorsal fin. This dolphin, designated FB-1 (for freeze brand number 1), was seen 34 days after tagging in a group of 100–150 spinner dolphins far to the north of its tagging site, off the Kohala coast between Kawaihae and Upolu Point (fig. 43). This is the northernmost sighting of any of our identifiable animals, including those identified by scars and marks. Two weeks later it was resighted in

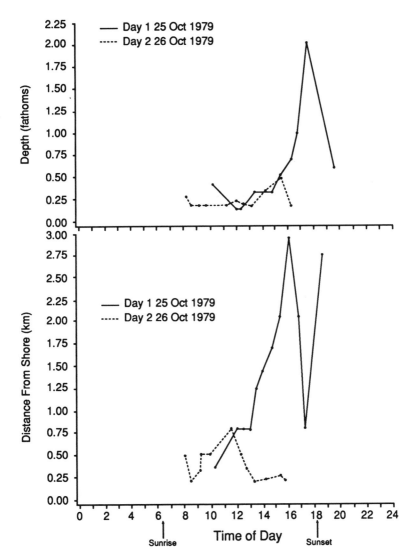

Figure 44. Movements of RT-1 relative to water depth, time of day, and distance from shore.

Kealake'akua Bay. On both occasions its dorsal fin rototag and freeze brands were clearly visible (see Appendix B).

Three dolphins were radiotagged (figs. 44, 45, and 46). None of the tags lived up to our expectations, although much information was gathered during their relatively short periods of functioning. The radios transmitted for 1.3, 6, and 1.5 days, respectively. Although the small size

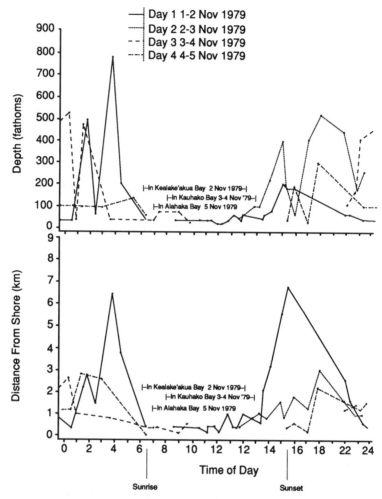

Figure 45. Movements of RT-2 relative to water depth, time of day, and distance from shore.

of these transmitter packages relative to those we have used previously (Würsig 1976, Irvine et al. 1981, Norris and Dohl 1980a) was no doubt advantageous in terms of reduced drag in the water, the system suffered from short range.

All three dolphins moved out over the relatively deep island slope with large schools of variable size and moved back and forth along the shore in the dark, sometimes over moderately deep water. Surprising distances were sometimes covered during these movements; in composite, this involved essentially the entire lee coast of the island of Hawaii.

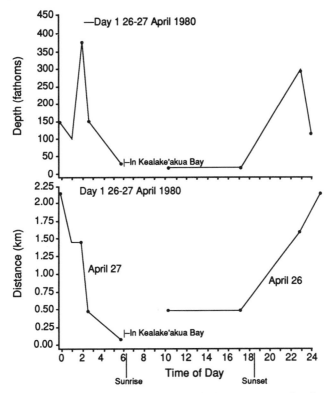

Figure 46. Movements of RT-3 relative to water depth, time of day, and distance from shore.

After a night of such back-and-forth travel, landfalls were made as the animals edged toward shore near dawn. The landfall chosen appeared to relate more to the vagaries of nighttime movement than to a directed choice toward a given rest area or cove.[5]

Radiotagged dolphin number 1 (RT-1), a female captured in the morning off Wawahiwaa on October 25, 1979, quickly rejoined her school following release at 10:05 as it headed steadily southward (fig. 29). The school of about 100 animals, which contained the naturally marked dolphin named Four-Nip, milled near the mooring area at the port of Kailua. The school was at first quite scattered, but it then coalesced as it headed southward in a slow zig-zag pattern at 11:45. By 12:30 the level of aerial activity in the school began to increase. This was followed by a general movement offshore from Kahaluu. At 16:24 the

5. Depths are given in the normal nautical *fathom* as taken from the charts we used, which equals 6 ft or 1.8m.

school turned and began traveling rapidly northward approximately over the 100-fathom (183-m) contour. At Honokohau Harbor, the school headed offshore again, swimming rapidly and with much aerial activity. Following a crew change, we tracked RT-1 to about 9 km (5.6 mi) off Makalawena at 19:25 where we lost the signal and returned to shore because of building seas.

The animal was relocated at 08:06 on October 26, in a very active school of about 80 animals headed south toward Hoona Bay. All age classes of dolphins were seen in the school. By 10:52 the school was headed south through Kailua Bay. The school continued moving steadily south nearshore, passing Keauhou at 12:20. They soon reached Keikiwaha Point at the northern cusp of Kealake'akua Bay, where they milled briefly. At 14:46 they entered the bay where another group of spinners, including the naturally marked individual named Finger Dorsal, had spent much of the day. All of the dolphins moved actively out of the bay together, moving steadily northward parallel to shore. At 16:02 they all milled north of Keikiwaha Point, and at 16:35 the school moved due magnetic west toward the setting sun. After we lost the signal from the tagged animal, we followed the school visually to about the 100-fathom contour, where they went into spread formation. They then coalesced, following the contour, and moved northward along the coast.

Thus, this animal and its school had moved more than 30 km along the coast in less than a 24-hr period and had moved offshore, presumably to feed, at two widely separated points. Its school did not occupy an onshore rest area during the day, although it slowed for a time and appeared to engage in rest, and its school coalesced with another school in the process of going to sea toward dusk.

The second radiotagged dolphin (RT-2) was followed for 6 days. A female, she was captured at 08:50 on November 1, 1979, in Hoona Bay. She spent the first day and night in the northern part of the study area, but then moved south of Kealake'akua Bay where she spent the last four days of her traverse. In the daytime, she and the school(s) with which she was associated stayed close to shore (but not in rest coves) and at night moved into deeper water. This pattern is shown in figure 30, which shows that only 9 of the 29 daytime positions (31%) were outside the 100-fathom contour but that 14 of 20 (70%) nighttime positions were in deeper water.

Following release at 09:04, the dolphin rejoined its school as it moved rapidly southward near shore. The school slowed as it reached Honokohau Harbor and then continued slowly into Kailua Bay where it milled with scattered subgroups (now totaling about 60 dolphins) until 13:05 when the dolphins moved slightly farther offshore. At about 13:30, the school retraced its path, and by 15:35 its broad rank formation was headed west of Keahole Point. As the school reached the 100-fathom contour it went into a spread formation of scattered subgroups. By 17:00, 9 km (5.6 mi) west of Makalawena, the school reached its farthest distance from shore. By this time, it had coalesced and showed much aerial activity among the now closely knit subgroups that dove and surfaced in subsynchronous fashion.

At this point, we returned to Honokohau Harbor for a crew change. We found the school again off Makolea. It moved south off the coast throughout the night. By November 3, 1979, RT-2 was in a school of 60–80 dolphins, and at 07:00 they entered Kealake'akua Bay. The school slipped quickly into a rest pattern and moved slowly over the rest area until 13:30 when, with a flurry of aerial activity, they left the bay, moving southward. By 14:45 they were over the 100-fathom curve off Honaunau, still headed south in scattered subgroups. They turned northward again in the late afternoon until they were off Kealake'akua Bay and then headed southward in the dark. By 22:00 they were off Honaunau, again in 400 fathoms (730 m) of water. At this time, the school appeared to be quite scattered. Hydrophone recordings produced abundant click trains and whistles. Our tracking vessel, Nai'a, returned to Kealake'akua at 24:00.

We found from an aerial survey plane in Kauhako Bay the dolphin again at 12:47 the next afternoon and followed it to the south of our field station at Kealake'akua. By 22:00 it had returned offshore of Kealake'akua Bay and was engaged in long dives at the same place we had seen it the night before. The signal strength of our radio was declining until we had to approach within about 1.5 km (1 mi) to receive it. At this point, the dolphin began a long traverse to the south that took it 9 km (5.6 mi) north of Milolii by 03:40 on November 4, 1979, before turning north again. This point is only a short distance from the southern tip of the island of Hawaii. Therefore, when we consider the movements of the first tagged animal with this one, essentially the entire lee coast of the island was traveled during these offshore sorties.

RT-2 moved northward on November 4, 1979, now in the company of two well-known individuals, Four-Nip and Finger Dorsal. About half this school entered Kauhako Bay at 8:32, while the remainder milled offshore about 1 km (0.6 mi) to the south. They coalesced offshore again at 10:22, and most of the animals moved northward, while RT-2 and her associates milled in Kauhako Bay for much of the day.

The activities of RT-2 were concentrated in the southern study area for the remainder of November 5, and on November 6, the dolphin entered Kealake'akua Bay in a school with the original tagged and branded dolphin FB-1. RT-2 still had the radio transmitter on her dorsal fin, but one of the mounting bolts was missing (they are designed to dissolve). The dolphin was not located again.

Our third radiotagged animal, RT-3, a male, was captured from a school of about 35 animals off Kahaluu, just north of our base camp, on April 26, 1980. It first moved offshore (fig. 31), and then on the morning of April 27, it came into Kealake'akua Bay with a school of about 50 dolphins that included calves and what we now recognize as probable adult male subgroups. The school briefly left the bay at about 10:00 and returned again at 12:00 to spend the day in the bay; they left at 18:27 that evening, moving northward, whereupon the signal was lost.

Three visual sightings followed. First, RT-3 was again seen in Kealake'akua Bay in a school of about 50 dolphins; we could see that the radio was in place but missing its antenna. The second was on April 29, also in Kealake'akua Bay, but this time with about 80 dolphins, and the third was on May 1 with an estimated 80–100 animals. In addition, dolphin FB-1 was sighted along 95 km (60 mi) of northern Kona, or the lee coast of Hawaii, while dolphin RT-2 traveled 63 km (40 mi) of coast (fig. 43).

The patterns of RT-3 were similar to the others. Like the other tagged animals, at no time did it seem an obvious part of a school subgroup. Instead it swam alone farther from its nearest neighbor than were other dolphins, and it did not swim with the same synchrony as the others. Nonetheless, its surfacing times appeared to match those of other schoolmates, as determined from the theodolite tracking. This phenomenon of radiotagged dolphins remaining near but not being an actual part of school subgroups has also been noted for bottlenose dolphins (*Tursiops truncatus*) by Irvine et al. (1982) and for dusky dolphins (*Lagenorhynchus obscurus*) by Würsig (1976). More recent work using extremely small radio tags seems to eliminate the effect (Würsig et al. 1991). This may indicate that the separation results from a tagged animal's inability to swim precisely in the same patterns as other subgroup members.

These radiotracked dolphins showed larger scale daily movements than we expected. However, the short duration of these radiotracking events did not allow matching the extent of movements we recorded over longer periods by use of scars and marks analysis, in which animals not only traversed the Kona coast but circled the island to the Hilo area.

THE QUESTION OF SCHOOL FLUIDITY

A high degree of fluidity in school membership is indicated by the following observations. None of the radiotagged dolphins continued to travel from day to day in schools of stable membership. Instead, the as-

Dolphin #	25 October	26 October
01		x
02	x	
09		x
13		x
45	x	
51		x
68	x	
70	x	x
77	x	
101	x	x
102	x	
107		x
116	x	
119	x	
122		x
123		x
124		x
128		x
179		x

Figure 47. Associations of radiotagged spinner RT-1 with other identified dolphins.

sociates of these radiotagged animals known from scars and marks came and went to such an extent that school fluidity within a large pool of dolphins was clearly the norm (figs. 47, 48, and 49). For instance twenty-eight identifiable dolphins were recognized swimming with RT-3 from April 27 to 30, 1980, but only four of these were seen in the same school with the animal on consecutive days (numbers 61, 69, and 93 on fig. 48). Though the lack of consecutive sightings might be partially a function of our not completely photographing each school, considerable effort was made to photograph all animals at each sighting. Several of the naturally marked dolphins that were seen once and not reidentified were among the most distinctive animals in our catalogue, and it is unlikely that they would have been missed, either by eye or in our review of the hundreds of slides involved. In addition, fluidity was evident because school size changed frequently in the course of all tracks.

Occasionally, we were able to watch schools mix or change composition. For example, R-1's school swam south of Kealake'akua Bay during the late afternoon of October 26, 1979, and met a school already in the bay. The schools merged and left the bay together shortly thereafter, and after moving northward as far as Maihi Bay, headed offshore together. Schools were also seen mixing and splitting during the night when the dolphins were moving along the coast in scattered schools. This nocturnal behavior can explain the changes in composition of schools that we observed during the day in the rest coves.

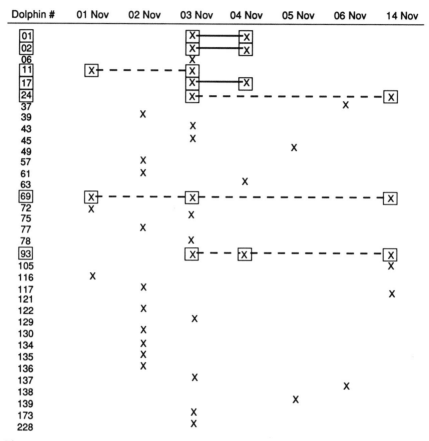

Figure 48. Associations of radiotagged spinner RT-2 with other identified dolphins.

The radiotracks give no hint of the relationship of dolphins to putative populations well offshore in Hawaii where one observer (Wayne Perryman, NOAA, pers. comm. 1980) reports seeing them. We have not seen such populations during a number of traverses of deep offshore Hawaiian waters.

Nonetheless, the radiotracks do fill in many gaps in our understanding of the spinner dolphin's daily cycle and show that the associates of a given dolphin offshore are as fluid as we had perceived them to be inshore (see chap. 6). They challenge our preconceptions of what a spinner dolphin school consists of because the numbers of associates of each tagged animal continually fluctuated. Clearly the participants in spinner schools shift frequently both inshore and offshore. Therefore, is the entire population on the island of Hawaii the effective population unit for

Dolphin #	27 April	28 April	30 April
02		x	
03		x	
06		x	
11		x	
17		x	
24		x	
26		x	
43		x	
49		x	
52	x		
61		[x]————————[x]	
63		x	
69		[x]————————[x]	
74		x	
94		x	
129		x	
145			x
151		x	
153		x	
164		x	
166		x	
167		x	
169		[x]————————[x]	
170		[x]————————[x]	
172		x	
173		x	
174		x	

Figure 49. Associations of radiotagged spinner RT-3 with other identified dolphins.

these dolphins? Or does it possibly involve more animals from still farther offshore? These key questions cannot be answered by our data, although we will allude to these questions again when we discuss our results from natural scars and marks analysis.

The daily movements of the three radiotracked dolphins were quite similar in several respects. Daytime movements were generally made inshore, with animals either occupying a rest cove such as Hoona, Kealake'akua, or Kauhako bay or simply moving slowly north and south along the nearshore coast, often reversing course sometime during the day.

During the afternoon, the animals moved offshore and showed the little observed spread formation before coalescence into more coherent feeding schools. Their movements toward dusk were often close to due magnetic west, precisely toward the setting sun, and they usually went as far as 100–1000 fathoms (183–830 m) depth. At night they tended to move along the coast in about this same depth range. Figures 44, 45, and

46 depict the relationships among time of day, water depth, and distance from shore for the three radiotagged animals.

When using bays, the animals tended to enter them in the morning, often early, and to leave well before sunset. The largest offshore excursions and passages through the greatest depth consistently occur in the evening hours. Dolphins never remained in deep offshore waters continuously throughout the night. Instead, their movements took them back and forth over deep water escarpments, and at times back into shallower inshore waters as they moved along the coast. Thus, these radiotracks, though short and few in number, did tell us much that was new to us, especially about the extent of movements in a single night and the fluidity of the schools a given dolphin occupies.

We move now to describe the aerial behavior of spinner dolphins, including the curious behavior called *spinning*. We show that it is just one of several aerial patterns that help define the activity state of a school.

FIVE

Aerial Behavior

Kenneth S. Norris, Bernd Würsig, and Randall S. Wells

The "trademark" of the spinner dolphin is its aerial behavior. An experienced cetacean naturalist can identify the species about as far away as it can be seen by its remarkable spinning leaps. But the spinner is not the only species of dolphin that performs characteristic leaps. Other species such as the common dolphin (*Delphinus delphis*), the right whale dolphin (*Lissodelphis borealis*), the beaked whales (Family Ziphiidae), the bottlenose dolphin (*Tursiops truncatus*), the rough-tooth dolphin (*Steno bredanensis*), the Pacific white-sided dolphin (*Lagenorhynchus obliquidens*), and the dusky dolphin (*Lagenorhynchus obscurus*) all have aerial patterns characteristic of their group. The mysticete whales also perform leaps, often called "breaches," which also can be characteristic of a given species. The possible function(s) of such cetacean aerial behavior have remained uncertain, although both scientists and laymen alike have made many speculations about them.

Herman Melville in *Moby Dick* (1851, p. 105) may have been speaking of spinner dolphins when he wrote of "the huzzah porpoise," judging by his description of their aerial behavior. He describes it in this way:

> This is the common porpoise found almost all over the globe. The name is of my own bestowal; for there are more than one sort of porpoises, and something must be done to distinguish them. I call them thus, because he swims in hilarious shoals, which upon the broad sea keep tossing themselves to heaven like caps in a Fourth-of-July crowd. Their appearance is generally hailed with delight by the mariner. Full of fine spirits, they invariably come from the breezy billows to windward. They are the lads that always live before the wind.

Mark Twain (1851, p. 392) may have written of them, too, in his chronicle of travels entitled *Roughing It*. His impressions were more om-

inous than Melville's, as he traveled amidst a school of what were probably spinners off Honaunau, less than 10 km south of our Kealake'akua
base camp (spinners are by far the most common dolphin in this inshore
area).

> . . . we dashed boldly into the midst of a school of huge, beastly porpoises
> engaged at the eternal game of arcing over a wave and disappearing, and
> then doing it over again and keeping it up-always circling over, in that way,
> like so many well-submerged wheels. But the porpoises wheeled them-
> selves away and we were thrown upon our own resources.

By the twentieth century, scientists began to think more scientifically.
Why do dolphins carry out aerial behavior? Why is it so ubiquitous? Why
does it seem species specific? The first scientific analysis of spinning was
done by Hester et al. (1963), who photographed spinner dolphins on the
eastern tropical Pacific tuna grounds and traced body postures from a
sequence of moving picture frames recorded during a single leap. The
resulting diagram clearly showed the kinesics, or body movement, of a
spin (see fig. 56). It showed how the dolphin, by twisting its neck and
body while in air, is able to perform as many as about four rotations during a 1.25-sec leap that can take it about 2 m into the air. These authors
tested the idea that spins might be used to dislodge remoras. They concluded that this could not be the complete explanation because spinning
animals often do not carry such ectoparasitic fish. They also rejected
courtship displays as an explanation because spins were performed by
both immature and mature animals of both sexes, a point with which we
concur. Even very small juveniles spin.

CATEGORIES OF AERIAL BEHAVIOR

Norris and Dohl (1980a) analyzed spinner dolphin aerial behavior and
concluded that there were at least seven aerial patterns exhibited by the
species. All but one of these patterns, the arcuate leap (fig. 50), involved
slapping the water upon reentry along with production of a boil of bubbles at the water surface and a plume underwater upon reentry. These
classes of aerial behavior roughly arranged in increasing order of activity
are nose-outs (and fluke-outs), tail slaps, back slaps, head slaps, arcuate
leaps (and salmon leaps), tail-over-head leaps, and spins. Note that the
salmon leap is a pattern seen only in very rapidly moving schools. It is
possible to subdivide aerial behavior further and to erect more classes
than we list here. But we think that such overclassification only serves to
obscure the relationships observed.
    The following analysis of spinning resulted from one cruise devoted
solely to studying the aerial patterns of spinner dolphins, in which an at-

Figure 50. A spinner dolphin making an arcuate leap.

tempt was made to analyze the various behavior patterns themselves, their location of occurrence in schools, and the context of their occurrence. This concerted effort was followed by many separate observations of these aerial patterns at sea and finally by many hours of observation of a captive school in which we could dissect the occurrence of the patterns, the details of their performance, and their collateral effects (sounds produced or effects on the behavior of schoolmates). In time, the form and context of aerial behavior began to fall into predictable patterns, which are presented here.

Using the incidence of these various patterns at any given time, the surface observer can assess the activity state of a school (most simply, the speed of locomotion) with some accuracy. Since most social patterns seen in dolphin schools are related to activity state, the surface observer can make a modest assessment of expected events below the surface by quantifying the aerial patterns. For instance, the various patterns found among socializing dolphins are also ordered in relationship to activity state, as is the level and kind of vocalization.

The daily pattern of movement and school state also correlate closely with the details of aerial activity. Typically, aerial activity was seen with high frequency in open water outside Kealake'akua Bay as dolphins swam into the cove in the morning. Once schools came in past the inner headlands of the bay (Manini Point), aerial activity subsided rapidly. It was seldom seen during the rest period unless some event, such as the passing of a tour boat or a water skier, caused a brief flurry of surface activity. In the late afternoon, the dolphins became active again, and aerial patterns were often abruptly evident, marking the cessation of the rest period with some precision (fig. 26). Once rest was ended, the dolphins generally increased their surface activity until it reached a high point about 10 hr after initial entry into the bay (table 3). Those groups that stayed in the bay for a briefer period than average tended to con-

TABLE 3 Frequency of Aerial Activities Seen in Kealake'akua Bay*

Aerial Activity	Frequency (Animals per hr)	Peaks of Activity
Nose-outs	0.072	Morning
Tail slaps	0.240	Morning/afternoon
Flips	0.240	Morning/afternoon
Head slaps	0.240	Morning/afternoon
Salmon Leaps	0.360	Afternoon
Side and Back slaps	0.480	Afternoon
Spins	0.600	Several peaks throughout the day

*Data are standardized by animals per hour. Most peaks were diurnally bimodal, with one peak in the morning and one in the afternoon. Peaks of activity more than two times as high as other daily peaks were scored for just that high part of the day, while peaks closer together were scored for both morning and afternoon. Arcuate leaps were infrequently seen and are not included.

dense this pattern, but when such records were averaged, they also showed the least amount of aerial activity during the approximate midpoint of their stay.

Our process in selecting the eight patterns described here was first to watch the dolphins for a long period, attempting to gain a "gestalt" of the entire process, and then to let the real modal patterns emerge in our minds, rather than to catalogue immediately how many classes it was possible to erect. This point is not trivial since behavioral literature is rife with long lists of subdivided behavior, often divided so finely that functions are lost in the "noise." For example, this is commonly done by those who create "ethograms" for a given species. At any rate, the following descriptions are of aerial patterns that we saw being produced by many different animals again and again.

NOSE-OUTS AND FLUKE-OUTS

Nose-Outs

The earliest indication of arousal in a resting school is often given by one or two dolphins thrusting their rostra from the water as they move slowly at the surface. We call this a *nose-out*. Sometimes the snout is splashed back against the surface as the animal snaps its head downward. On one occasion, three dolphins were seen slowly "sparring" with nearly vertical rostra. The behavior lasted for an estimated 7 sec.

Nose-outs (along with fluke-outs) are the lowest activity level aerial pattern in our list (fig. 51). They are frequently seen among quietly socializing animals just before or after the rest period. Their appearance at

Figure 51. A spinner dolphin side-slapping in a school in which nose-outs are especially evident. Hookena Bay.

the end of a rest period is often the first sign that the school as a whole is waking.

Fluke-Outs

The *fluke-out* is a very passive pattern that we do not consider as true aerial behavior, but because the pattern is so distinctive, we mention it here. It is sometimes seen in quietly moving rest schools. An occasional animal may literally surface vertically, tail-first, thrusting the tail stock and flukes into the air before subsiding again. Such animals may emerge to about the level of the umbilicus and hold still or wriggle the tail in the air for up to 8 sec.

SLAPS

Tail Slaps

Tail slaps are slaps of the tail flukes made in either the normal or inverted position relative to the water surface (fig. 52). They produce a loud percussive noise that sometimes can be heard in air for many meters away from a school. They are sometimes given in long trains that we have termed *motorboating* because the resultant sound resembles

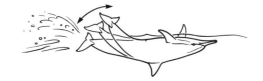

© 1991 Jenny Wardrip

Figure 52. Top, the posture of a spinner dolphin making a tail slap; bottom, a head slap.

the noise of a very slowly moving motorboat. Some of these trains of tail slaps last for as long as 15 sec and involve 20 or more individual slaps. Motorboating by dusky dolphins has been associated with feeding. The dolphins apparently keep surface-herded fish from escaping laterally by circling around the fish school while tail slapping (Würsig and Würsig 1980).

We have observed the production of such slaps by both captive and wild dolphins with little or no forward movement involved. The animals simply rocked up and down longitudinally, their tails moving out of the water as their heads moved down. After the tail emerged, it was slapped sharply down against the water surface and the pattern was repeated. Some of the longest motorboating series we observed were made by inverted animals slapping the dorsal surface of their flukes against the water.

Single slaps are thought to have a signal function for many odontocete species, ranging from sperm whales to harbor porpoises (Norris and Prescott 1961). Such slaps may possibly signal danger, such as when a human observer disturbs a school, or they may precede synchronous dives by dolphins in a school. We saw them from our theodolite station when no human observer was near the animals.

Clearly, most tail slaps in the spinners we observed are unrelated to danger. In most cases they occur in slow-moving, undisturbed but not

© 1991 Jenny Wardrip

Figure 53. A back slap.

resting schools. They are often seen when behavioral state is changing. No precursor behavior has been noted.

Head Slaps and Back Slaps

As a dolphin moves slowly along, it may emerge from the water as far as about mid-body and then slap its anterior belly or back against the water. It can do this in any rotational orientation (figs. 52 and 53) creating *head slaps, back slaps,* and *side slaps.* These slaps are patterns of moving schools. They seem especially abundant during overall school acceleration and most frequently occur in the direction of travel. Their splash projects water forward in the direction of travel as the animal moves along.

We saw no precursor behavior to these slapping patterns underwater in dolphins in captivity. A captive dolphin performing slaps simply swam partially out of the water using rapid and very short amplitude strokes of its flukes. When one-third to about three-fourth of the dolphin's body was out of water, the animal twisted or flexed so as to strike the nearest part of its body against the water upon reentry, producing a smacking sound. To the ear, these sounds seemed much less intense than those generated by spins or tail slaps.

LEAPS

The impression given by all the traveling slaps discussed so far and the leaps (which follow) is that they mark the direction of movement with their elongate reentry splashes. Taken together, they allow the surface observer to form a precise instantaneous estimate of the activity level of a school. The dolphins' activity level seems to be determined by a group process, so it might be more accurate to say that, for them, such signals are likely to be related to the synchrony of behavioral state.

Figure 54. A salmon leap.

Arcuate Leaps

The clean, arcing leaps made by rapidly moving dolphins (fig. 50), are called *arcuate leaps*. As in other dolphin species, they seem to relate to improving the efficiency and speed of locomotion (Au and Weihs 1980, Hui 1992). That is, by leaping, the dolphin is able to take one or two tail beats with the tail still in the water while the anterior body is in the much less dense air. This allows the animal to take advantage of the much reduced drag upon its airborne body.

Such leaps are seen in a variety of circumstances in aroused schools. Dolphins "surfing" or simply traveling in large swells often pour from the water in such leaps. A school may bound away from the source of a momentary disturbance in a series of arcuate leaps. Such a leaping school usually emerges from the water as a series of subgroups that probably represent these same formations underwater. When we have observed dolphins exiting from a rest cove in the late afternoon in a last dash to sea, locomotion typically consists of a series of arcuate leaps separated by brief periods of rapid swimming.

Unlike most other aerial patterns, arcuate leaps seem incapable of producing noise or bubbles upon reentry. Arcuate leaps are performed by dusky dolphins while they are herding fish. The dolphins come to the surface to breathe, overshoot the surface in a arcing leap, and the use the weight of their body in the air to help propel them to depth (Würsig and Würsig 1980).

Salmon Leaps

A more active aerial pattern is what we have termed a *salmon leap* because it looks like a salmon leaping up rapids or falls—slightly arched and stiff-bodied, usually falling back on its side. The salmon leap, which usually takes the animal completely out of water, is typically seen in fast-moving schools such as those that move out to sea after rest (fig. 54).

Figure 55.   A tail-over-head leap.

Tail-Over-Head Leaps

The most athletic leaps of all are the *tail-over-head leaps*. They have not been observed underwater and hence the presence or absence of precursor behavior for this pattern remains a question. In the tail-over-head leap, the animal bursts from the water in a high, arcuate leap and literally throws its tail over its head, usually accompanied by a spiraling trail of water (fig. 55) At the end of the leap, the animal slaps the dorsal surface of its flukes and body smartly against the water as it reenters tail first. Sometimes in very active schools, a dolphin will combine a spin with a tail-over-head leap and yet still contrive to make an audible slap upon reentry. Such leaps are seen by the human observer as a bewildering mélange of flashing flukes, flippers, fins, and body.

SPINS

Spins are energetic patterns often performed in a series by a single animal (fig. 56). In one such sequence, fourteen spins in quick succession without stopping were performed by an individual. It is usual to observe sequences of four or five spins by a single animal. It is easy to tell that such a sequence is produced by a single animal because each dolphin performs the pattern somewhat individualistically. Some animals spin while nearly vertical, and others more nearly horizontal, reentry patterns vary, and so on. Furthermore, such sequential spinning has been observed to occur at a single general place in the school. Finally, spin series are usually performed with declining activity level. Frequently, the

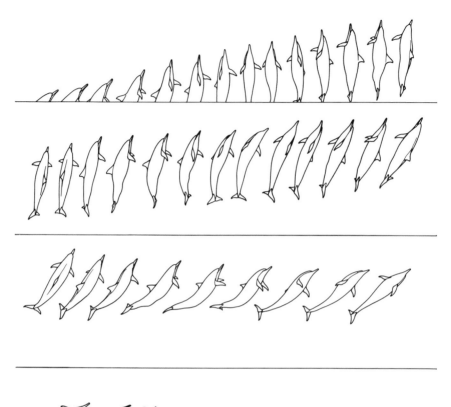

Figure 56. Sequential drawings of spin, as seen above water, including the boil of water left upon reentry.

last attempt may fail to take the dolphin clear of the water, and it may not complete the last revolution.

Dolphins spin in any part of the school, for as we observed the behavior, we failed to define a sector where spinning was most frequent. During the day, the pattern seems most common at dusk, and from observation of a captive school, it is probably very frequent during the night.

The spin has a well-defined underwater precursor pattern that involves both stereotyped body movement and vocalization (fig. 57). Our

observations of precursor behavior were made on a small captive school held at Sea Life Park Oceanarium on Oahu. A dolphin about to spin moved slowly a few meters below the surface in a rather rigid horizontal orientation. It bent both the head and tail down and up together in a rapid whole-body flex. At the same time, rapid slicing motions of the pectoral fins were made and the dolphin emitted a series of barks. Sometimes these motions were so rapid that they appeared as blurred vibrations, increasing in frequency. None of these movements tended to propel the animal forward with any efficiency. If the animal were close to the bottom, the movements rubbed the genital region rapidly over the

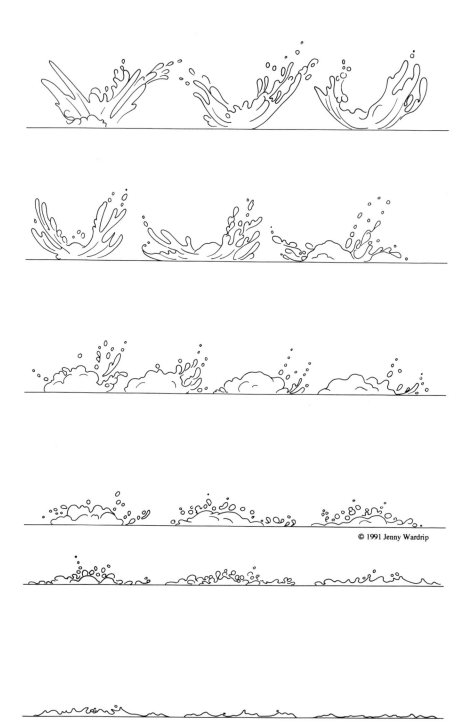

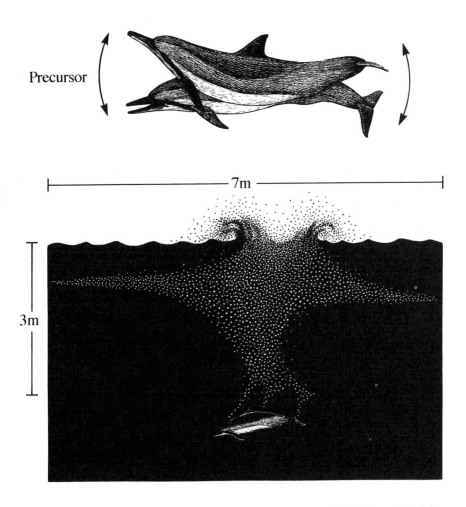

Precursor

7m

3m

© 1991 Jenny Wardrip

Figure 57.   The spin as seen below the surface. Top: Precursor behavior as seen at Bateson's Bay. Bottom: Bubble trail left by a dolphin sinking after a spin.

substrate. If the animal were in open water, these rapid jerky movements were followed by a rapid dive downward to about 2–3m (~7–10 ft). Then the dolphin arced upward, starting at an angle of about 70° relative to the surface. With ever-increasing amplitude of fluke strokes, until they were wide and powerful, the animal burst through the surface head first. When all of the dolphin except for the flukes and a few centimeters of the tail had left the water, the flukes flashed sideways at what looked to be a nearly 90° twist. It is suspected that this twisting of the flukes is an effect, not a cause, as the dolphin threw the airborne part of its body into a twisted pattern, causing the body to twist and rotate. But this is uncertain because we could not watch above and below the water simultaneously. Both high energy click trains and the short bark series just mentioned typically accompanied spin precursor behavior.

Upon reentry, the animal hit the water on its back and side, rotating rapidly (fig. 57). In the captive school, a loud, percussive crack of sound was usually heard at this point. The rotating dolphin then sank, stretching out full length. Its appendages and body drew a distinctive plume of bubbles down with it. After the spin was complete, this plume typically drifted and extended in length as much as an estimated 4m (13 ft) before it dissipated. Such bubbles were strongly entrained by the head, flippers, and dorsal fin. As many as one and three-fourths rotations occurred underwater after reentry before the animal righted itself and swam away, often to repeat the pattern again. The bubbles produced by this reentry persisted for a considerable period of time, in one case (fig. 57) about ten times as long in duration as the leap itself. Thus, these bubbles should make an excellent temporary echolocation target. From our observations of spins by captives, after the leap and its reentry noise are over, there seemed to be plenty of time for another dolphin in the school to wheel around and echolocate the bubble plume, thus determining the location of the spinning animal in the school.

Leadership, or dominance do not seem to be functions of the spin. In fact, as Hester et al. (1963) noted, the opposite seems to be true. All age classes spin, including very young animals. Norris and Dohl (1980*a*) noted that in a captive school, a new introduced dolphin not yet accepted into the group of more established dolphins was a frequent spinner. They go on to say (pp. 832–833):

> The more spread out a school, the more frequent spins seem to be. In feeding schools, which were the most dispersed of all formations [they had not observed the even more dispersed spread formation] spinning and other high-energy behavior occur almost constantly.

PATTERNS OF OCCURRENCE

It is clear to us that spinner dolphins engage in a variety of aerial patterns that are distinct and that, if quantified, would fall along an activity continuum. Once our classification was perceived and checked as a real recurring set of patterns in nature, it also became clear that such patterns vary regularly according to the time of day (table 3). Most aerial activity recorded in our daylight observations occurred in the morning or afternoon, and the type shifted regularly throughout the day. Thus,

TABLE 4   Mean Distance Traveled and Aerial Activity per Hour
of Dolphins Entering, Resting, and Leaving Kealake'akua
Bay for Two Sample Days.

Time of Day	Sample Size	Mean Distance Traveled (m)	Count of Aerial Activity
July 30, 1979			
6–7	18	414.1	0
7–8	23	68.7	3
8–9	37	138.1	27
9–10	34	77.3	6
10–11	26	53.6	4
11–12	20	77.4	0
12–13	21	78.2	0
13–14	31	165.3	20
14–15	06	386.4	44
February 2, 1980			
9–10	09	92.2	0
10–11	32	183.5	24
11–12	21	125.5	3
12–13	11	164.9	27
13–14	37	65.9	8
14–15	22	71.9	0
15–16	38	383.8	0
16–17	19	688.7	20

deep in Kealake'akua Bay, the patterns typical of the rest period—the fluke-outs, nose-outs, and tail slaps—all occurred mainly in the mid-day hours and at low frequency per animal.

Those aerial activities that occurred most often per animal tended to take place in the afternoon when the animals had aroused from rest. These were the active patterns—spinning, tail-over-head leaps, arcuate leaps, head slaps, other body slaps, and salmon leaps. The latter were only seen in fast-moving schools.

An enumeration of aerial behavior seen throughout two sample days, one in winter and the other in summer, corroborates this general pattern (table 4). Arriving animals moved rapidly into the bay, slowed for 3–4 hr from mid-morning to mid-afternoon and then became active again. In these two samples, aerial behavior was low early in the day, rose, and then all but ceased during the slowest locomotion, only to rise sharply again as the school sped up after the rest period was over (see fig. 26).

FUNCTIONS OF AERIAL PATTERNS

We propose that these aerial patterns serve at least two major functions. First, we suggest that the most athletic patterns, such as tail-over-head leaps and spinning, may serve to dislodge ectoparasitic remoras and perhaps other ectoparasites. Second, we propose that the classes of aerial behavior that produce noise and bubbles may be parts of redundant sequences of behavior that, taken in their entirety, provide short range omnidirectional markers in active schools. Thus, these contribute to defining both the school envelope and the positions of its members to other school members as they move along, often out of visual contact with one another. That is, with a number of animals in aerial behavior bouts throughout the school, any school member may be able to detect the disposition of the school envelope by listening for precursor and reentry sounds or by echolocating the resultant bubble trails.

Remora Dislodgement

We doubted the possibility that spins and tail-over-head leaps were involved in remora removal, as had been suggested by Hester et al. (1963), until we examined our high speed photographs of spinning and leaping dolphins. Of 79 photo series examined, 35 showed one or more remoras attached to a spinning animal. Since remoras are not that frequently seen on spinner dolphins underwater, we believe they do not occur in the population as a whole with nearly the frequency we saw in these photographs, although their abundance remains unmeasured. It is also only conjecture that the spin or subsequent reentry can actually dislodge such fish. It is worth noting that one of us (R. S. Wells) has recorded a similar high incidence of remoras attached to leaping bottlenose dolphins in Florida waters.

If we look at the remora question the other way around, another dimension is added to the argument. That is, do remoras *seek out* spinning dolphins and hence account for the rather high incidence of occurrence of leaping dolphins with remoras attached? This, would seem possible, given the few remoras one sees under water. Why would they do this? Perhaps by hitching a ride on a leaping dolphin the fish is able to enter and surprise schools of prey without the need for stealth. When one considers the buffeting a remora must take when clinging to a spinning dolphin it suggests that the fish somehow benefits on other terms. Could it be that the remora is able to surprise prey by being transported into their midst by clinging to a spinning dolphin?

Spinning cannot, however, always be involved with ectoparasite removal. Only a modest fraction of spinner dolphins seen at sea have remoras in attendance. Captive dolphins that did not carry remoras continued to spin in episodes that roughly matched the abundance pattern of these behaviors at sea. That is, these captive animals entered daytime rest periods and high energy aerial behavior ceased, but spinning behavior was observed throughout the remainder of the day–night cycle. Spinning and other active aerial patterns were very common at night in these captive dolphins. The implication is that a diurnal pattern of spinning occurs with or without remoras being present.

Sound and Bubble Production

The second proposal is that the sounds and/or bubbles made by aerial behavior and its precursors might produce acoustic markers serving to define the deployment of a school. This idea remains somewhat circumstantial. Can dolphins hear the reentry sounds of a spinning dolphin? We think this is possible, although Watkins and Schevill (1974) did not record the sounds of spinning with an underwater array when animals passed relatively nearby. However, we have repeatedly heard distinct, even rather loud sounds from the impact of spinning in captive animals, and at sea have sometimes heard slapping sounds both above and below water from nearby spinning dolphins. In our view, the reentry slap may be one part, perhaps the least important part, of a redundant string of signals that define a larger behavioral event that as a whole, can ensure accurate identification of the behavior by nearby dolphins, instead of being a unitary event that could easily be lost in noise. Conversely, the slap could merely be an adventitious sound associated with the production of a bubble plume and not of functional significance.

These views seem concordant with a number of facts about spinner dolphins and seem to violate none of our observations. The following information is related to this proposal.

The vocal emissions of dolphins all seem to have some level of directionality and therefore seem not to be especially suited to the task of school envelope definition (see Au 1980). However, the slaps of reentry and especially the bubble trail target are expected to produce a target that can be sensed from any direction. Whistles are much less directional than clicks. Nonetheless, when one makes an acoustic recording of the passage of a school many kinds of sounds can be heard until about the time the school is abeam of the observer. As the school passes the recording station sounds diminish and soon are all but gone, even though the school is still close by. The least directional and longest ranged vocal sounds appear to be burst-pulsed signals, followed by whistles.

One of these relatively omnidirectional sounds, a short burst-pulsed bark, was noted to precede the spins of a captive dolphin. If it occurs in nature, it may serve to alert school members within hearing that a spin is about to take place. Just before a captive dolphin spun, such barks were given as the dolphin repeatedly flexed stiffly a few feet below the surface, moving about at different headings before the leap, perhaps serving to broadcast this preemptory sound as widely as possible. Such very brief barks may be best suited to an alerting function. They seemed not to be sustained long enough for another dolphin to do more than note their general sector.

With school members alerted to the general sector of a spinner, they may then seek more precise information. If the reentry slap can be heard, it should be reasonably omnidirectional, although it is not expected to propagate very far since it is produced at the surface where acoustic clutter is strongest. Thus, it might not propagate across the width of a school. We only occasionally heard the reentry sounds of spinning in our listening at sea.

The twisting plume of bubbles drawn down by the dolphin as it sinks (fig. 57) may be more important than the slap since it should allow precise localization by echolocation and should be detectable over a considerable range. Bubble plumes make especially excellent acoustic targets (Glotov 1962), as outlined in chapter 8.

OTHER CORRELATES OF AERIAL BEHAVIOR

If aerial behavior and its associated behavior serve to define the deployment of a school, a number of other correlates should occur.

1. Spinning should be found throughout the range of the species. Leatherwood et al. (1983) have recorded the behavior around the world.
2. Other dolphin species should have similar needs and should also show aerial patterns that could serve the same general purposes, as also seems to be the case. Aerial patterns vary from species to species, as if they might also have a species recognition function. If such recognition occurs, it is expected to take place underwater.
3. Aerial patterns should be concentrated during times when members of spinner schools must rely most heavily on acoustic contact. This is clearly true. Such patterns are very abundant in the dark and found in all actively moving schools. They all but cease during rest when schools are thought to be mediated visually.
4. Aerial patterns should occur throughout the school if they are to define its shape. Repeated casual observation showed no place or sector in a school in which spins were absent, and no area of concentration was obvious.
5. If aerial behavior is used for acoustic definition of a school, the various patterns involved in it should clearly be involved in sound or acoustic target generation.
6. Because of the possibility of confusion when two or more schools are close to one another, such marker sounds should probably not

propagate far beyond the confines of the school, as was found to be the case.

7. The rate at which such marker signals are produced should be economically produced and sufficient to define school deployment in relation to the speed of change of the dimensions of a school, but not excessive for the task.

8. The plume of bubbles drawn into the water by the rotating dolphin may be distinctive in shape (fig. 57). Because the animal sinks about 3 m (10 ft) below the surface clutter before it recovers and begins another leap, such an echolocation target should be widely available to alerted dolphins as a target from which precise range and bearing can be obtained.

9. The barks that precede the spin and perhaps the reentry slap may allow a school member to discriminate between the cascading bubbles produced by a breaking wave and those produced by a spin.

CONCLUSIONS

All the observations we have made fit the model that spinning behavior serves to produce a matrix of acoustic markers that define school shape as the dolphins move along, most of them out of effective visual range of one another. Schools, though fluid and moving as fast as 6–8 knots, change shape relatively slowly. The need for acoustic markers should be satisfied by modest levels of occurrence schoolwide, that is, a few markers or spins every few minutes, as has been observed.

In addition to the function of spins and leaps as a means of school envelope definition, they may also be part of the general social facilitation that goes on in dolphin schools. There seems to be a contagious aspect to spins and leaps that relates to the general arousal level of the school, as has also been found in dusky dolphins (Würsig and Würsig 1980). Thus, a single behavior may contribute to more than one aspect of school life, a not uncommon circumstance in animal behavior.

What might militate against the theory that aerial behavior produces markers of school movement? Spinning is obviously energetically expensive behavior to perform. Are the values we propose sufficient to explain its presence? All we can say is that it is crucial for the member dolphins to keep track of schoolmates under all conditions, and the difficulties of the process may be enough to justify the energetic cost of the behavior.

In the next chapter, our discussion moves on to consider how the populations of spinner dolphins seem to be organized in terms of movements of individual dolphins and their association patterns over time. These data were derived by studying a cumulative record of individually identifiable dolphins based on the natural scars and marks of their dorsal fins and backs. Such animals were identified from a photo dossier of all the dolphins we encountered over the entire course of this work.

SIX

Population Structure

*Bernd Würsig, Randall S. Wells, Melany Würsig,
and Kenneth S. Norris*

A species' place in nature is delineated by the descriptors of its populations, reproductive patterns, associations, and movements. Traditionally, the information used in population analysis of cetacean populations has come from analysis of dolphins killed for one reason or another. For the genus *Stenella*, studies of animals taken from the drive fisheries of Japan have contributed a great deal, as have studies of dolphins taken in the tuna seine fishery.

The commonly caught striped dolphin (*Stenella coeruleoalba*), an open water relative of the spinner dolphin, has been studied both in terms of its population parameters and its social structure (Oshumi 1972, Miyazaki 1977, Miyazaki et al. 1974, Miyazaki and Nishiwaki 1978). Another close relative of the spinner, the spotted dolphin (*S. attenuata*) has received attention both in Japan (Kasuya et al. 1974) and in many studies done with animals taken from the eastern tropical Pacific tuna grounds (see bibliography by Holbrook 1980 and recent review by Perrin and Gilpatrick, in press). We will return to some of these works later when we attempt to interpret our own results.

SCARS AND MARKS ANALYSIS

For living cetaceans, a method called *scars and marks analysis* has contributed most to the understanding of dolphin movements and school compositions (Würsig and Würsig 1977, 1979, Würsig 1978). This method consists of photographing each wild dolphin school encountered so thoroughly that the fin and mid-back of each animal is recorded clearly again and again (Appendix B). Then, because cetacean skin records scarring much as if someone drew on the animal with a stylus and because fin damage may persist for long periods, the resultant photographs can be compared over several years' time. In this way, a dossier of individual animals can be built up and a record of their movements and association patterns developed.

Because the time is brief during the surfacing of a school, such photography is accomplished at high shutter speeds (often to 1/4000 sec) using power winders that allow several frames to be taken on a single surfacing of a school. If the sexes of marked animals can be determined, the sexual correlates of these population features can also be assessed. However, for modestly dimorphic species such as spinner dolphins, this is difficult and was only occasionally accomplished in this study. Our newly constructed viewing vehicle, *Smyg Tittar'n*, now makes such determination routine because we can easily see the genitalia of nearby animals (see chap. 3).

A thorough scars and marks catalogue for the spinner dolphins of the island of Hawaii was the first order of business. Each time a new school was encountered, it was repeatedly photographed and its members catalogued. Over the term of the study, this allowed us to describe the site specificity of marked dolphins, their associations, their movements relative to the entire island coast, and the degree of coherence of their schools. Finally, it allowed us to make a rough estimate of the island population of spinner dolphins.

What we found from the scars and marks catalogue was a surprise to us. Instead of documenting coherent schools of dolphins that had integrity over time, we found most of the spinner dolphin associations to be very fluid. The composition of the schools that came almost daily into rest coves changed radically from day to day.

Instead of a "population" of dolphins living around the island composed of a recognizable suite of animals, the island dolphin population appeared to be open to an unknown reservoir of dolphins and in a continual but not total, state of flux. This high degree of fluidity is shown by the association patterns of marked dolphins with radiotagged animals RT-1, RT-2, and RT-3 (figs. 47–49). Considering the large number of dolphins identified in schools containing these radiotagged animals, very few associates were seen with them on consecutive days. Nonetheless, our scars and marks catalogue does show some associations of dolphins within schools that appeared over and over again (figs. 58–62). It also shows that certain animals were most often associated with certain stretches of coast. This overall pattern of loose structure is of special interest since it may indicate a fission–fusion society with similarities to the societies of some social primates and other social mammals (Altmann 1967).

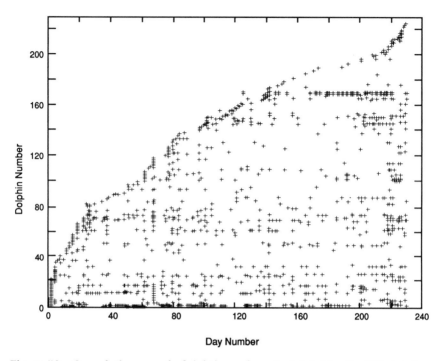

Day Number

Figure 58. Cumulative record of sightings of marked dolphins for 230 sighting days, along the Kona coast of the island of Hawaii, May 1979 to October 1980 and June 1981.

While persistent association patterns emerged among certain dolphins, such patterns were far from a pervasive element in the school structure we uncovered. Only a few associations seemed to be long lasting (figs. 63 and 64), while associations by age and sex were more evident. This fluidity contrasts to the much more stable association patterns seen in shore-dwelling bottlenose dolphins (Wells 1978, Shane 1980).

OCCUPANCY OF REST COVES

In the rest coves, occupancy was clearly to some extent opportunistic and highly variable both in the number of dolphins seen on any given day and the identity of individual dolphins. Nonetheless, each resting area seemed to exhibit a reasonably definable upper limit of numbers of dolphins that it would hold. This was different for each different resting area and apparently related to the area of usable resting grounds, specifically the areas of sandy bottom (fig. 65). The Honokohau–Kiholo area had both the largest schools and the greatest sand area.

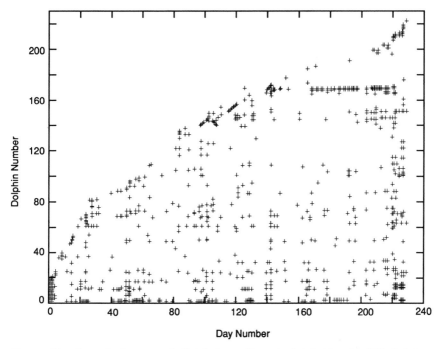

Day Number

Figure 59. Cumulative record of sightings of marked dolphins for 230 sighting days in Kealake'akua Bay.

INDIVIDUAL IDENTIFICATION

During the 18 months of the continuous study from May 1979 through October 1980 and during two weeks in June 1981, 224 individual dolphins along the entire Kona coast were identified from photographs (fig. 58). Of these identified animals, 32 were determined by body marks and were rarely resighted. The remaining 192 dolphins, whose markings were recorded from the dorsal fin and the adjacent back, represent the main body of our identification work. Of these animals, 66 were sighted only once. Of the 126 dolphins sighted more than once, a few were seen repeatedly.

A total of 36 dolphins were sighted ten or more times each. These most frequently identified individuals provided the majority of data relating to long term associations and movement patterns. However, many of the dolphins that were sighted less often were readily identifiable from clear photographs; they simply appeared infrequently in our study area. Therein lies a dilemma—we cannot be sure why a given individual was infrequently present in our photo log. These animals conceivably could have been missed because of the subtlety of their recognition

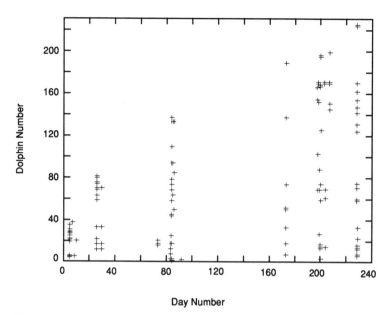

Figure 60. Cumulative record of sightings of marked dolphins south of Kealake'akua Bay.

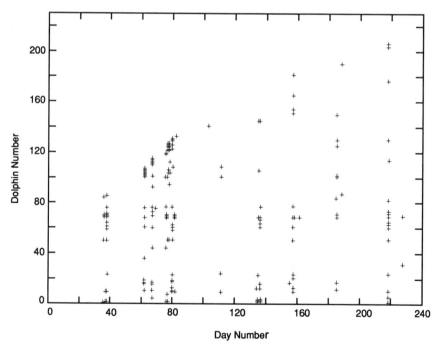

Figure 61. Cumulative record of sightings of marked dolphins north of Kealake'akua Bay, including Kailua Bay.

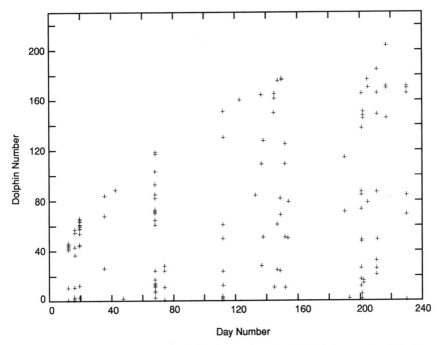

Figure 62. Cumulative record of sightings of marked dolphins north of Kailua Bay.

marks (although some of them had obvious markings), they could have avoided the photography team more than the more commonly noted animals did (although we made attempts to photograph entire schools), or they could simply have seldom been present inshore where most of our photography was done.

The latter possibility seems the most likely. Evidence from our radiotagging studies (chap. 2) shows that a significant population of dolphins appears not to enter rest coves each day but rather stays a modest distance offshore, patrolling back and forth over the island slope during the day. The frequent occurrence inshore of certain marked animals could also represent some aspect of population segregation in which a particular part of the social complex of the larger dolphin school tends to come ashore, while other parts tend to stay offshore.

Finally, some bias doubtless exists in our records because scars and marks do change with time (fig. 63), and some may have changed sufficiently to become unrecognizable. Judging from the constancy of most marks, this was probably a minor factor in our results.

Figure 63. Some well-marked and frequently sighted dolphins. Top, dolphin no. 2 (Four-Nip), April 11, 1980; bottom, dolphin no. 1 (Finger Dorsal); opposite, dolphin no. 17 (Low Notch).

ISLAND USE PATTERNS

To obtain information on differential use of different parts of the Kona coast of the Big Island, we divided our study area into four sectors for analysis: (1) Kealake'akua Bay, (2) south of Kealake'akua Bay to (and including) Kauhako Bay, (3) north of Kealake'akua Bay to (and including) Kailua Bay, and (4) Keahuolu Point to Kukio Bay, the northernmost sector of our sample area (see fig. 62). The occurrence of marked dolphins in these four areas is shown in figures 58, 59, 60, and 61. This record does not include samples from the Makalawena area, north of Keahole Point, where our records are too sparse to indicate patterns of occupancy.

There was a total of 1233 sightings of marked dolphins—795 in area 1, 119 in area 2, 134 in area 3, and 185 in area 4. The differing number of sightings for the various areas seems primarily to reflect survey effort and not differential use by dolphins. The average number of sightings per dolphin was 6.4, ranging from 1 to 69 sightings per individual.

We prepared a computer program that calculated the frequency of sightings in relation to the number of total sightings by area for each dolphin. The results allowed us to compare relative frequencies of occurrence by area, standardized for sighting effort. The 36 dolphins seen ten or more times demonstrated that dolphins do indeed spend more time in some areas than in others. Ten of these dolphins were seen mainly in Kealake'akua Bay and to the north of it. Six were seen

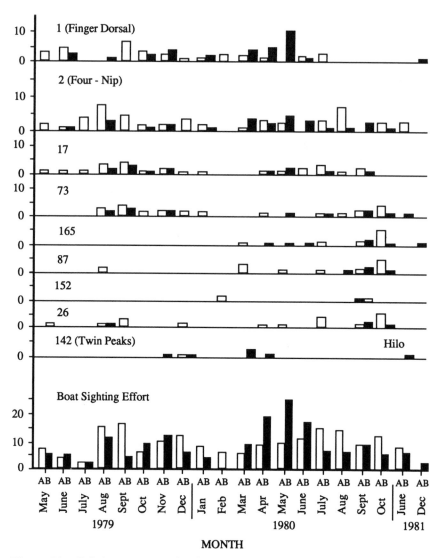

Figure 64. Sightings per month of nine selected dolphins in Kealake'akua Bay (shaded), and outside the bay (unshaded); monthly sighting effort is at the bottom.

throughout the south sector but not north of Kailua Bay. Eight individuals were seen mainly south of Kealake'akua Bay and seldom in the bay or to the north of it.

Only 6 of these 36 dolphins were seen with about equal frequency throughout the study area. Time analysis showed us that this overall occurrence was due to shifts from one area of general occupancy to another. For example, dolphins #11, #12, #13, and #14 were first seen in Kealake'akua Bay, but they shifted to the Keahole Point area to the north when not present in Kealake'akua Bay (fig. 64).

ASSOCIATIONS

We also analyzed association patterns among individual dolphins. Dolphin #1 (Finger Dorsal) and #2 (Four-Nip; fig. 63) were seen together on 26% of their total sightings (#1 was sighted 44 times and #2 69 times), throughout the year and throughout the study area. But their association within a given school appeared quite fluid, that is, they did not obviously swim with one another. This is exemplified by a series of sightings in Kealake'akua Bay in September 1979 in which #1 was observed in the bay on September 4 and 5, both were observed in the bay on September 8 and 11, and #2 was observed alone on September 15. Although there is a slight possibility that these dolphins were together on days when only one was recorded, both dolphins are highly distinctive and can be identified from both sides. It is unlikely that on good photographing days, such as these were, that one or the other was missed.

Dolphin #1 (Finger Dorsal) was first sighted in Kealake'akua Bay on May 1, 1970, by Thomas P. Dohl. During our study, the animal was seen in Kealake'akua Bay through July 1980, but he disappeared and was not seen again until he was photographed by Shannon Brownlee in December 1980. Then, during a brief survey effort in June 1981, the dolphin was again missing. It is instructive to follow this often-sighted animal in some detail. As figure 64 shows, Finger Dorsal was quite often seen outside of Kealake'akua, and in fact, he was found in all of the observation sectors studied here. During May 1980, he was most often seen at Makalawena at the farthest northern reach of the Kona population. Then he seemed to disappear altogether from the region.

Tight associations among dolphins were observed, but when viewed against the large bulk of resightings, they were rare. An example is the pair of dolphins #17 and #73, which were almost always sighted together. They were both present on the Kona coast except during February and March of 1980. Such tight associations were not always pairs. Dolphins #165, #87, #152, and #26 formed a distinct group that seemed to develop during our work. After September 16, 1980, they were always seen together in larger schools, while before this date their association was only sporadic. Our difficulty in sexing most dolphins prevented us from determining if these associations were of age-related or sex-related groups, although work on other dolphin species (e.g., bottlenose) suggests that the latter might be the case (Wells et al. 1980, Wells 1991).

MOVEMENTS

The sighting log indicated that individual movements can involve the entire coast of the island of Hawaii or can be on a much smaller scale between rest coves at different parts of a single sector of the island. The longest movement we recorded was for the very distinctively marked dolphin named Twin Peaks (#142). This dolphin was seen five times off Makalawena at the northern boundary of the Kona population (from November 1979 to April 1980) and once in Kealake'akua Bay (December 1979). No other sightings of it were recorded, even though the dolphin was easily distinguishable from either side at a distance. Its fin had been bifurcated, leaving only two protruding points of tissue. On June 16, 1981, this animal was photographed on the opposite side of the island of Hawaii from Kona, near Hilo Bay, about 180 km from Makalawena.

These records show (as substantiated by Norris and Dohl 1980a) that the large spinner dolphin schools are labile assemblages with smaller units of greater cohesion within them.

ACCESSION RATE

As figures 59 through 62 indicate, previously unsighted but identifiable dolphins appeared in our log at a rather high rate throughout the study. Once in the log, such animals were sometimes sighted over and over. An example can be seen in figure 60 when dolphins #16 and #1 appeared in the log and then remained present until nearly the end of the study period.

The accession rate of new dolphins to our log shows that at first, newly recognized animals were rapidly entered in the log, as would be expected in sampling an unexamined population. By the time the log included about 80 animals (on approximately day 30), the rate of new entries began to fall off and through the remainder of the study stayed at about 1.4 new dolphins per day. The curve showed no sign of reaching an asymptote by the end of the study.

At first, the variation in accession rate of the curve could easily be explained by variations in our sighting effort, either because we concentrated our observations on one area of coast for a time or because there were changes in the basic effort of accession itself. The relatively constant rate of accession that occurred later suggests that the population of dolphins we were sampling was large relative to the school numbers we observed and that its boundaries fell well beyond the boundaries of the coastal coves where we spent most of our time.

POPULATION PATTERNS AND SEGREGATION

The number of dolphins frequenting the shore appear to vary widely throughout the year. Our data suggest movements between the "off shore reservoir" and shore rest coves of some magnitude. A steady stream of offshore animals seems to make its way into shore rest areas. Once in occupancy, an animal or a group of animals sometimes returns again and again over many months and sometimes disappears altogether.

We support these conclusions with two observations. First, many clearly marked animals appeared only briefly in our sighting log. Second, aerial sighting data from surveys done under good sighting conditions revealed far fewer dolphins than some surveys done under poor conditions. In fact, the numbers of dolphins sighted per aerial survey varied from a minimum estimate of 59 dolphins per flight to a maximum of 515, and variations in weather were clearly not great enough to explain this. These numbers are based upon our best estimates from the air and, when possible, from counts taken from photographs made from the aircraft.

There seemed to be no recognizable differences in the accession curve of new dolphins for the different sectors of the Kona coast (figs. 58–62). We gained the impression that cove use is to a degree opportunistic based partially on where a nighttime school finds itself as dawn approaches and probably partially on knowledge of the shoreline and what coves it offers.

In terms of the sector of shore use, some dolphins seem to visit a given coastal sector from time to time and some are close to being resident. Many inshore dolphins reappear daily for a time in one part of the coast with a decreasing probability of being found later an increasing distance away from that core area. This suggests that nighttime tracks of feeding schools may tend to focus on a part of the total coast rather than upon all of it, leaving the dolphins over and over again in the same general sector of coast as the daytime rest period approached. This seemed true even though one radiotracked school of dolphins took a very long traverse along almost the entire Kona coast.

When the data from the flight series and the radiotracks were combined with the sighting and resighting data, the outlines of a possible population structure for Hawaiian spinner dolphins could be proposed. A population involving both inshore and offshore parts is postulated. Recall that not all tracked dolphins come ashore every day; some remain offshore to cruise along the coast during the day. The very different resighting rates of individual dolphins are suggestive of this. This seems clearly to be a real phenomenon rather than one based on the difficulty of recognizing individual scars and marks.

It may be that a segregation within the larger dolphin school takes place, with certain subgroups tending to stay offshore while others tend to use the coastal coves during the day. Both of our most frequently sighted dolphins (#1 and #2) were males, and #2 (Four-Nip) was often noted traveling with juveniles. Four-Nip has been sighted frequently in shore coves in a new series of observations in 1991, which have now revealed him as a very old male. It will take further study to determine if this indicates a tendency for subgroups, perhaps including alloparents and younger dolphins, to rest ashore while adult subgroups cruise the coast.

A POPULATION MODEL

Our model of spinner population dynamics involves shifts of the entire population during seasonal development of the stormiest island weather, with the dolphins generally seeking the lee side of the island. It also involves a large, coalesced offshore nighttime school that fragments and enters rest coves in numbers regulated by the size of the rest area available to them. Establishing a rest school seemed to involve "shopping for coves." On several occasions, if a given cove was occupied by a dolphin school, we either saw a new school enter and meld with it or saw such newly arrived schools turn away at the mouth of the bay when it was already occupied and travel elsewhere.

We have no idea of what their social arrangements might be like in the dark, except to say that nighttime schools appear to be large. Perhaps they involve the coalescence of the smaller schools from the coves, although this is only an impression developed by traveling through a nighttime school with our vessel floodlights on.

We suggest that the shoreline island of Hawaii has an approximate carrying capacity of dolphins seemingly determined by the size and distribution of usable rest areas. It seems logical to assume that the size of the larger population, including those who seem to stay offshore, is determined primarily by available food supply at the island and only secondarily by the school numbers that occupy coves.

We have not attempted to quantify the flux of spinners in our study with the spinners that seem to live far offshore of the island of Hawaii or near other islands in the chain, although it might be possible to produce a model to estimate this.

The question of what a spinner dolphin family unit might consist of and how such units function within the 1000 or so other dolphins we estimate to be present at any one time in these seemingly fluid assemblages is an important one since it relates to questions of kinship and gene exchange rates. Do we deal with patterns that could preserve kin relations? Or are effective family boundaries blurred by the fluidity we observed and by the apparently promiscuous mating patterns we report here? It is difficult to imagine maintenance of a tight familial unit in such a society. In fact, the functions of such a unit seem to be subsumed by the school in its various permutations. But, because of our lack of knowledge of flux with the putative offshore group, we simply leave the question open.

We have clearly not perceived all of the descriptors of spinner dolphin school composition. It is true that we could often see differences among the various rest schools in terms of age group composition. On occasion, the schools that came to Kealake'akua Bay would include numbers of juveniles and mother–young pairs, while other schools that entered rest coves were without such groups and seemingly composed largely of adults.

COMPARISONS TO OTHER POPULATIONS

It is instructive to consider what is known of the population dynamics of two other members of the genus *Stenella*, both of whom live in oceanic populations in the same general regions as the spinner dolphin. These are the striped dolphin (*S. coeruleoalba*) and the spotted dolphin (*S. attenuata*). Like the spinner dolphin, both have populations that live in the open eastern tropical Pacific. Although they move considerable distances during seasonal shifts in tropical water masses, both are considered residents of the tropics. Also, both have other populations that are migratory in circumferential oceanic current systems. For example, all inhabit the Kuroshio Current and are fished off the shores of Japan.

The association between the spinner dolphin and the spotted dolphin in the eastern tropical Pacific is a particularly strong one, indicated by the fact that they are frequently caught together in the same seine hauls. The striped dolphin, however, seems not to associate closely with other dolphins (Miyazaki and Nishiwaki 1978) and is often found much farther away from the tropics than the other two, swimming into temperate water masses. The spinner dolphin and the spotted dolphin have shore-dwelling populations, while only the spinner seems to rest in coves or lagoons. All three species seem to exhibit population structures that are at least broadly similar. All have been studied in some detail with regard to populational features. Much is known about the reproduction and recruitment of the spotted and spinner dolphins on the eastern tropical Pacific tuna grounds (see Holbrook 1980), and these data are reviewed in chapter 9.

The striped and spotted dolphins have been studied by Japanese cetologists who were able to examine in great detail whole schools taken in the Japanese drive fishery. These extensive school samples allowed examination of many features of the life of *Stenella* schools that have been otherwise unavailable to scientists. For example, plausible sequences of behavioral maturation, in which a young dolphin passes through a series of social patterns and associations, have been constructed for the striped dolphin by looking at the totality of many such captured schools (Miyazaki and Nishiwaki 1978).

These dolphins were captured from the migratory populations of the species taken while the dolphins moved in the Kuroshio Current and its marginal waters northward in spring months (centered in April–May) and southward in fall and winter (centered in October–January). The schools move together as a loosely defined front that, with reference to a given point on shore, builds, peaks, and declines. The front is composed of a great number of schools. This movement seems to "spread out" the population structure into something like its component parts. Instead of having all age- and sex-related associations melded together into single schools, as seen in the Hawaiian spinner dolphin schools, such subgroups sometimes appear in discrete form in the captured schools. For example, we were able to discern juvenile subgroups within Hawaiian spinner schools at times, but such subgroups may swim by themselves in the Kuroshio Current schools of the striped dolphin.

In a subtle way, we could see the same effect in Hawaii when parts of the offshore nighttime school came into rest coves. It was then easy to see that some schools included mother–young pairs, others juvenile groups, and still others adults of both sexes.

Miyazaki (1977) and Miyazaki and Nishiwaki (1978) were able to define features related to sex and age composition from 45 such schools, including nearly 6000 animals. From this sample, they attempted to reconstruct the dynamics of the population of dolphins from which the schools came, including the course of social arrangements throughout a dolphin's life. This effort assumed that the schools they studied were units of some degree of permanence. They also assumed that, as a dolphin was born and grew, it progressed through schools that emphasized nurture and juvenile patterns and then into schools in which adults and adult patterns were emphasized.

This is the same assumption with which we began our work on spinner dolphins, but in time, it became obvious that it was not entirely correct. Instead, because we sampled our population through time (while the Miyazaki and Nishiwaki sample represented a single point in time for each capture), we were able to perceive the high degree of mixing of adult schools that they could not see. Miyazaki and Nishiwaki (1978, p. 113) perceived this difficulty to a degree:

> There are a variety of modes of fetuses and newborn calves seen in Fig 21.
> It can be presumed from this figure that individuals of this species do not

stay in a certain school, they may have mobility, if they were stable there would be more periodical modes of fetuses and newborn calves.

The Japanese dolphin schools were caught by fishermen using four swift (20-knot) drive boats that left port near the Izu Peninsula before dawn, located schools of dolphins, formed an arc behind them and drove them toward shore, where they were guided into a cove by additional vessels and finally encircled with a net. The dolphin schools are taken from the inner margin of the northward-moving Kuroshio Current, whose nearness to shore changes greatly. This nearness to shore seems to influence the age and sex composition of the schools the fishermen catch. This is an important point because it indicates that the total structure of the larger migrant striped dolphin schools probably extends beyond the reach of the fishermen, whose capture range is determined by the length of daylight. The fishermen probably sample just a part of the population that moves by.

The schools the scientists examined ranged from 25 to 2136 animals. Unless the schools were small, only a portion of the dolphins could be sampled by the scientists before operations of the fishery took the remainder away. The authors do not state how they selected the dolphins for sampling, and thus some concern exists about possible bias in their sampling procedure, an important point when most of the data derived from the work relate to the statistics of sex and age compositions of the captured schools. However, we will assume that such sampling was random or that any effects were unimportant to their conclusions.

The various schools, as revealed by the catches, showed variable age- and sex-related modes that suggested to Miyazaki and Nishiwaki (1978) a regular lifetime progression of a given dolphin through a series of age- and sex-related association patterns. In the Hawaiian population we studied, some parts lived seemingly anchored to a single island. Because of this, such proposed patterns of change are more difficult to observe than in the Japanese spread schools and are made still more difficult by our problem of visual determination of sex, and often age, of the dolphins we observed.

Miyazaki and Nishiwaki divide the 45 schools they examined into a series of classes based on age, sex, and sexual state relationships. They recognized juvenile schools, adult schools, mixed schools, mating adult schools, nonmating adult schools, mating mixed schools, and nonmating mixed schools. To some extent, this is a classification based on modal composition and sexual state rather than on sharp differences among schools. In other words, their conception of school structure could be an artifact of classification. A mixed school, for example, might include all age classes but have a preponderance of adults. The other categories, such as resting condition or mating schools, are similarly modal and definitional. For example, a nonmating school is defined as one emphasizing middle and late stages of pregnancy.

Juvenile schools tend to be the most sharply delineated of these school types. Typically, juveniles were found traveling without mothers and young and without the presence of parturient or suckling females, but usually with a small number of apparently late juvenile and adult males. Although some of these schools were caught during October when testis weight was highest in other schools, it was less on average in the males associated with these schools. Perhaps males not engaged in the reproductive events of the larger school travel with these juveniles.

We saw such juvenile subgroups in the Hawaiian schools, although they were typically embedded in but distinct from a larger school of other age and sex classes. We also observed male consorts in these subgroups, including our two most frequently sighted marked animals, Finger Dorsal and Four-Nip. Because the association of these two animals extended over a long period, it seems possible that we were seeing a demonstration of a social role instead of a simple maturational state, as indicated by testis size.

Miyazaki and Nishiwaki (1978) include a graph (their fig. 8) that shows the diurnal change of school size in the striped dolphins. The largest schools they sampled (500–2327 animals) were all encircled shortly after first light, from 5 to 8 A.M. For spinners, this might correspond to sampling the coalesced nighttime feeding schools before they fractionated and moved into rest coves and descended into rest. Two huge juvenile schools (636 and 903 individuals) were included in the Japanese records, as were mating and nonmating adult and mixed schools. It is worth noting that the number of individual dolphins involved in these open water migratory fronts is greater by an order of magnitude than the spinner groups with which we worked.

The Japanese work, like our own, was mainly focused during the daylight portion of the diurnal cycle. This means that if striped dolphin schools coalesce at night, which is suggested by the early morning captures of large schools, many of the school compositions indicated by Miyazaki and Nishiwaki may be transitory assemblages. The various subdivided daytime schools may come together in the dark only to fractionate again the next day, perhaps in mixes as different as those we recorded from day to day in Kealake'akua Bay. If striped dolphins are like spinners, we would expect this to be the case.

Nonetheless, Miyazaki and Nishiwaki (1978) do demonstrate important internal structural features of an open water dolphin assemblage. They show that tendencies exist within schools to gather in age classes and in groups related to mating and parturition. They provide evidence that a given young dolphin can expect to pass during its life through a series of association patterns and roles within a population, even though there may be much fluidity to the associations they make as they travel.

Of all categories in the larger dolphin society, the juvenile period seems the best defined in their study. If the function of these subgroups is like that for other mammals, this period serves as a time of imitation, instruction, and practice of adult cultural patterns (Gentry 1974). Its importance is underscored by its length in dolphins generally and its influence in lengthening the life cycle of the species in question.

Although not discussed here, the population dynamics of the remaining species of the genus, *S. attenuata* (the spotted dolphin), has also received a good deal of attention. It is the major species caught in the eastern tropical Pacific yellowfin tuna seine fishery. It is also taken occasionally in the Japanese drive fishery, where entire schools have been studied in detail (Kasuya et al. 1974), just as Miyazaki and his colleagues have done for *S. coeruleoalba.*

ABUNDANCE

Our study does not allow us to resolve the dynamics of the Hawaiian spinner population nor to provide a very precise estimate of numbers frequenting the island of Hawaii. The number of acquisitions of new animals to our scars and marks catalogue continued to increase throughout the entire study, indicating that we had not catalogued an entire or discrete population. Many clearly marked animals moved in and out of our data set. Even so, some dolphins appeared to be residents nearshore and were seen over and over again, some for as long as a decade or more. Even these often changed their areas of shoreline occupancy. Our work revealed a just-offshore population of spinners at night apparently larger than the total occupying the coves ashore during the day at any one time. However, we documented no population farther offshore among the islands, even though our trips by vessel to the island took us through offshore waters, nor can we provide any evidence of movement across the deep channels between islands.

Nonetheless, our impression at any one time was that we were dealing with a local group of animals near the island shore of about 1000 in number. This impression came from three sources. First, our flight records let us make actual counts, and allowing for the inevitable inaccuracies of estimation from the air, this is what we estimated.

Second, we prepared a sample of 20 good sighting days from our vessel *Nai'a* during which we believed our photographic record was complete for the schools we encountered, both in the bays and farther offshore. We felt that we were able to identify all recognizable animals in these schools. When we calculated the percentage of recognizable animals in these schools, it was on average about 20% (14–36%). By the end of the study, we recognized 192 individuals by markings visible from the surface. Using this number, we approximated the minimum number of dolphins within the population that frequents the shore of the island of Hawaii at 960 animals (192 × 5).

Not all these dolphins seem to frequent the Kona coast rest coves on a continuous basis. Instead, the entire population seems to move around the island relative to the weather, either entering coves or presumably staying in the just-offshore group. Also, there was clearly a flux of animals in our records, so this figure is an order-of-magnitude impression only.

Third, if you add up all the numbers of dolphins seen from the coves on the lee side of the island during any day of occupancy, it usually falls in the range of about 400–700 animals. Allowing for others in the near-shore waters, we get a total number on the order of 1000–2000.

Future work with island spinner dolphin populations such as ours will benefit if better estimates of the percentage of animals remaining offshore versus those occupying shore coves can be made. One successful approach to obtaining a direct count (which we did not perform) might be to select a time when the dolphins are concentrated on the Kona coast, radiotrack dolphins all night until daylight, and then follow and count their associates in offshore schools at the same time that an aerial count of dolphins within rest coves around the island was being made.

PERSPECTIVES

Other questions remain. Why should some dolphins remain offshore while others rest in island coves? Are we imperfectly seeing a daily population segregation in which, for example, certain age classes of spinners such as juveniles tend to come ashore to rest along with other parts of the school that may somehow be involved with them?

Do the often-sighted adult males Finger Dorsal and Four-Nip serve as alloparents in a division of labor within the larger spinner school? Certainly, both were often seen in the company of juveniles and both were frequently sighted nearshore over a considerable period of time. Such sightings were in fact highly disproportionate in number compared to sightings of other well-marked animals. If such an alloparental role is part of the behavioral suite of Stenellid dolphins, it may explain why Miyazaki and Nishiwaki (1978) found that in eight out of nine juvenile striped dolphin schools they sampled there was a significant cadre of adult males (but only a few adult females) present. The answers to these questions lie in more detailed observations than we were able to perform.

SEVEN

The Visual Domain

Kenneth S. Norris, Randall S. Wells, and Christine M. Johnson

One frequently reads of dolphins described as acoustic animals as if the sense of vision were unimportant to them. This is far from the case. Except for two riverine species, *Platanista gangetica* and *Platanista minor*, which are nearly blind (Purves and Pilleri 1973–1974), dolphins are equipped with excellent sight. In fact, dolphin vision is heavily involved in events both inside and beyond the school. In a number of ways, it is adjusted to the special environments in which dolphins live. To fully appreciate these visual environments, the observer must venture below the surface. These environments are very different from those we experience in air.

We have seen that the spinner dolphin divides its day into two phases—a short, visually mediated period when the animals rest and a much longer portion during which acoustic emissions are constantly evident. Even at night, however, sight is surely important. The visually mediated rest period begins around midday and extends into the afternoon. It is a time of tight, cohesive, slow-moving schools. The remainder of the diurnal cycle is spent in active and sometimes scattered or widely spread schools of evident subgroup structure. Even then, as long as the sun is in the sky, we expect vision to play a crucial role. Bioluminescence at night must also provide important visual cues for school coordination, prey capture, and predator avoidance.

We wondered what these two sensory worlds were like for the spinner dolphin. In this and the following chapter we discuss our findings about both their visual and acoustic domains.

Spinner dolphins live most of their lives near the sea's surface. They surface to breathe often, usually less than 1 min intervening between breaths, although on long feeding dives they can stay below the surface for as long 4 min or more. Thus, for Hawaiian spinners, daytime activity is largely carried out in well-lighted, reasonably clear surface water. We found this environment to be a visually difficult one for humans, but one to which spinner dolphins have become adapted in interesting ways, particularly in dealing with high light levels that fluctuate widely and rapidly and with the effects of suspended particles that scatter light (Lythgoe 1979).

DIURNAL LIGHT CYCLE

McFarland (1986, 1991) and Loew and McFarland (1990) have summarized knowledge of the lighted environment of the surface sea. These discussions reveal a very different world in the nearshore sea than in the air just above it, especially with regard to light intensity changes over time. McFarland (1986, p. 390) says:

> In any given body of water, at least near the surface, light intensity varies during each day over a dynamic range of approximately 8 to 9 decades of magnitude. The most rapid change in intensity is associated with the twilight periods following sunset and preceding sunrise. . . . Differences in the rate of change of light intensity throughout the day are considerable. Near high noon, in tropical seas, light intensity changes slowly at about 0.1% per minute. During twilight, light intensity can change approximately 50% per minute. As dark clouds suddenly obscure the sun, light intensity can decline by as much as 75% over a few seconds. Even more pronounced are the changes in light intensity produced by the passing of surface waves, which can effect changes well in excess of 200% per second.

LIGHT SCATTERING

Not only does sun intensity change rapidly over short periods of time but the amount of light penetrating below the surface causes scattering to fluctuate widely as well, especially nearshore where particulate matter is abundant in the water column. The effect is to allow excellent lateral vision for many meters early in the day when the sun is low, but to create greatly reduced visibility near noon when the angle of incidence is closest to 90° and more light penetrates the sea surface. Where we worked, the water was generally rather clear, especially offshore where the halo of planktonic life that gathers around the islands and in the outwash from streams washing off the land are less evident than close to shore.

Under the worst visual conditions, when the sun was high over the rest bays and the water was cloudy with plankton and sediment, we estimated that we could see laterally about 15 m (50 ft) or sometimes less. Usually we could see twice that far, especially before the sun was high and its light had not begun to scatter to a significant degree underwater. Because concentrations of light-scattering particles drop off sharply with distance from shore this effect is much less evident in open water. This effect of light-scattering particles produces what Lythgoe (1966) has called the *veiling brightness*, and it generally tends to be greatest in nearshore waters.

This type of large-particle scattering is called *mie scattering* (Gates 1962), which refers to the aggregate reflections of particles suspended in water. In air, particles larger than about 1μm in diameter scatter light to form the white of smoke or steam. In water, light scattered by particles this size is responsible for the bluish haze that gradually extinguishes vision at a distance.

Another sort of scattering often called *molecular* or *Rayleigh scattering* is also important at sea. In this type of scattering, molecules hit by light reemit it. The scattering effect is inversely proportional to the fourth power of the wavelength, which results in blue light scattering about ten times as well as red light (Gates 1962). This results in both blue sky and blue water. Down-welling light is *backscattered* toward the surface, creating the deep blue of open ocean water when one looks downward. The same backscatter occurs laterally and creates the blue backdrop of life in the open sea for a swimming dolphin. In addition to these scattering effects, dissolved chemicals in seawater also contribute to the color of water at a given location, giving specific tints that vary from place to place.

At Kealake'akua Bay, a lateral view was typically into a blue haze; only in shallow water did the tint verge upon greenish aquamarine. The view of a dolphin school against such a backdrop was of animals sharply in focus near the observer but fading in definition and ultimately matching and disappearing into the blue backdrop at the limits of vision. Near this limit, dolphins appear as bluish-gray shapes, their patterns and outlines made indistinct by scattering. The white pattern components of their skins could be seen much farther than the darker components because the contrast of white dolphin skin to the ambient environmental hue is emphasized at a distance until it can literally "glow" in relation to its bluish background (Madsen and Herman 1980). We came to know these as "flash patterns," which were useful at the extreme limits of vision. Even beyond the limits where body outlines were sharply outlined, these distant light patches and their associated movements could signal such behavior as courtship or play.

This effect seems to relate to the complex question of perceived brightness in a surrounding dark field (see Jameson and Hurvich 1961). If the visual response curve of retinal receptors is most sensitive in one wavelength region while environmental light is mostly of another wavelength, the effective contrast of such pattern components can be en-

hanced. This is probably the basis of the frequently observed, almost luminous light patterns of odontocetes when viewed against the blue
background of the sea. The whitish pectoral fins of humpback whales,
the white lips of several dolphin species, and the pale belly patterns of
many dolphins do indeed flash out with luminous brilliance underwater
as compared to dark pattern components (see fig. 3.7 in Madsen and
Herman 1980).

Conversely, the dark backs of dolphins disappear quickly into the blue
distance. Animals traveling away from the observer and not flashing
their whitish pattern components first become overlain with blue and
soon become invisible to the human observer and perhaps also to the
dolphin observer.

FLICKER

The underwater photic environment of the near-surface sea presents
another problem to dolphins living there—*wave-produced flicker*. Flicker is
produced because the surface of the sea is seldom truly still. Two sorts of
waves perturb the surface almost constantly, even on very calm days.
These are gravity waves and capillary waves. *Gravity waves* are the familiar large-scale sea wave oscillations that can travel very long distances
and may intersect, cancel, or reinforce one another (Van Dorn 1974). In
open water, such trains arrive at a given locus from different directions
and throw the sea surface into complex and sometimes large-scale motion. Nearer shore, headlands and the rising bottom may refract and order them into the familiar trains of breakers we see crashing on a beach.
The other much finer scale waves that disturb the sea surface are *capillary waves* or surface ripples (McFarland and Loew 1983), which are local
phenomena induced by surface wind. Both wavelength and frequency
differ between these two wave states. Gravity waves pass at much less
than 1 Hz to a few Hertz, while capillary waves may flicker by at about
10 to 50 Hz or more.

At the surface, these waves create trochoidal surfaces of differing dimensions that serve to focus light down into the water various distances
(depending upon the focal length of the wave surface), producing the familiar light beams that one often sees playing down into the dark water
of clear tropical or offshore seas. We see them because such beams reflect from tiny particles in the seawater.

Gravity waves can create focusing surfaces that can reach many
meters across, while capillary waves are much smaller scale phenomena.
The focal plane depth of such a passing wave surface increases as the
wave period increases. Thus, long period waves will play light deeper
into the water and focus it far below the same effect produced by a cap-

Figure 65. Complex trochoidal patterns on a dolphin near the surface, in condition of low surface turbulence.

illary wave. Typically, because the sea surface is shaped by a complex of waves, a range of focal depths exists at any given moment (fig. 65). If there is an object below the surface of the water, it will be illuminated with a shifting pattern of bright and dark regulated by the sea surface contours above and the object's depth.

Not only does each of these waves produce this pattern of light and dark but the variations of intensity may be great and can result in intense light flashes directed downward (Schenck 1957). The result is an environment of flickering light rising and falling at different frequencies and amplitudes with moving and intersecting patterns of light and dark falling upon swimming animals (fig. 66). Spinner dolphins, as well as all other near-surface pelagic marine life, spend a fair part of their lives in this difficult visual environment. A seldom-appreciated aspect of this environment is that the focusing effect of the sea surface may produce flickers of light that can be several times as intense as the sunlight overhead (McFarland and Munz 1975).

To our eyes, such flicker may appear as a moving pattern (as just described), or depending upon the speed of flicker, it may fuse into a single brightness. Hecht et al. (1934, p. 243) describe these features of the flicker curve as follows:

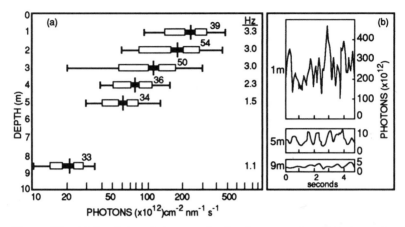

Figure 66. (a) Statistics of the magnitude of near-surface light intensity fluctuations as a function of depth in the lagoon at Eniwetak Atoll, Marshall Islands, shown are mean irradiance, standard error of the mean, standard deviation and range of measurements, numbers above line are coefficient of variation, and (b) examples of data recordings at three depths. The vertical axis is irradiance, horizontal axis is time (from Loew and McFarland 1990).

The critical frequency at which fusion of rhythmically produced illumination takes place depends for its value on a variety of factors. The most effective of these is the intensity of illumination.

Hecht and Verrijp (1934) add the following:

> In the intermittant stimulation by light the outside agent alternates abruptly between zero and a particular maximum intensity. When the frequency of this alternation is sufficiently low, the difference between these two extreme conditions is completely perceptible to sensation: the brightness during the light period is maximal, and during the dark period it is zero. As the alternation frequency increases, the two sensations become less sharply delimited in time and less clearly separated in intensity: the light period loses brightness, and the dark period gains in brightness. The more frequent the alternation, the less is the difference between the successive sensations; and when the frequency is sufficiently high the difference between successive sensations vanishes so that the outside fluctuating light appears continuous.

By varying the intensity and repetition rate of a flashing light source presented to a subject, a response curve can be constructed for a given species. The space under the curve defines the conditions for seeing individual flashes of light. Outside the curve, the flashes fuse into a brightness. McFarland and Löew (1983) have suggested that pelagic species are visually adjusted to the frequencies of flicker, often in relation to the spatial frequency of their pattern. One measure of this in temporal terms is the *critical fusion frequency,* or CFF, of a species. The CFF is a point on the flicker curve of a given species that describes the highest rate of flicker that it can perceive as a flashing light instead of a continuous light. McFarland and Loew show that the highest CFF values seen in fishes occur in diurnal epipelagic species such as atherinid fishes, who live just beneath the sea surface in an environment where both flicker rate and contrast are greatest.

Spinner dolphins penetrate this surface environment hundreds of times a day to breathe. However, contrary to what one would expect from watching them above the surface, our underwater observations suggest that most spinner dolphin movement takes place at about 2–30 m (~7–100 ft) depth. At such depths, flicker rate has slowed and contrast against the background is greatly reduced as compared to shallower depths. To a human observing below the water, flicker is not very apparent until the dolphins move close to the surface.

Nonetheless, the pattern of dark and light lines on dolphins is clearly evident on photographs and stop-framed videotapes taken while the dolphins swam 2 m or more below the surface. Clearly, in this circumstance, the human visual system fused the majority of such flicker into a constant brightness while the film and video did not.

McFarland and Loew (1983, p. 181) extend their interpretations of flicker to the evolution of vertebrates in general with these words:

> Are the close relationships between spatial and temporal vision evidence of an ancient time reference within which the physiological properties of the eyes of both invertebrates and vertebrates evolved? We think so. For example, the first vertebrates evolved in shallow tropical Cambrian seas . . . and coexistent with them were ancient crustaceans, arachnids, and molluscs. All must have experienced flickering light with frequencies not unlike those of tropical seas today.

To our knowledge, no flicker curve or CFF value has been determined for any dolphin. The CFF values for humans differ for rods and cones, with the curve for rods peaking at 30–40 Hz and that for cones at approximately 60 Hz. A dolphin's complex body patterns of dark and light can only be seen in detail if they pass the eye of an observing dolphin at a rate well below the CFF. This seems normally to be the case with dolphins, where the rate of passage of the pattern components of one dolphin in the eye of its neighbor should fall in the general range of about 0.5–3 Hz.

The importance of such flicker effects to oceanic dolphins might be profound. They could relate to the dolphin's ability to read the body patterns of neighboring animals and thus to coordinate school movement amid the clutter of flashing patterns of light. From our underwater viewing vessels, we humans could clearly make out every feature of the passing dolphins' patterns. To give an idea of our capabilities at discrimination, we could examine scars in detail and make an assessment of whether or not they had been produced by sharks. We noticed some slow-moving flicker, but it did not prevent such assessment. The dolphins we observed were deep enough that they represented low contrast subjects swimming against a dark blue background.

We look with human eyes. What might dolphins see? While the CFF relationships are a general function of the vertebrate eye (Loew and McFarland 1990), the specific responses of dolphins probably differ from those of humans so that full understanding of these relationships must wait.

The mammalian response to flicker at maximum contrast sensitivity is about 10–15% of the CFF. For a dolphin to use the patterns of a schoolmate to their best advantage, their movements should approximately match this range, which in fact they seem to do. That is, the swimming movements of dolphins in a school are relatively slow, about 0.5–3 Hz, which may make them maximally discriminable by schoolmates.

EFFECTS OF LIGHT INTENSITY ON SKIN

While the eye can average out the flickering environment of the surface sea, a dolphin's skin cannot. In other words, the focusing effects of surface flicker constantly play high intensity light across a dolphin's back as it swims in the daylight surface sea. This not only heats the animal but subjects it to heavy irradiation. It is no wonder that clear water dolphins are uniformly dark dorsally. It is also not surprising that pale captive dolphins moved to clear water, as we found when moving captives from California to Hawaii, will quickly darken.

REFRACTION

Snell's window, a refraction effect, is a clear, circular "window" or disc in the sea surface subtending an angle of 96° over the head of a dolphin traveling below the surface. Its characteristics are determined by the different refractive indices of air and water, the angle of the sun relative to the sea surface, and the depth at which the animal swims. Through this window, the animal can look out into the subaerial world, but beyond the edges of the clear disc, the sea surface appears as wavering, silvery mirror through which it cannot see.

The effect of ocean wave trains upon Snell's window is an interesting feature of the dolphin's visual environment. A dolphin swimming near the surface in a regular train of swells cannot see up into the passing swells until they reach the proper angular constraints of Snell's window. Then, the window will be partial depending upon the animal's position and the angle of the sea surface. Some prey fish such as flying fish, halfbeaks, needle fish, and others have been seen swimming in such trains of swells where they may have some degree of protection from predators simply because they travel in the "hump" of water that takes them out of the line of sight of a predator who does not occupy the swell with them.

Dolphins may use Snell's windows during rough weather to locate appropriate places to surface and breathe where they can avoid tumbling water during the course of a breath. We have seen such windows ourselves from our underwater vessel and noted how simple it should be for a dolphin to locate "good breathing water." The turbulent areas were marked by bubbles and the smooth water by a mirror-like surface.

THE DOLPHIN VISUAL SYSTEM

Dolphins possess a *double pupil* (Herman et al. 1975) that may relate to the shallow water in which they live. A curtain-like extension of the upper iris margin, the *operculum*, comes down over the pupil in bright light leaving only two tiny pupils in each of the two lower lateral quadrants of the eye (fig. 67). The function of the operculum is to form two very small pupils, each smaller than can be produced by the circular reduction of pupil size in the human eye. These two pupils give the dolphin several advantages. First, the dolphin eye can "stop down" to two tiny pupils in the high intensity environment of the surface sea. These small pupils may help avoid the problem of focusing very intense light on a single point of the retina. Instead, two points of lesser intensity can be used. A second advantage is that the pupils in spinner dolphins are on the lower quadrant of the eye, which, in a normally swimming dolphin, generally point slightly downward in the direction of the bluish water beneath the animal rather than directly into the flickering glare of the surface.

The double pupils have a third possible advantage. As we will see, the visual fields of the two spinner dolphin eyes have at best a very modest field of overlap and hence the opportunity for binocular vision and the depth of field it brings is scant. Computer analysis of eye movements of two *Tursiops* showed binocular correlations between the eyes less than 10% of the time (Dawson et al. 1972, Dawson, 1980). Also, a nerve degeneration study of *Tursiops* has shown complete crossing of the optic chiasm by nerve fibers from the eyes and hence no evidence of binocular vision (Jacobs et al. 1975).

Figure 67.   The iris of a Pacific white-sided dolphin (*Lagenorhynchus obliquidens*) in bright light, showing paired pupils in a single eye.

Instead, could these two pupils, working together in the same eye, by their minor overlap and by motion of a subject across the retina through the two pupils, impart a measure of depth perception in a dolphin? Madsen and Herman (1980) have suggested a mechanism for obtaining depth of field from a single eye. They postulate that two fixed foci may exist, one specialized for close-up vision with a far point of 1 m (3 ft) or less, and a second focal point for distance vision with a near point of 2.5 m (8 ft) or greater. Herman et al. (1975) provide experimental support for this view and suggest that the double slit pupil may be its cause. The double pupil may also assist vision throughout the large visual field of the dolphin, allowing focus both forward and to the rear simultaneously.

A final benefit of the double slit pupil is that it allows dolphins to have sharp vision both above and below water during strong daylight (Herman et al. 1975). This ability exists even though the water-adapted cornea of a dolphin in air should produce gross myopia (Dral 1972). The cause of this effective above-and-below water vision is a matter of debate, but it may be due to the *pin hole effect* in which the tiny pupils themselves act to focus light instead of the lenses or corneas. Because the pupil is so small, it actually collimates light to the eye.

VISUAL PIGMENTS

McFarland (1971) has investigated the visual pigments of a number of cetacean species, including the spinner dolphin. The expectation that the sensitivity peaks of these pigments will match the color of the predominant light in a given habitat is not exactly realized. He comments (pp. 1073–1074) as follows:

> Recently it has been proposed for fishes that visual pigments are often not precisely coincidental with the predominate wavelengths of a given habitat, but rather, are offset somewhat to enhance visual contrast of objects viewed against the luminous aquatic background. . . . Lythgoe's suggestion offers an alternative explanation to the "sensitivity hypothesis." . . . Cetaceans, perhaps more than fishes, offer several advantages for the evaluation of λ_{max} position and its significance to underwater vision. For instance, as indicated, cetaceans do inhabit a variety of distinct photic environments and this distribution correlates broadly with the λ_{max} of the photopigment in each species. λ_{Max} is the measure of the point of maximum absorption of a visual pigment by wavelengths of light.
>
> Most cetacea are not restricted to a single photic environment, however, but show daily movements that drastically alter both the intensity and quality of ambient light they encounter. For example, bottlenose dolphins and spinning porpoises off Hawaii freely move from inshore to offshore waters during the day where sub-surface light is bright and varies from greenish to bluish in hue. . . . In addition, they dive to depths of 100 m or more where light of solar origin is attentuated and more monochromatic than at the surface. . . .
>
> Since cetacea possess predominately, if not exclusively, a pure rod retina and a single rhodopsin . . . they cannot utilize multiple visual pigments to maximize both sensitivity and contrast.

More recent inspection of the dolphin retina (Dawson 1980) suggests that it is indeed *duplex,* meaning that it is composed of two classes of light-sensitive cells. This indicates that contrast enhancement could be based on receptor cells, as well as on the visual pigment absorption maxima. That is, if the two classes of photosensitive cells possess pigments with nearly the same wavelength of maximum absorption, they may serve to enhance the sensation of contrast. Conversely, if the wavelength of peak absorption (λ_{max}) is the same, no contrast enhancement will result.

VISUAL SIGNS, SIGNALS, AND VISUAL FIELDS

The maneuvering of schooling and flocking animals appears to be ordered by a *sensory integration system,* or SIS. Each group member may use this system to receive and transmit environmental information throughout the group by means of movements and sign stimuli (Norris and Schilt 1988) (See also chap. 13). Such a passage of information may vary in speed from waves of influence passing seriatim between school members at up to 8 m/sec in some fish (Radakov 1973), to the passage of visual information that is essentially instantaneous provided animals do not interfere with one another. Acoustically passed information is slower, but still far exceeds the speed of any predator.

Both such sensory integration and the social processes of a spinner dolphin school appear heavily dependent upon various sign stimuli provided by the animal's pattern and body form, although our evidence is mostly inferential from behavior we have watched. For instance, several details of a dolphin's pattern may well be useful in the transmission of information about changes in body posture or attitude. By closely watching each other, spinner dolphins should be able to predict how a schoolmate will move moments before it does. But before describing examples of such putative sign stimuli, we first need to assess the sensory fields of spinner dolphins so that we can perceive how movement affects them.

Visual Fields

We define *sign stimuli* as fixed body patterns involved in visual message transmission and *signals* such as structural or pattern features involving movements, sounds, or smells as integral parts of their message transmission. The functioning of these must be dependent in part upon the remarkable visual field of the spinner dolphin, which includes vision ahead of the animal, 180° to each side, and large fields above and below the animal (fig. 68). If overlap between the sides of these fields exists, it seems to be modest, both in front of the dolphin and directly over its tail.

Such fields are apparently enhanced because the dolphin eye can be moved in and out within its socket to a degree by engorgement or drainage of the opthalmic rete behind the eye (Galliano et al. 1966). In this way, spinner dolphin eyes can be raised on prominences above the general body contour or they can be drawn in flush with the contour of the head. The eyes of spinner dolphins are situated at approximately the widest point of the head, allowing the animal maximum opportunity to look forward and backward.

We attempted to describe the spinner dolphin visual field by measuring the planes in which the pupil of the eye was visible from in front, behind, and below the animal. To make these measurements, we took a captive spinner named Kehaulani out of the water at Sea Life Park on Oahu and measured her visual field with a large goniometer built for the test. Visual angles relative to the body axis were measured using the outermost surface of the cornea as the apex of an angle described by

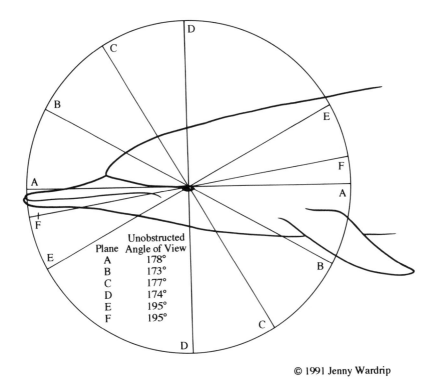

Figure 68.   Visual field of a spinner dolphin at six planes as determined by goniometry.

lines extending from that point to the first point where the dolphin's visual field was obstructed by its own body. These lines define the visual field of the dolphin in a given plane. The small angles in planes A–D are primarily a function of the supraorbital ridges that are located above and extending just anterior to the eye. They tend to limit vision above and ahead of the animal in regions covered by the echolocation field. Larger angular fields were found in planes E and F of figure 68.

In this spinner dolphin, the possibly slightly binocular forward field of vision above the level of the jaw did not appear to be as strongly developed as we have noted it in the bottlenose dolphin. An observer directly behind a spinner dolphin can clearly observe both corneas simultaneously, even when the eyes are flush with the contours of the body. But an observer in front of the animal, looking down from above the jaw, can detect both corneas only when the eyes are protruded, presumably by retial engorgement. Thus, spinner dolphins have available almost complete visual fields around their bodies both behind and in front. The slight impairment created by the supraorbital ridges is avoidable by slight movements of the animal relative to the object it is viewing. With the early warning system of echolocation functioning, no major sensory impairment is expected to occur.

Whole-Body Signals

The distant flashes of reflected light from the bodies, especially the bellies of dolphins, seen at the limits of vision are what we refer to as *whole-body signals*. We have no doubt that these signals, combined with their associated motor behavior, are distant indicators of social events in the school. Not only are body postures signaled by this means, but indications of speed, number of animals involved in a behavior pattern, and age of participants (juvenile versus adult) may be indicated. We learned to recognize three whole-body signals while observing underwater from the viewing vehicle *Maka Ala* and from film studies taken from the same vehicle. These are belly-tilts, tilt-aways, and belly-ups.

Belly-tilts (a tilt of the belly of one dolphin toward another) (fig. 69) and *tilt-aways* (the opposite rotation) are visual displays effective even when very subtly expressed. Even a slight tilt of the belly of one dolphin toward another, representing an axial rotation of only 5°, was predictive of continued and increased interaction between two animals. In a tilt-away, the slight tilt of the dark dorsal surface of one animal toward another predicted a cessation of association. Such rotations produce clear signals of dark or light to the observer below water (see chap. 14). *Belly-ups* occur when one dolphin swims near another in the inverted position. They seem commonly to be invitations to genital contact (see later). Although poor visibility was a great source of frustration to observers in the *Maka Ala* viewing chamber, the experience at such times of being able to see nothing but a glint of white from the belly of an inverted or tilted animal emphasized the visual importance of these patterns.

In our observations, few belly-tilts were given without another animal in the immediate visual field (10%). The visual nature of these tilting patterns is emphasized by the countershaded pattern of spinners and the near-vertical direction of incoming light when the animals are some distance below the surface. In fact, the patterns give rather precise angular information about the sender to a receiving animal. Belly-tilts occur during the meeting of two animals, during prolonged pair-swimming, during partner switches, and as a final display before the separation of two animals in a behavioral bout. Considering the limited movement available to these streamlined animals, such gross eye-catching movements are probably relied upon as general signals of recognition and acknowledgment.

An incident seen from the *Maka Ala* that involved tilt-aways is of special interest. It involved an animal rubbing its rostrum on another's tail

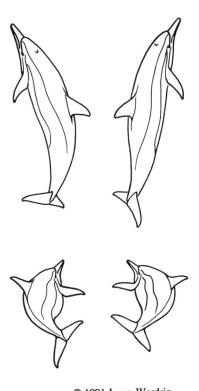

© 1991 Jenny Wardrip

Figure 69. A pair of spinner dolphins belly-tilting. Top, tilt-away;
bottom, tilt-toward.

stock. The recipient animal tilted-away, righted itself, and swam rapidly
away. Often in our motion picture analyses, we have seen tilt-aways occur
within triads, with the recipient animal being left as the other two animals engaged in a caressing bout. Such tilt-aways seldom elicited overt
aggressive behavior from a recipient. The tilt-away may be an act that
hides the white belly of one animal from another and/or places the pro-
truberent dorsal fin between one animal and the next.

    Belly-tilts and belly-ups seem to be linked in sequences that may lead
to overtly sexual behavior. We note, however, that our assignment of a direct reproductive function is our own, and perhaps may not be that of
the animals involved. We came to feel that throughout much of the non-
estrus year, much apparently sexual behavior is in fact primarily communicative and not a direct part of the reproductive process. However,
belly-tilts do seem at times to be involved in tests of sexual receptivity be-

tween heterosexual pairs. When we compared behavioral frequencies and serum hormone concentrations for the Sea Life Park captive spinners over a 14-month period, we found a significant relationship between the frequency of simultaneous or rapidly alternating belly-tilts (ventral presentations) and the occurrence of high concentrations of reproductive steroid hormones (Wells 1984) (also see chap. 9).

From the standpoint of signal "power", the belly-tilt is a shorter and less powerful posture than the belly-up (fig. 70). Only about one-fourth (22%, $n = 211$) of sequences in which at least one belly-tilt occurred evoked other belly-tilts or belly-ups. But nearly half (46%, $n = 103$) of sequences involving at least one belly-up involved posturing by the partner. Nongenital contact was more often associated with belly-tilts (48% of tilt sequences, $n = 103$) than with belly-ups (27% of belly-up sequences, $n = 211$). As might be expected, belly-up sequences often lead to genital contact (24%, $n = 211$). Conversely, in over half (56%, $n = 103$) of sequences involving genital contact, there was first belly-up behavior as a sort of foreplay. It is interesting to note that in every case ($n = 14$) where belly-tilting adults made genital contact, they had previously turned belly-up. But to put the belly-up pattern in its proper context, we note that in 76% of the 211 sequences with belly-ups, there was no genital contact involved. It is important to note that observations of the captive school at Sea Life Park Oceanarium showed that genital-to-genital contact was associated with high levels of reproductive hormones (see chap. 9), while these other patterns were not.

These three whole-body signals are probably very important in instantaneous integration of the social state of an entire school. The importance of such signaling is underscored by the relationship between the average water clarity in the habitat of a given dolphin species and the extent of the *flash elements* of its body pattern. The murkier the water in which a species lives, the more completely its entire body is involved in such flashing. Spinners and their dolphin allies from the ocean sea typically have tripartite patterns of a dark area on top, a gray panel on the sides, and a pale underside, along with considerable complexity in pattern structure in the various sectors of the body.

Coastal shelf animals, living in water with lower visibility, are often boldly patterned with dark and light (see Heyning 1989). Examples are Dall's porpoise (*Phocoenoides dalli*), the killer whale, (*Orcinus orca*), Commerson's dolphin (*Cephalorhynchus commersonii*), the spectacled porpoise (*Australophocaena dioptrica*), and the hour-glass dolphin (*Lagenorhynchus cruciger*). Murky water species may use the entire body to produce flash signals. The pure white belukha (*Delphinapterus leucas*), the pale humpback dolphins of the genus *Sousa*, and the various pale river dolphins come to mind. Even the ubiquitous bottlenose dolphin (*Tursiops truncatus*

© 1991 Jenny Wardrip

Figure 70. A spinner's pectoral fin and how it may signal an incipient change in direction.

and allies) has pale populations in murky water situations and darker, more contrasty animals in open water situations.

The Pectoral Fin and Associated Patterns

In contrast to the large white pattern components, which can be used in signaling at a distance, dark pattern components on an otherwise light-colored body should be optimal for signaling between nearby animals. The pectoral fin and flipper stripe, which runs from the anterior insertion of the flipper to the eye and angle of the gape (Mitchell 1970), of spinner dolphins together provide an example. These elements act as a single dark signal unit that contrasts clearly against the white of the throat and belly. Because the flipper stripe is a fixed pattern and the dark fin can be moved in various ways, together they form a unit that can instantly indicate incipient movements of one dolphin to its neighbors (fig. 70). All a neighboring dolphin has to note is whether or not the diagonal black band formed by these units together is in a broken or straight line. When the dolphin is cruising forward without deviation from a given course, the diagonal of the flipper stripe is unbroken and straight. When the animal is about to dive, however, the pectoral fin moves and may precisely indicate the amount and kind of incipient movement that a neighbor can expect. Thus, it seems that this complex pattern component may be important in allowing coordination within the school during maneuvering.

The normal echelon position of dolphins swimming in a school places the eye of one neighbor directly adjacent to the pectoral fin pattern of another where it fills much or all of its visual field, perhaps allowing the almost instantaneous synchrony of a school we have observed. In this position, a tiny movement of a pattern element should be perceptible by a neighboring animal. If the animals were farther apart, reaction speed would be reduced as perception of change became more difficult.

This pattern complex may have other uses besides maneuvering. For example, during aggressive displays, the pectorals are sliced back and forth in a rapid scissoring movement. When one dolphin faces another, as in aggressive encounters, such movements slice the dark fin over the immaculate white of the belly.

The pectoral fin notch located at the posterior insertion of the fin also may have special signal value. When the pectoral fin is in the "relaxed position," the notch is hardly evident, being all but covered by the inner posterior lobe of the pectoral fin. But when the fin is thrown forward, a roundish white extension of the belly-white is exposed as the fin moves forward or away from the body. This specific pattern arrangement is found widely in odontocetes (including *Cephalorhynchus, Lagenorhynchus, Orcinus,* and *Delphinus*) (Norris and Dohl 1980*b*), perhaps because there is wide utility for one cetacean to know about the incipient movements or emotional state of its neighbors, as told by the state of these pattern components.

Other Body Patterns

The body pattern of spinner dolphins is roughly composed of a dark gray cape, a lighter gray lateral field, and a white ventrum (Mitchell 1970, Perrin 1972*a*). The margins between these fields are sharply demarcated in the spinner dolphin and frequently in other odontocete species. They may be enhanced by a darkening of the margins between panels in some species such as *Lagenorhynchus obliquidens* and in the spinners of some regions (Perrin et al. 1979). They may even be speckled at these margins as is seen in the whitebelly spinner.

We suggest that the three major longitudinal pattern components may allow dolphins to determine instantaneously the degree of rotation of an adjacent neighbor. If the rotation of the back is toward a neighbor, the dark cape increases in area at the expense of the belly field. The belly-white covers a larger area if the animal rotates the other way. Such attitudinal pattern relationships may be used as part of the overall system of cues that allow school synchrony during dives, turns, and other movements. They would therefore be integral parts of the sensory integration system of a school, as described in chap. 13.

The closer dolphins come together, the more effective such lateral fields should be for signaling. The enhanced margins of these lateral fields may promote precision in the estimation of body attitude and speed of movement. The 1-m (3-ft) focal length of the bottlenose dolphin eye mentioned by Herman et al. (1975) may be related to close inspection of such patterns by adjacent schoolmates. This distance does indeed approximate the usual distance we have observed between adjacent swimming dolphins in active schools.

None of these arguments, in our view, negates the possibility that the tripartite pattern of most spinners may also serve as countershading, or as a shield against high levels of radiation. Most animal adaptations enter multiple functional domains, having been produced by the adaptive process of evolutionary approximation in relation to fitness of the entire animal.

The Rostrum

The dark rostrum of Hawaiian spinner dolphins may act as a pointer during school maneuvers (see chap. 13). In any case, we were able to predict the direction of turning movements of a school by watching the rostra of dolphins initiating a turn.

The snout tip of adult Hawaiian spinners is jet black in contrast to the white of the jaw tip of adult spotted dolphins (*Stenella attenuata*), although this dichotomy is muted in the eastern tropical Pacific. When these two species swim together, as they habitually do in the eastern tropical Pacific Ocean, it is easy to separate species at a glance. It seems that the jaw tip could be a sign that serves for species recognition and could also emphasize the aggressive posture of one animal toward another (see chap. 14). When such dolphins are facing one another, they lower their snouts and arch their backs in an aggressive posture. The effect is to lower the dark jaw tip over the immaculate white of the belly, thereby emphasizing the posture.

Locomotory Signs

If one watches dolphins from behind, as schoolmates frequently do, one can see that the diagonal patterns of the flanks extend downward onto the tail stock and thus contrast the tail and its movements against the belly-white. For a following dolphin, these pattern elements probably provide precise information about the degree of excursion of the tail, its rate of movement, and hence the speed of the lead animal. They also emphasize the bending of the tail stock and hence the intent to dive or to change course.

Sexual Pattern

At the posterior end of the white belly of spinners, the lateral field dips downward and ends in a pair of dark brush marks extending forward onto the belly-white, just above the genital region. The ventral margin of the caudal keel may be light-colored and also extends forward into the tapering white belly. At this juxtaposition are located the anus and the genital slit. As is true in many cetaceans, such pattern marks serve to pinpoint the positions of these anatomical features. Both sexual and mother–young encounters are doubtless mediated to some degree by them. Nonetheless, sexual dimorphism in spinners is modest. It is exhibited mostly by thickening and muscularity of the tail stock, a moderate postanal hump, and a dark erect dorsal fin in adult males, and associated behavior, and by the absence of these features in adult females.

Having outlined these features of the visual domain in spinners dolphins, we now move to the other great sensory modality in their lives—sound.

EIGHT

The Acoustic Domain

Shannon M. Brownlee and Kenneth S. Norris

Whenever spinner dolphins are active, they are noisy. Rest, at least when spinners enter bays, can be a time of almost complete silence. Some schools also must rest far at sea and these are not considered here. As the dolphins traverse slowly back and forth over sandy bottom, and if the school is of moderate or small size, only sporadic low level clicks and occasional whistles are heard. Overall sound abundance correlates generally with activity level, being low in low activity states such as rest and the slow-moving part of zig-zag swimming and high when schools are moving more rapidly.

We also found the noisy times of a spinner's 24-hr cycle were easily divisible, both by overall emission rate per animal per unit time and by the range of sound types. When dolphins arouse from rest and begin their daily zig-zag swim, their vocal emissions oscillate between the extremes of nearly complete silence and vociferous sound emission. Toward the end of this period of oscillation as the dolphins make their traverse out to sea at dusk, a crescendo is sometimes heard that we jokingly called the "Yugoslavian News Report."[6] It is a complex interplay of several sound types that was as unintelligible to us, who do not speak slavic languages, as any foreign language radio station would be.

During the nighttime feeding traverses along the island slope, our occasional listening and recording showed that dolphins continue to be very vocal. The percentage composition of emitted sound types changes again, compared to daytime recordings.

Our attempt was to describe the broad daily outlines of spinner dolphin acoustic emission. This meant typifying sound types and how fre-

6. Note that there is no true Yugoslavian language, only Serbian, Croatian, Slovenian and so on.

quently they were emitted for each of the recognizable parts of the 24-hr behavioral cycle. As these cyclical patterns become clearer (and they proved to be quite regular), we began to think about their broader behavioral context. How did they seem to be used by the dolphins? This was a much more difficult exercise than the simple description of sounds and their frequencies of occurrence, and we regard the hypotheses we put forward as tools for future thinking, not firmly grounded fact.

We say little about the details of recorded sound exchanges among individual dolphins since our focus here is on the sound emission of schools. A whole realm of acoustic study that may unravel much of the detail of school structure and function lies ahead in the analysis of individual spinner sounds, as Tyack (1986, 1991) has begun to show.

THE ACOUSTIC ENVIRONMENT

Before describing our findings, it is useful to consider the acoustic environment of a dolphin school and the limits and opportunities it provides for the use of sound in both the communication and echolocation of spinner dolphins. A concise way of thinking about the underwater acoustic environment and the problems faced by a phonating or listening dolphin is to consider the terms in the *sonar equation* (Urick 1975). This is an expression used by designers of human-made sound navigation systems (sonars) to predict how such gear will work in a given environment and to quantify the factors that might influence performance. These terms describe the circumstances of social communication as well as those of echolocation.

Urick gives the sonar equation, which involves two-way transmission, as follows:

$$SL - 2(TL) + TS = NL - DI + DT$$

where SL = the source level of a projected signal
  TL = transmission loss between source and target
  TS = target strength
  NL = noise level (ambient)
  DI = receiving directivity index of sonar system
  DT = detection threshold

(Note that dolphins use systems in which much of their economy is achieved by using a highly directive source, unlike some sonar systems that broadcast a signal widely and pick up directionally.)

The following sections discuss the terms in the Urick sonar equation from the standpoint of wild dolphin acoustics.

Source Levels and Beam Characteristics

Dolphins are known to have considerable control over the characteristics of their click trains, including source levels (Norris et al. 1967, Murchison 1980, Au et al. 1978, Mackay and Liaw 1981, Au et al. 1985). The bottlenose dolphin, at least, is able to emit trains of quiet clicks or to increase them to the remarkable measured intensity of 228.6 dB (re 1 μPas 0.92 m), which is near the finite limit of sound where further energy begins to produce heat instead of sound. These brief sounds are both emitted in very tight beams (3-dB beam width at 100 kHz, 9°) (Norris and Evans 1967, Au 1980) and received in a similar tight cone of reception (Au and Moore 1984). The bottlenose dolphin can send and receive sounds ten times as broad as the range of human hearing (Johnson 1967, Popper 1980).

Distortion of the Sound Path, Transmission Losses, and Bubble Effects

Because dolphins spend most of their time close to the surface, they face special sound propagation and backscatter problems. These can be produced by the refractive and reflective effects resulting from a number of oceanographic features such as the layering of water masses of different temperatures or salinities and the boundaries of internal waves. Doppler shifts and multipath propagation may be important at sea, perhaps produced from wave trains moving in various directions relative to the phonating animal.

Sound is bent at interfaces in amounts described by Snell's law because of propagation velocity changes from medium to medium. Local oceanographic complexities may even produce "shadow regions" where sound does not penetrate because it is routed around a volume of water by such refraction, or by "acoustic clutter" such as that produced by bubbles (Clay and Medwin 1977).

Perhaps the most important effect from the standpoint of phonating dolphins is the acoustic behavior of small bubbles, which can be very efficient scatterers and producers of sound. Clay and Medwin (1977, pp. 194–201) describe their scattering effect this way:

> At resonance, the scattering and absorption cross-sections of a typical bubble at sea are of the order 10^3 its geometrical cross-section. . . . During stormy periods, the water within meters of the sea surface contains enough bubbles to alter the speed of sound by several meters per second. . . . A sound beam propagating through bubbly water attentuates and backscatters.
>
> To understand the amplified acoustical cross sections [of such bubbles] it helps to realise that the bubble at resonance is effectively a "hole" of very

low acoustic impedance compared to that of the water medium. This hole distorts the incoming acoustic field over a very large volume surrounding the bubble.

The underwater bubble plume produced by a breaking wave entrains bubbles of various sizes and these have various fates. The smallest bubbles quickly dissolve and the largest ones quickly rise to the surface. The ones in between (about 4–60 μm in diameter) may persist in acoustically significant populations for as long as 5 hr (Gavrilov 1969) in the top 10 m (33 ft) of the sea. Glotov (1962) found that the largest number of such bubbles per unit volume have a radius of 60 μm and have a resonant frequency of 54.2 kHz (Clay and Medwin 1977), which is about the median frequency range found in dolphin echolocation click trains.

It is for these reasons that we have wondered if a spinning dolphin, which draws a plume of bubbles down into the water as it reenters after its aerial sortie, might be producing a recognizable acoustic target that would allow schoolmates to assess its position. By assessment of the aggregate of all spinning animals during a short period of time, a school member could determine the approximate dimensions and disposition of the school envelope as it moves along. Since the frequency composition of echoes from such a plume will change with the size of bubbles making up the plume, an echolocating dolphin could possibly gain an idea of the recency of a spin. It seems likely that both the dolphin reentry plume and a breaking wave would produce good echoes, but we did not have the equipment to measure this. It would clearly be interesting to do so.

In addition to providing a topography of acoustic reflection and scatter, clouds of oscillating bubbles are also sound emitters (Medwin and Breitz 1989, Longuet-Higgins 1990), contributing heavily to the production of ambient sea noise centered in the high audible–low ultrasonic region.

Target Strength
Spinner dolphins feed on rather small, deep-scattering layer prey (12–50 mm total length for some squid and shrimp recovered from stomach contents) (Norris and Dohl 1980a) (see chap. 12). Many of the fishes spinners are likely to catch have swimbladders containing gas bubbles capable of such strong backscatter that a single small fish can make a strong echolocation target (Clay and Medwin 1977). Marshall (1954) lists about one-third of the mesopelagic fish fauna on which spinners feed as having swimbladders. This includes myctophids, stomiatoids, gonostomatids, hatchet fish, trichuiroids, and some melamphaeids. Entire schools of closely spaced organisms may produce low frequency targets useful for location of food at some distance.

Ocean Noise

All dolphin acoustic navigation and communication is done against a backdrop of ambient ocean noise. This noise results from several sources, including the anthropogenic sounds of ships and other machinery, biological noise from the many phonating organisms in the ocean, waves, seismic activity (such as underwater eruptions), density currents, wind, rain, bubbles, and thermal noise (Tolstoy and Clay 1966). Ocean noise occurs both as discrete sounds if the listener is near enough to a source or a hiss of broadband noise integrated from many sources that is a background over which all other sounds must be heard.

The amount of this noise varies markedly from place to place, even within the daily routes taken by spinner dolphin schools. The dolphins must both navigate and communicate in the face of such noise, or in acoustic parlance, they must deal with signal-to-noise ratio. Dolphins are capable of extracting meaningful signals from considerable levels of such noise.

In the shallow water environments frequented by the Hawaiian spinner dolphin, recordings are usually made against a random background of sharp broadband clicks thought to be produced by small, bottom-dwelling snapping or pistol shrimp (Family Alpheidae), forcibly closing their chelipeds. These sounds extend into the low ultrasonic range and can sound much like single dolphin clicks (Ritzmann 1974). To the ear, they sound in aggregate like the sharp crackle of cellophane being crumpled or bacon frying. Pistol shrimp are shallow water animals, so this part of the acoustic background will decline in deeper water. Here, water noise from whitecaps or tumbling waves, or motion noise produced by the recording hydrophone and boat being moved through the water, may produce a masking background against which dolphin sounds must be detected.

The problem of dolphins being able to hear through the water that constantly rushes past their ears was somehow solved at some point in dolphin evolution. All we know is that measurements of their locomotion indicate a partially laminar flow past their bodies (Lang 1966). Motion noise is a near-field effect involving the bulk movement of water past a listening device (van Bergeijk 1966) and hence of very short range. If a smooth, nonturbulent transition between the moving surface of the animal and the relatively still water beyond has been achieved in dolphins, even only over the points of entry of sound into their bodies, they should be able to hear better than we can usually hear with our mechanical devices.

Dolphins are clearly adapted perceptually to extract information from the background of ocean noise. Both their click trains and whistle emissions seem to be conformed to maximize the animal's opportunity to do this. The clicks are given in long, repetitive trains having a repetition rate that is idiosyncratic to a given emitting animal, and whistles are usually signals that change frequency in a rather stereotyped pattern over time. Because the dolphin must hear some part of its own click trains or whistles as the sounds are emitted, it may perform autocorrelations between outgoing sounds and returning echoes, both in terms of the frequency composition of the individual clicks and their repetition rates and the frequency composition of whistles through time.

Most dolphin species seem to emit a click and wait for its echo before emitting the next signal. This is thought to allow direct comparison over many clicks of such click–echo pairs. By use of expectancy that an echo will occur at a given time, they can extract the echoes from deep in the background noise. The first experiment to show this relationship was done with *Tursiops truncatus* (Norris et al. 1967), but note that the belukha or white whale (*Delphinapterus leucas*) does not always place echoes in the interspaces between outgoing clicks (Turl and Penner 1987). Echo placement after an emitted signal allows the assessment of time–separation–pitch, a psychophysical phenomenon produced by the repetition rate apparent as a pitch of a given frequency related to the distance from a target (Nordmark 1960). Both the repeated trains of clicks and the slight variations in the rate at which they are emitted may allow an animal to pull the echoes of its own sounds out of noise.

Whistles, too, seem conformed to minimize the signal-to-noise problem for a listening dolphin. Typically, they are frequency modulated signals given in a series, and each dolphin is thought to have its own signal called a *signature whistle* (Caldwell et al. 1973). Tyack (1986) has shown that a second dolphin may open a communication exchange with a whistling dolphin by mimicking its whistles. Thus, a communication channel between two animals out of visual contact can be established. Each animal in such an exchange should be able to pinpoint the other by expectancy of a particular sound contour. Because the signals are given in rhythmic series, it can expect "information" to occur at a given time in any communication exchange. Since many dolphins may be whistling at the same time, a sort of "acoustic map" of the school can be produced for its members.

Detection Thresholds

Dolphin audiograms done for *Tursiops truncatus* (Johnson 1967) show that for this species, dolphin hearing extends to about ten times the upper limit of human hearing, or slightly more than 150 kHz. Sensitivity is also great, with maximum levels of -50 to -60 dB (re 1 µPas) in the region between 40 and 70 kHz (Popper 1980). No information of this sort exists for spinner dolphins, although the frequency composition of their clicks suggests that their hearing is at least roughly similar to that of the bottlenose dolphin.

SOUND EMISSION PATTERNS

Here we attempt a statistical comparison of the abundance of the various kinds of spinner dolphin sounds emitted during each of the various behavioral states we have recognized throughout the 24-hr cycle. Because we could not simultaneously listen and observe underwater, the record does not allow us to do a meaningful analysis of more specific circumstances of these sound emissions.

However, we do know the daily behavioral sequence of spinner dolphins clearly, and we know that this sequence is repeated from day to day without basic change. This gave us an unusual opportunity for correlation between the diurnal patterns of behavior and the sound emissions of a wild cetacean. Most other research that has been attempted on correlation between behavioral state and sound emission (e.g., Taruski 1979) has lacked the clear behavioral baseline available to us. At this writing, for cetaceans, only the work of Tyack (1976) on Argentine *Tursiops* schools and the studies of Ford and Fisher (1983) and Ford (1984) on killer whale dialects have developed solid correlations of this type. Thus, while our work represents a step forward in assigning sounds to the broad classes of behavior seen during a diurnal cycle, it says little about the specific uses of such sounds. Toward the end of this chapter, however, we go beyond the limits of these data and erect an hypothesis for others to test. It describes how we think some sounds, especially whistles, might function in a dolphin school.

Data Gathering

Our data gathering was based on two simultaneous assessments. First, we assigned each recording to a given behavioral state in the daily cycle, and we are confident that our assignments are correct, that is, that we were listening to dolphins during rest, waking, zig-zag swimming, feeding, and so on. Our work over the many months of the study showed these states to be predictable and repetitive.

Second, to calculate the number of phonations per animal per unit time, we first estimated the number of animals in the schools that we were recording and then made an estimate of the percentage of the school that was being recorded. This was an approximation made by first attempting to make recordings with the dolphins approaching the hydrophone and simultaneously making a count. If a school passed by during recording, as it frequently did, we attempted to estimate the number of dolphins coming toward the recording hydrophone versus those that had turned away or those that had swum past it. For this analysis, our data were culled to be loud, clear records from approaching schools. In our recordings, it was typical for sounds to be abundant and loud as a school approached the hydrophone and to fade as the school passed by, even if the retreating school was not far away when listened to. Thus, we challenge the common contention that both dolphin whistles and burst-pulsed sounds are both nearly omnidirectional. Measurements at sea are needed.

To arrive at a figure for the number of animals, we first estimated school size by attempting a count of the animals present. This involved watching the school for a time to determine the synchrony of surfacing of the various subgroups, estimating their numbers, and summing for the school as a whole. We made repeated estimates until we were reasonably confident of the approximate number of animals involved. These estimates were at times correlated with estimates from shore, the cliff, and the airplane. Such comparisons allowed calibration from these different vantage points.

Then, when the hydrophone was lowered, we spoke a running commentary into a hand-held cassette recorder, describing the movements, headings, and distances of animals from the hydrophone as the recording was made. This allowed us later to select segments of tapes in which the preponderance of animals clearly moved directly toward the hydrophone and were within good listening distance (about 70 m or less). Undoubtedly such visual estimates contain considerable error since animals underwater may not all have faced our hydrophone. Factors in our favor are that most dolphins in a school typically do move in the same general direction and that our recordings were long enough that minor movements by individuals should have averaged out. At any rate, some very clear correlations emerged between frequency of sound emission and behavioral state (fig. 71). This is probably as much a tribute to the magnitude of the acoustic differences we sought to define as it is to our success at standardizing.

Conversely, our ability to make such clear discriminations waned when we sought to typify less well-defined behavioral states, such as when comparing two different transitional stages in which the level of arousal was similar. For example, our ability to discriminate acoustically between schools entering rest coves in the morning and awakening schools after the rest period is doubtful.

Since we were unable to make useful visual estimates of school size in the dark, our estimates for feeding schools were made during the relatively few times when the pattern was seen at dusk. To accomplish this,

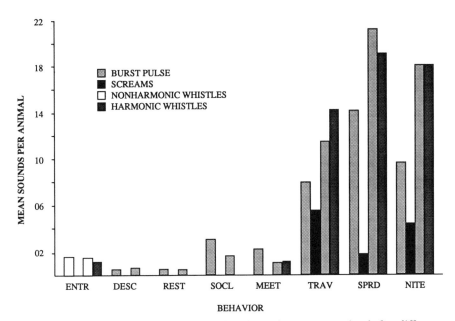

Figure 71. Mean number of sounds of various classes, per animal, for different classes of behavior.

schools were followed from rest in Kealake'akua Bay and watched through spread formation until the subsynchronous diving that typifies feeding schools was seen.

Classes of Sounds

Before describing our correlations, we must first outline the classes of sounds used in our analysis. Spinner dolphin sounds are presently indistinguishable from the sounds of other oceanic dolphins such as other species of *Stenella,* and *Delphinus* that feed on small scattering-layer fishes, shrimps, and squids (for a possible exception, see Thomas et al. 1983). All these species emit broadband clicks, pure tone whistles with harmonic structure, and a diverse repertoire of burst-pulsed signals.

To the listener, burst-pulsed signals include an array of very distinct sounds and invite application of names such as quacks, blatts, banjo twangs, barks, chuckles, or even lowing sounds like those of cattle (Busnel and Dzeidzic 1966). We avoid using most of these classes here, except among ourselves in the field, but we do define the two clear kinds of burst-pulsed signals that can easily be discriminated instrumentally on the basis of physical structure. These we call repetition rate signals and multiple transient signals.

Repetition rate signals are those in which the human listener hears the repetition rate of click trains as a frequency or pattern of frequencies. Analysis breaks these down into very rapid trains of clicks. Dolphins appear to be adapted to separating clicks out as individual signals from such trains at much shorter intervals than humans can (about 100 μsec) (Bel'kovich and Dubrovskiy 1977). Thus, it is not clear what the dolphin hears, even if a human listener can hear the envelope of such sounds and can make the presumption that the animal is responding to the same phenomenon. Such repetition rate signals may have envelopes and contexts that suggest social use.

Multiple transient signals are those in which many overlapping individual transients make up the sound. These cannot be resolved into simple click trains.

Spinner dolphin clicks are extremely brief, rapid rise time transients that have energy from the low audible range to at least 65 kHz, which was the upper limit of our recording system. There is perhaps no essential difference, in terms of the means of generation, between click trains and repetition rate signals except for the speed of emission. In the literature, the slower click trains (up to about 700 clicks/sec) are generally assigned an echolocation function, while signals of higher repetition rate are considered social in function. This assignment is a gratuitous human one based mostly upon the fact that our ears begin to fuse clicks into a tone representative of the repetition rate at about 700 clicks/sec (Nordmark 1960). This functional assignment must clearly be viewed with caution. It is not unlikely that, for the dolphin, both sociality and echolocation are functions of both classes of sounds.

Whistles are predominantly pure tone emissions, some with and some without evident harmonics and some that grade between the two states. They typically last from 0.1 to 1.5 sec, although most cluster near 0.7 sec duration. In our records, most whistles swept up in frequency (61%). The screams we recorded were rather narrow band signals that nonetheless could not be classed as pure tone signals. A discernible bandwidth of about 1 kHz was evident (figs. 72 and 73 and table 5).

Burst-pulsed signals appear to be the most complex spectral class of spinner dolphin sounds. They last from 0.2 to 4.0 sec (mean = 0.5 sec), with energy from the low audible range to the upper limit of our system. Most range between about 5 and 60 kHz. Our analytical instrumentation did not allow us to subdivide this class of sounds because the waterfall display apparatus we used had a scan rate too slow to record all component clicks and many were missed altogether. We had to be content to lump them in our statistics. It is probably fortunate that these sounds appear to be primarily social, and hence our analysis of them may relate to the general prevalence of sociality in spinner schools.

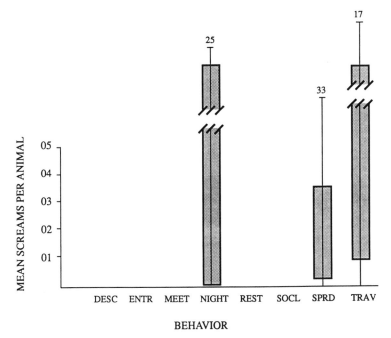

Figure 72. Mean number of screams per animal, per 5-sec period, for different classes of behavior.

Analysis of whistles, however, is direct with the waterfall display. As discussed earlier, our attempts to allocate whistles into discrete categories began to fail as sample size increased. The boundaries between what we at first saw as discrete kinds of whistles began to blur as more whistles entered the data base. If spinner whistles are like those of other dolphin species, the effect may be caused by the following features. First, it has been shown that dolphin whistles have structures that are specific to individual animals (Caldwell et al. 1973, Tyack 1986). It is also true, however, that such signature whistles may be modulated according to the emotional state of the individual, thus carrying both information about a signal source and the animal's state. Large opportunistic samples of whistles from schools of varying composition where the sounds cannot be assigned to a known individual quickly become unclassifiable. This was the case in our recordings.

Sounds and Behavior

Sound emission per animal varied widely both in types of component sounds and in the frequency of their occurrence through the 24-hr cycle. We used discriminant analysis to show that a given behavior period

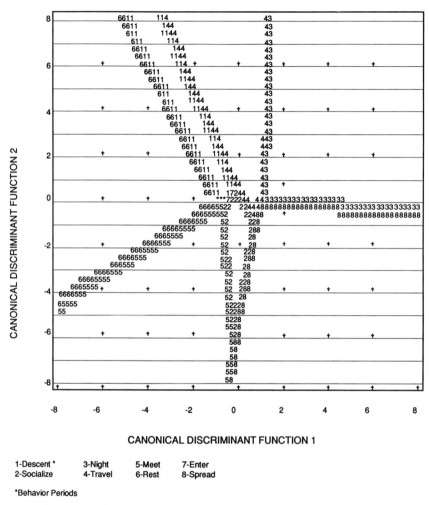

Figure 73. Two-dimensional plot of multivariate discriminate functions for various classes of sounds, by behavior period.

could be typified by the frequency of clicks, burst pulses, screams, whistles with harmonics, and pure tone whistles emitted per dolphin during 5-sec samples (fig. 73).

It was simple to discriminate among samples from active periods such as traveling, spread and feeding schools versus samples from less active periods when dolphins entered rest bays, descended into rest, and swam quietly over the shallow sand patches during rest. We suspect that a major class of information that these signals transmit is simply activity state.

TABLE 5 Statistics for Spinner Dolphin Whistles

Parameter*			Behavior Periods				
	ZigZag (11) 1	Enter (25)	Meet (23)	Travel (124)	Spread (92)	Night (114)	Cumulative
Begin freq. (kHz)	12.77 ± 0.68	12.06 ± 1.17	10.85 ± 0.26	11.12 ± 0.81	8.42 ± 0.49	10.50 ± 0.78	10.42
End freq. (kHz)	12.45 ± 1.51	17.86 ± 0.86	15.08 ± 0.95	14.55 ± 0.71	16.74 ± 0.80	16.98 ± 0.94	15.85
Duration (sec)	0.45 ± 0.13	0.72 ± 0.13	0.81 ± 0.13	0.64 ± 0.05	0.79 ± 0.10	0.65 ± 0.05	0.68
Number of breaks	1.09 ± 0.86	0.84 ± 0.50	0.17 ± 0.16	0.22 ± 0.10	1.39 ± 0.51	0.26 ± 0.14	0.57
Number of inflections	1.55 ± 0.66	1.80 ± 0.53	1.78 ± 0.57	1.61 ± 0.24	2.42 ± 0.72	1.18 ± 0.21	1.69

*Breaks refer to abrupt shifts in frequency or short stops in emission of a whistle. Inflection points refer to changes in direction in contour of whistle. There are no statistics for some periods (such as rest) because there were too few whistles in recordings. Numbers in parentheses after behavior indicate number of occurrences.

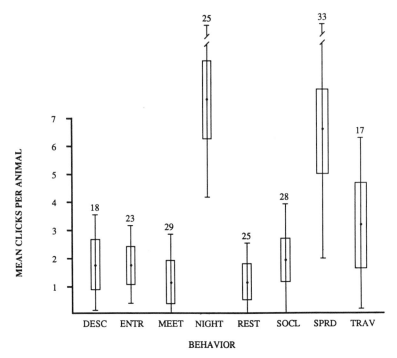

Figure 74.   Mean number of clicks per animal, per 5-sec period, for different classes of behavior.

Our difficulty in discriminating among similar states may reflect the need of the members of a dolphin school to assess the activity states of others, thereby promoting behavioral synchrony throughout the school.

Zig-zag swimming, which follows the rest period, was a time of behavioral oscillation between an active and a resting state and hence difficult for us to typify except by its pattern of extremes. When listening to zig-zagging dolphins, one hears much vocalization during the active sorties followed by a decline in their occurrence as the dolphins slow down and begin to mill. This pattern is repeated, sometimes over and over again. Milling dolphins in the midst of zig-zag episodes often remain quiescent for highly variable periods (a few minutes to an hour or more) before beginning a new oscillation into activity. Finally, the active periods reach a climax and the dolphins begin their concerted rush out to sea, leaping and phonating. The zig-zagging period itself is sometimes short but sometimes lasts for a matter of hours in the afternoon.

Although dolphins click frequently during all active periods, they are nearly silent during rest (fig. 74). At the opposite extreme, our records show that they may emit up to seven clicks per animal per 5-sec sampling

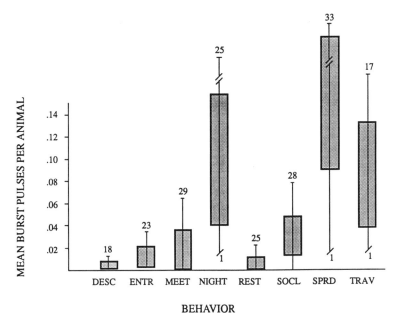

Figure 75.   Mean number of burst-pulsed signals per animal, per 5-sec period, during various classes of behavior.

period when the school is traveling or feeding. In our sample of quiet periods, dolphins emitted no screams and only a few whistles or burst-pulsed signals. The number of burst-pulsed signals per unit time rose dramatically as animals began to travel offshore, and these sounds were even more abundant during spread formation at dusk (fig. 75) (see Powell 1966 for similar behavior in the bottlenose dolphin).

Occasionally two schools met in Kealake'akua Bay. To our surprise, such events did not seem to be marked by distinctive levels or kinds of sounds. The relative and absolute probabilities of occurrence of sound classes remained indistinguishable from recordings of active schools entering the bay and from occupant schools in which social patterns such as caressing were prominent. Once again, the broad features of these sound emissions seemed to reflect activity state of the school and not just specific behavioral events. Perhaps this is not surprising, given the apparently fluid nature of schools. It is also probable that if recognition occurred between two meeting schools, the event may have happened at a greater distance than we expected (because dolphin signals propagate so well) and hence may have been missed by us. Or it may be that there is simply no need for recognition and hence the silence we monitored.

Our findings concerning whistle emissions are similar to those of Steiner (1981) who analyzed Atlantic spinner dolphin whistles. He found that such whistles generally sweep upward in frequency, last between 0.02 and 2.4 sec, and typically do not have multiple inflection points. That is, neither of our records shows sinusoidal patterns of frequency modulation.

Spinner dolphin whistles have a clear contagious aspect. Our tapes of zig-zag swimming, spread, or traveling schools frequently contain overlapping clusters of whistles. Periods of up to 4 sec without whistles are often followed by times when seven or eight whistles overlap (table 5). This clustering of whistles was sometimes especially notable in schools at the end of zig-zag swimming as high speed travel out to sea began. Then, many school members began to whistle in rhythmic series producing interweaving tangles of whistles that create an ululating effect.

The abundance of whistles increased during the travel that follows zig-zag swimming. Whistles were very evident during nighttime feeding periods (fig. 76 and 77 and table 6), when they were 14 times more abundant on average than during rest. Whistles reach maximum abundance during times when school members are widely spread. Closely similar whistles are often recorded in regular series, probably emitted by the same animal. We expect that these represent the signature whistles of Caldwell and Caldwell (1965). The whistle contour bears characteristics of a single emitting animal. In our records, we did not feel confident about identifying individuals beyond assigning a single whistle series tentatively to a single animal. This was done by examining the conformation, length, frequency range, and repetition pattern (rhythmicity) of a given whistle series. We could have been confounded in such assignment if mimicry was involved, as it most likely was.

The Caldwells' reported that an individual dolphin emits a specific whistle contour during its adult life. Perhaps dolphins need such identifiers because they often find themselves out of sight of one another at night, during dives, or in murky water. Therefore, a common need exists for them to identify communicants in an exchange by other than visual means. At such times, they cannot use physical appearance or expression as we frequently do to define a communication channel. Tyack (1986) showed that these signatures can be a way in which pairs of animals can engage in a specified exchange through mimicry. That is, if one animal emits its characteristic whistle and another then engages it in a whistle bout by mimicking the first animal's whistle contour, the exchange becomes specific to the two animals instead of being a general whistle response by the entire school, although the rest of the school must listen in.

TABLE 6 Statistics from Multivariate Discriminant Analysis of Spinner Dolphin Sounds Recorded from Various Behavior States.*

Discriminant Groups	C1	Function 1	Function 2	P (1) 2	P (1) 3	V(1) 4	V (2) 5
1. Descent	Screams	0.002	1.081				
2. Socialize							
3. Night	Nonharmonic Whistles						
4. Travel		0.688	−0.168	0.01	0.05	91.8%	6.13%
5. Meet							
6. Rest	Burst pulse	0.753	−0.781				
7. Enter							
8. Spread	Clicks	0.825	0.313				

*Behavior periods (labeled groups) were differentiated on the basis of sounds per animal emitted during each period. Four discriminate analyses were run: in the first, each behavior period was considered a single group; in subsequent analyses, behavior periods were lumped.

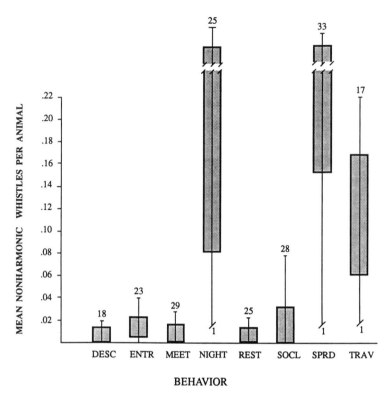

Figure 76.   Mean number of nonharmonic whistles per animal, per
5-sec period, for different classes of behavior.

Pure tone whistles and whistles with harmonics are emitted at all
times of day. While their abundance does not vary more than about two-
fold over the day, such whistles become much more abundant at dusk
and into the night. We counted average whistle emission rates at five to
ten times more abundant at dusk and during darkness than during daytime. This is what one would expect of a whistle-based communication system that supports social cohesion when animals cannot see one another.

Screams are wholly nighttime sounds. They can be heard in recordings as schools begin to travel out to sea at dusk, and they are abundant
at night. We have no clue as to how they are used by the dolphins, except
to say that when they are emitted other signal emission is also high, indicating a high degree of alertness (fig. 72).

Click trains are emitted during all classes of behavior, and there is a
pattern in the frequency of their occurrence. Emission rate is low during
all daytime patterns, being least frequent during rest. It is our subjective
impression that click trains emitted, during quiet periods are themselves

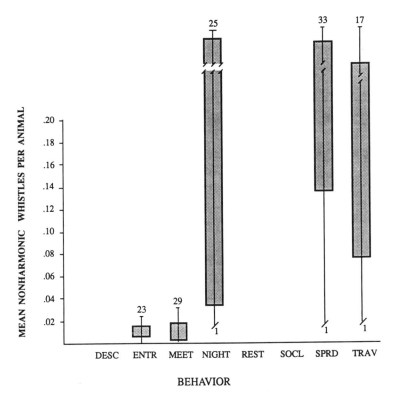

Figure 77. Mean number of harmonic whistles per animal, per 5-sec period, for different classes of behavior.

of low level compared to the loud trains frequently heard in the dark. One hears a quiet "muttering of clicks" during these times rather than the insistent sharp trains of more rapidly moving animals.

When we estimated school size and then counted clicks for a measured time during a school's travel out to sea, click rate was found to double relative to rest. Once the school reaches deep water and goes into spread formation, click abundance remains very high and may increase further. However, because schools may merge and visibility becomes a problem, absolute counts are not easily estimated.

By use of discriminant analysis, it is sometimes possible to test the association of certain sounds with a given behavior period. Night and day associations are clearly separable, while active daytime activity is difficult to divide into clearly different patterns. (fig. 73).

This apparent pattern of changing click train abundance fits the need for acoustic navigation and food finding in dark water. The reader should bear in mind, however, that if clicks have a communicative function, it also fits that model.

INTERPRETATIONS OF ACOUSTIC SIGNALS

It is difficult for the scientist listening to schools of dolphins moving out of sight below the water surface, or invisible in the dark, to say what the recorded signals might mean. But it is valuable to erect a theoretical framework to focus questions and to sharpen future observations. The remainder of this chapter covers the interpreted functions of the various acoustic signals, including whistles (in general), signature whistles, and burst-pulsed signals.

Whistles: The Phatic System

We propose that whistles are the basic organizational and structural signal class that regulates school organization and function, as Tyack's (1986, 1991) work seems to support. We suggest that whistles given by dolphins scattered throughout a school can define the limits and disposition of the school. Variations in how whistles are emitted can carry graded or analogic information about emotional state, level of alertness, hierarchy, the presence of food or danger, and similar information. These variations can carry both information and disinformation. In short, we propose that whistles function as what Jakobson (1960) called a *phatic system* of communication. In such a system, whistles are providing an open communication channel, akin to an open phone system to which all can listen, and that the modulation or temporal patterning of the whistles allows transmission of a variety of context-specific information. This definition overlaps the more commonly used term *contact call,* but we believe it also extends the concept usefully beyond the simple location of another animal in time and space. Such a signal system can be modulated in intensity, frequency, and frequency pattern through time to produce a complex system of great potential information-carrying capacity (Theberge and Fall 1967, Bain 1989).

The data about dolphin whistles that led us to this proposition came from several sources. First, captive studies have established that for the few species that have been studied, dolphins emit whistles typical of an individual (signature whistles) and that modulations indicative of emotional states can be superimposed on the basic signal (Caldwell et al. 1973, 1990). Second, while our animals whistled in all behavior periods, whistle abundance was much greater when animals could not see one another. Third, there is a clear contagious aspect to the whistle emissions we recorded. Whistles frequently occur in clusters as one would expect if they were signal response systems, and they clearly involve mimicry (Tyack 1986). Such contagion is greatest when the dolphins become fully alert after rest and in the dark. Fourth, whistles may be emitted or not emitted in relation to the status of a given dolphin in a hierarchy (Caldwell and Caldwell 1968).

Finally, whistles are clearly much less directional and can be heard farther than clicks normally can. Subjectively, when one approaches a distant school, the listener first hears whistles and burst-pulsed signals, and only later, clicks. For these reasons, whistles seem to be the acoustic signal most suited to supporting an open channel available to all school-members.

Classification of Message Types

It is useful to examine Jakobson's (1960) communication model further, although the terms he used have not become generally applied in the modern animal behavior literature. He described four modes of signaling: *referential signals,* which refer to specific things in the environment; *emotive signals,* which indicate the emotional state of the signaler(s); *vocative signals,* which are assertive signals directed at the receiver; and *phatic signals,* which provide the open communication channel for discourse of various kinds that we have just described and which we feel contains the other classes of signaling embedded in it.

Once the phatic matrix is established, the other kinds of messages may pass over it as modulations, mimicry patterns, or responses to the basic signal. We see this scheme as describing what might be occurring when spinner dolphins whistle and chorus.

Mimicry of another's signature whistle has a special place in such a system (Tyack 1986, Caldwell et al. 1990). By use of mimicry, a schoolmember can enter into a specified exchange with another animal anywhere within the school. If other animals enter the exchange and a chorus begins to develop, information about general school state may be carried to all participating or listening members. If the chorusing becomes general, relational information between respondents may be a school phenomenon, rather than simply that between two animals. Aspects of hierarchy could emerge. Such chorusing can then bring an entire school into a particular emotionally based motivational and relational state.

We propose that such a phatic communication system is a basic means allowing dolphins to organize their schools when vision will not serve. If this model is correct, the repetitive dolphin whistling we have repeatedly recorded performs a number of functions simultaneously. An unmodulated signature whistle by one animal may carry an analog message that a human might interpret in these words: "This particular dolphin is here, and all is well." The place, the individual identity of the dolphin, and its state are carried by a single signal.

The emotional message is carried by modulations of the carrier whistle. For example, the whistle of a frightened animal may quaver, its pitch may rise, or it may be repeated in rapid succession.

When a sender is mimicked, a specific channel is opened to which not only the respondents but all may listen. When several dolphins chorus, school alertness and cohesion can be indicated by reaction times of one animal to another's whistles. Precedence of the whistle of one animal over that of another may be indicative of dominance versus subordinance in a whistling pair.

The messages between two respondents may have a managerial aspect in which more than one kind of organizational message may be carried to various listeners by a single exchange (Owings et al. 1986). This might depend on their own position or status or sensory capability, such as position in a hierarchy or the need for a new state of organization.

We interpret the fluctuating occurrence of contagious whistles during zig-zag swimming as partially social facilitation, with dolphins testing each other's alertness until the school is primed for its transit out to sea. From the aggregate of such signals from all schoolmembers, the disposition and state of an entire school can be carried to all members. Finally, the phatic system is probably the acoustic substrate upon which the sensory integration function of a school (Norris and Dohl 1980 *a, b*) can be carried out whenever visual mediation is impossible (see chap. 13).

Signature Whistles

Most of the evidence for signature whistles has come from studies of captive dolphins. Caldwell and Caldwell (1965) found that 99.4% of emitted whistles from a series of captive dolphins were signature whistles, repeated over and over *ad nauseum*. This constancy of whistle type proved so great that the Caldwells first erected a one dolphin–one whistle hypothesis, feeling that their dolphins were possibly incapable of other whistle production. But these workers later discovered that dolphins have much capability at modifying the basic signal (Caldwell et al. 1973). A later study with mimicry of machine-made sounds (Richards et al. 1984) showed that dolphins have extremely refined capability at modifying these and other sound emissions. Caldwell et al. (1990) suggest that each individual may have individually distinctive patterns of changing whistles with context.

The transmission of secondary messages superimposed upon the basic whistle carrier is known to occur. For example, when in fear or pain, dolphin whistles sometimes show sharp breaks or steps in frequency, much like a human voice cracking in times of stress. Such stepped sounds have been recorded from harpooned dolphins (Busnel and Dziedzic 1968). Also, during birth, an animal that normally emits a whistle composed of a single rising and falling tone (or loop) may link loops together into a continuous series of several loops (Dreher 1966). Such multilooped whistles seem to attract schoolmates and to cause the general excitement level of the school to increase.

Social Facilitation

It is worth emphasizing that the basic phatic signal we propose, the whistle, seems to remain constant when there is no emotional or relational information to carry, as the Caldwells found with captives. For the phatic system to work, there must be a level of invariance in the basic signal upon which modulations can indicate changing states. It is not surprising that in captivity, where changes in emotional state usually lack the context of a school, the basic phatic signal may be repeated over and over without basic change or modulation.

We feel that zig-zag swimming and the sounds that typify it are parts of a group process of transition from one behavioral state to another. In this way, the group affirms that its acoustic channels are functional and its members synchronous as the school sets to sea.

There are parallels of such group testing from studies of mammal herds and the vocalizations of flocking birds. The movements of jackdaw flocks seeking roosts in the evening (Lorenz 1952) are cyclical, and changes in direction occur in a manner strikingly reminiscent of the behavior of dolphins during zig-zag swimming. The pseudopod movement of hamadryas baboons (*Papio hamadryas*) seems to serve much the same function (Kummer 1968). Piñon jay flocks circle and call before determining migratory direction, and the development of behavioral synchrony and preparedness of gelada baboons is communicated through chorusing (Richman 1980).

Perhaps the most striking parallel is found in the "pep rallies" of Cape African hunting dogs (*Lycaon pictus*). Estes and Goddard (1967) describe the twittering calls of these canids made prior to going on the hunt. As each animal awakes, it joins its packmates in a chorus of twittering and touching muzzles prior to going on the hunt. (Recall that caressing is prominent during the active phases of spinner dolphin zig-zag swimming.) The intensity of the chorus rises and falls as the dogs make false starts as if to move off and then come back to their sleeping grounds. Finally, the entire pack moves off in unison.

During the hunt, the dogs use calls that apparently locate one another spatially. Estes and Goddard (1967) suggest that the pep rallies serve to inform individual pack members of the readiness of the entire pack to hunt, to excite and alert the pack, and to reaffirm relationships among pack members. We extend this idea by suggesting that in group-dwelling animals such as hunting dogs or dolphins coordination be-

tween members is crucial to a degree we humans may find hard to understand. All of these features probably take place during the chorusing
of zig-zag swimming. In the fluid spinner dolphin society, such choruses
may also reacquaint individuals with the whistles of others they should
expect to hear that evening and reaffirm hierarchical relationships. In a
fission-fusion society in which membership is fluid, such coordinatioin is
probably vital.

Burst-Pulse Signals

Now let us examine burst-pulse signals. In primates, canids, and dolphins, burst-pulse signals seem to have similar contexts. Spectrally
rich growls, purrs, buzzes, barks, and mews are used as emotive and
vocative signals among individuals (Rowell and Hinde 1962, Fentress
1967, Green 1975, McCarley 1975, Lehner 1978, Petter and Charles-
Dominique 1979).

Spinner dolphin burst-pulse signals are extremely varied. They range
from almost private sounds passed between courting adults, to peremp-
tory barks or loud "bangs" that seem to be used as aggressive signals in
a captive school. Such sounds sometimes travel long distances at sea. We
have listened to burst-pulsed signals issuing from a school traveling
against the cliff of Kealake'akua Bay, as we recorded from a craft drifting
at the entrance of the bay more than 1.6 km (1 mi) away. If burst-pulse
signals are emotive and vocative in context, we would expect to record
them in greatest frequency during socializing periods when animals are
in active physical contact, rubbing bellies, chasing, and exhibiting "sexual patterns," as is the case.

Other animals exhibit a wide range of such sounds during greeting
events. For example, chimpanzees use such signals to indicate recognition and degree of friendliness when they meet (Marler and Hobbett
1975).

Our data indicate that daytime patterns in general and rest in particular are times of reduced social activity compared to night patterns, and
acoustic emission frequency is reflective of this difference. Thus, although much social activity occurs during the day when the animals are
not resting, burst-pulse vocalizations are even more prevalent at night.
Our several day-long observation sessions of the captive school revealed
that acoustic and behavioral activity are at very high levels throughout
the dark hours. Data from other delphinids support the proposition that
burst-pulse signals are emotive and vocative in nature. Tyack (1976)
found that bottlenose dolphins emitted the greatest number of burst-
pulse signals during milling periods when the animals also exhibited sexual and play behavior.

Our conclusion from observations of both captive and wild schools is that all daytime acoustic patterns are to some degree depressed compared to levels of these patterns at dusk or at night. Perhaps this may not relate so much to alertness as to the availability of visual communication during the day.

Beyond the attempts we make here to provide a general classification of acoustic events during the repetitive diurnal sequence of spinner dolphin behavior lies the whole domain of specific contexts and uses of individual sounds. Such interesting matters await more specific studies than we have attempted here.

NINE

Patterns of Reproduction

Randall S. Wells and Kenneth S. Norris

As every behaviorist knows, a first order of business in beginning to understand a wild animal society is to define its reproductive patterns. This chapter examines spinner dolphin reproduction from two viewpoints. First, it outlines the patterns of reproductive seasonality found in Hawaiian spinner dolphin populations and the factors that seem to regulate these patterns. Then it outlines the reproductive hormone production of individual Hawaiian spinner dolphins and relates the results to social patterns. What little we know about reproductive behavior in wild spinner dolphin schools is covered in chapter 14, "Social Behavior," while here we take up the physiological and populational aspects that have recently been summarized (Perrin et al. 1984).

We were at first puzzled by what looked like sexual behavior in Hawaiian spinner dolphins occurring all year long. Then we began to discriminate true reproductive behavior, which was strongly seasonal in occurrence, from social caressing, which occurred without evident peaks all year long. This understanding pointed to our next step, which was to define the reproductive cycle of Hawaiian spinner dolphins in seasonal terms. This included reviewing the work of others on the seasonality of births in oceanic populations of spinner dolphins, as well as our own attempts to define the hormonal events of estrus and testicular change in Hawaiian spinner dolphins.

PATTERNS OF CETACEAN REPRODUCTION

Delphinid cetaceans are estrus mammals, bearing single, large young, able to swim quickly after birth but which were highly undeveloped socially. The fetus develops in one horn of a bicornuate uterus and is typ-

186

ically born flukes first. The newborn then receives a considerable period of parental care (Slijper 1966, Harrison 1969). Gestation generally lasts about 10–16 months in various species (Perrin and Reilly 1984). We note that some degree of seasonality of mating and calving has been demonstrated for every delphinid that has been studied in any depth, even in tropical species. Such seasonality in cetaceans is associated with broad peaks of recrudescence and decline of the testicular tissue of males, and sharp periods of sexual readiness, and ovulation in the females (the estrus cycle) (Mackintosh 1965, Hohn et al. 1985).

Gestation and Development

Perrin et al. (1977a) estimate 10.7 months as the gestation period for eastern tropical Pacific spinner dolphins, an average birth length of 77 cm, and an average length at attainment of sexual maturity of 165–170 cm (Perrin and Henderson 1984). Lactation appears to be the most variable component of the reproductive cycle. It has been estimated for whitebelly spinners as lasting from 15 to 18 months (Perrin and Oliver 1982).

The Male Cycle

The testicle weight to body weight ratio of cetaceans can sometimes become very high, among the greatest recorded for mammals (Harrison, et al. 1972, Kenagy and Trombulak 1986). Such large testes are associated with the breeding season. At other seasons, they decline to modest resting levels. Perrin and Henderson (1984, p. 424) have described such testicular variation in oceanic spinners of the eastern tropical Pacific as follows:

Average weight of testis (with epididymis) at the onset of spermatogenesis is 85.1 g, based on a logistic fit, for the whitebelly spinner. . . . In the northern whitebelly spinner, a mode of large-testis-weight animals appears in February, centered around 700–800 g. The mode persists through May (although possibly retreating slightly); appears in June centered around 400–500 g; moves out to 600–700 g again in July–August; and all but disappears in September through January. The months of peak testis weight (and, presumably, peak fertility and breeding) are February and July–August.

They also note (p. 424) that "a sizeable proportion (>20% at the upper end of the body-length size range) of northern whitebelly spinner testes weigh over 700 g (up to 1354 g in the sample), and all of these examined histologically have had at least 'some' sperm in the epididymis."

Barlow (1984) notes that "some males [of *Stenella*] always have active sperm production" and that, presuming that females are sexually receptive, conception can occur at any time of the year. This matches what we know of the appearance of newborn spinners in the Hawaiian population.

The usual correlate of a large change in testicle size in schooling or herding mammals is the performance of rapid multiple matings timed to the sexual readiness of females, for example, as seen in flocks of domestic sheep where a single ram can "cover" a dozen or more receptive females in a very short time.

The Mating System

The mating systems of group-living mammals can usually be described as polygynous, polygynandrous, or polyandrous. The seldom-used term *promiscuous* indicates an absence of choice. In a polygynous system, a male may attempt to "possess or sequester" several females for exclusive mating (LeBoeuf 1974). In polygynandrous mammals, males mate with females who are sexually ready and both males and females can have multiple partners. Polyandrous mammalian groups, such as packs of African wild dogs (Frame et al. 1979), may contain only one or two sexually mature females who bear the young for a troop. Such mammals, unlike dolphins, may simultaneously bear multiple young.

Interpreting the dolphin system at our present state of knowledge requires caution. It is possible that paternity for spinner dolphins could be limited internally by the female. For example, if a vaginal plug is produced after an initial mating, all subsequent copulations may be ineffective. Vaginal plugs, or at least what seem to be such plugs, have been reported by Harrison (1969) for the closely related genus *Delphinus.*

Which of these arrangements is used by spinner dolphins? The polyandrous system (except for serial polyandry) is impossible because dolphins bear single young. Although we have noted more than one male mating within a short time with a single female, only one could be the father.

The polygynandrous alternative seems to describe best what we have seen for the Hawaiian spinner dolphin. Certainly when females come into estrus, this is likely to be detected by males, and these females appear to be preferentially sought out. But if the tangles of dolphins (*wuzzles*) we describe in Chapter 14 are mating groups, the question of whether or not choice is involved remains obscure. We note that many matings have been observed outside such groups and that both male and female choice may occur there. The great seasonal change in testis size just described, in which the peak weight coincides with a mating peak, and the short female hormonal peaks fit the polygynandrous model.

Mate Selection

It may be difficult for males to sequester or guard females within the confines of spinner schools, in the usual arrangement seen in polygynous animals ashore. Nonetheless, Connor (1987) has reported that coalitions of male bottlenose dolphins (*Tursiops*) at Shark Bay, Western Australia, have been seen cutting out and herding oestrus females. Although we, too, saw such coalitions of males in Hawaiian spinner dolphins (chap. 14), they swam within the school envelope and we did not see herding behavior. Instead, we saw mating in the midst of the school without any evident attempts by school members to segregate out from the main group.

The Female Cycle

The ovaries of delphinids bear the marks of ovulations, perhaps permanently, in the form of corpora albicantia that can be counted to relate ovulations to age. Benirschke et al. (1980) suggest that spinner dolphins are spontaneous ovulators. Ovulation is not, however, a linear function of age, but instead a curvilinear function that reflects a reproductive decline, sometimes slight, in older females (Perrin et al. 1976). Pregnancy rate and calving interval for the whitebelly spinner have been determined by Perrin and Henderson (1984) to be 30.0–32.8% and 3.0–3.3 years, respectively.

Reproductive Seasonality

With one exception, studies of the seasonal occurrence of births in oceanic spinner dolphins show "diffuse seasonality" (Barlow 1984). The exception is the eastern spinner dolphin, which is the spinner population most heavily impacted by the tuna seine fishery; its populations had been reduced to an estimated 17–25% of prefishing levels by 1983 (Smith 1983). Eastern spinners show a single broad annual peak of calving, centered in late spring and early summer (Barlow 1984).

The closest oceanic population of spinner dolphins to the Hawaiian Islands, the whitebelly spinner, is much less impacted, its population having been reduced to an estimated 58–72% of prefishing levels by 1983. The frequency and seasonality of birth of whitebelly spinners based on examination of fetuses taken from netted dolphins shows a seasonally diffuse reproductive pattern with a slight bimodal tendency; a modest peak occurs in late spring and summer and another in midwinter (Barlow 1984, his fig. 7). This corresponds well to two seasonal peaks of testis weight reported by Perrin and Henderson (1984) for the same form.

Barlow (1984) points out that the unimodal peak of eastern spinner reproduction could be explained as a density-dependent response of a heavily exploited population in which the determinants of the peaks of calving were multivariate, at least involving variations in calving interval, gestation rate, and the proportion of sexually mature to resting females. Calving interval does in fact seem to respond to exploitation (Perrin and Henderson 1984).

DETERMINANTS OF REPRODUCTIVE SYSTEMS

Fowler (1984) has clearly defined the broad determinants of cetacean and other populational systems. He says (p. 373):

> Cetacean populations, along with the populations of other large mammals (and animals in general) are regulated through density-dependent changes in reproduction and survival. These changes seem to be expressed most commonly in the processes associated with recruitment (i.e., birth and juvenile survival) and to involve causes associated with food resources. Growing evidence indicates that social and behavioral factors are also important causal elements.
>
> Evolution, however, seems to have resulted in certain features of the dynamics of animal populations being somewhat independent of both the causal factors involved in density dependence and its modes of expression. There is an observed pattern in which the level of populations which produce maximum net growth in numbers is more closely related to the maximum rate of increase per generation time than to other factors. Spanning a broad spectrum of species types (fishes, insects, mammals, protozoans, and bacteria), this pattern involves maximum growth rates at high population levels for species with low rates of increase per generation time. Conversely, species with high rates of increase per generation time exhibit maximum growth rates at lower levels (often less than 50% of their carrying capacity).

Because they are large mammals, cetaceans generally show maximum population growth rates at high population levels, unlike most fish, which (because they typically spawn many eggs at one time) tend to show fastest population growth at lower percentages of their carrying capacity (Fowler and Smith 1981). As an aside, this is an important reason why it requires special care to apply management procedures derived from studies of fish to large mammal populations, such as those of the Cetacea.

Fowler et al. (1980) in studies with *Stenella* spp. have noted that density-dependent regulation is the norm. Fowler (1984) says that the relationship between population level and maximum population growth involves two levels of organization: trophic dynamics and direct evolutionary process. This is a major point we discuss in chapter 16 when we compare evolutionary arrangements in a variety of mammals.

Perrin and Henderson (1984) have examined the growth and reproductive rates of the heavily exploited eastern spinner dolphin and the less heavily exploited whitebelly spinner to see if the predictions of density-dependent compensation are occurring in the reproductive patterns of the two forms. They found instead that gross annual reproductive rates (proportion female × proportion of females mature × pregnancy rate) were closely similar in the two populations (about 8–10% in each case). They were unable to state the cause of this anomalous finding, which does not directly support the density-dependent assumptions of Fowler (1984). However, they did note that the two populations—both below their maximum net productivity level, with whitebelly spinner slightly below and the eastern spinner far below this level—may be responding in a curvilinear fashion that could mask a density-dependent response. In other words, the eastern spinner is so far below normal population levels that its density-dependent compensation system may have begun to fail. At any rate, though their work does not directly support density-dependent compensation, it need not be thought to invalidate the widely supported concept.

The highly dimorphic eastern spinner is reported to have smaller testis size than other races of eastern Pacific spinners (Perrin et al. 1977*a*). In mammals, such sexual dimorphism is generally inversely correlated with sperm competition (Kenagy and Trombulak 1986). This suggests that polygyny may be relatively more developed than in other races.

HORMONAL AND BEHAVIORAL CORRELATES
OF REPRODUCTION

The population dynamics-related aspects of spinner dolphin reproductive patterns are overlain upon the hormonal events and correlated changes in the reproductive organs of the dolphins that together define the basic reproductive patterns of the species. We now report on direct measurements of both seasonal patterns of reproductive behavior in the captive school of spinners at Sea Life Park Oceanarium on Oahu and the hormonal cycles of the same dolphins, as determined from a biweekly sampling of blood and observation of behavior. The work was published earlier as part of a large symposium volume on cetacean reproduction (Perrin et al. 1984), but because it is an integral part of this wild–captive study, we quote extensively from that paper here.

Because the captive school was very small, we were unable to reflect on possible schoolwide reproductive events with these data, but only upon the regular course of physiological and behavioral change that appears to occur in individual animals. It seems likely that had our sample been large and from wild animals, we would have found considerable variation among individuals in the timing of these patterns, if the diffusely modal pattern of seasonal reproduction we discussed earlier for the whitebelly spinner is any measure.

The occurrence of hormonal events and such aspects of reproduction as ovulation, implantation, and mating behavior are typically strong in mammals. For example, endocrine concentrations in rhesus monkeys (*Macaca mulata*) are directly related to ovulation events and correlate strongly with mating behavior (Wilson et al. 1982).

Radioimmunoassay techniques now allow such assessments for dolphins of both sexes using very small blood samples. Sawyer-Steffan and Kirby (1980) and Kirby and Ridgway (1984) have successfully applied these techniques to outlining the reproductive patterns of the bottlenose dolphin (*Tursiops truncatus*) and the common dolphin (*Delphinus delphis*). Here we combine the radioimmunoassay technique with a behavioral assessment. This includes the frequency of occurrence of some behavior patterns that seem to be related to reproduction. The search for a correlation was especially important to us because in other parts of our study, we were attempting to separate sexual patterns that appeared to have been co-opted into the social process from true sexual patterns. Patterns that closely tracked hormonal events were expected to indicate sexual function, while those that did not were probably more broadly social in context.

The literature of cetacean reproductive biology has a long history, especially in terms of the anatomy of reproductive systems. Only when captive odontocetes became common did behavioral observations begin to accompany knowledge of anatomy (Slijper 1966) and some aspects of physiology and behavior (McBride and Hebb 1948, McBride and Kritzler 1951, Tavolga and Essapian 1957, Tavolga 1966, Harrison 1969, Harrison et al. 1972, Caldwell and Caldwell 1972*a*, Tayler and Saayman 1972, Saayman et al. 1973, Saayman and Tayler 1977).

Bateson (1974) describes sexual behavior in spinner dolphins, as do Norris and Dohl (1980*a,b*). Puente and Dewsbury (1976) provide a detailed analysis of sexual behavior in the bottlenose dolphin. But measurements of hormonal relationships to other aspects of sexuality in dolphins remain scarce.

HORMONAL CYCLES OF HAWAIIAN SPINNERS

The results reported here were based upon a suite of blood samples taken from the captive spinner dolphin population at Sea Life Park from August 1979 through June 1981. Five dolphins were present at first (two males and three females), but from February 1980 on, the colony consisted of two females (Kahe and Kehaulani) and a male (Lioele).

The biweekly blood samples were accompanied with a 24-hr behavioral observation period either directly preceding or following sampling. Three hormones were measured by radioimmunoassay techniques: testosterone, sensitivity to 0.05 ng/mL; estradiol, sensitivity to 1.9 pg/mL; and progesterone, sensitivity to 33 pg/mL.

Wells (1984, p. 466) described the behavioral methods used in assessing reproductive and social behavior in this study as follows:

> Behavioral observations were made in conjunction with each blood sampling session. The behavior patterns of the colony were monitored for the first ten minutes of every half hour through the twenty-four hours preceding or following hormone sampling. An observer at the side of the observation tank narrated behavioral events and times of occurrence into a tape recorder. The durations of heterosexual pairings and frequencies of occurrence of all other behavior patterns considered to have sexual or social connotations were measured from transcriptions of the tapes. The durations of heterosexual pairings were measured as periods during which the dolphins swam and surfaced synchronously within approximately 2 m of each other, engaged in similar activities, and swam approximately side-by-side. Other behavior patterns were scored as one occurrence as long as the participants were together and engaging in the same activity. If the dolphins separated to terminate the activity or surfaced to breathe, and then engaged in the activity again, two occurrences were scored. The behavior patterns were further partitioned on the basis of the role of the participant. A dolphin was classified as "giving" the behavior if it initiated or was the most active participant in a behavioral sequence; the more passive participant or the individual receiving the attentions of the "giver" was classified as the recipient. Distinctions were sometimes difficult and some of the behavior patterns involved mutual participation. . . .
>
> The distribution of durations of heterosexual pairings and frequencies of occurrence of behavior patterns were compared relative to the concentrations of the reproductive hormones measured during the same period with a Kolmogorov–Smirnov Goodness of Fit test. . . . Testosterone concentrations were considered "high" if greater than 30 ng/ml, "intermediate" if between 8–30 ng/ml, and "baseline" below 8 ng/ml. High estradiol concentrations were those over 90 pg/ml, intermediate were between 50–90 pg/ml, and baseline were below 50 pg/ml (this categorization scheme follows that of Wilson [et al.] 1982). Progesterone concentrations over 3 ng/ml were considered indicative of ovulation, 1–3 ng/ml were intermediate, and below 1 ng/ml were baseline.

The Male Spinner

During the 14-month sampling period, testosterone levels of the male dolphin varied 60-fold. Serum testosterone levels ranged from less than

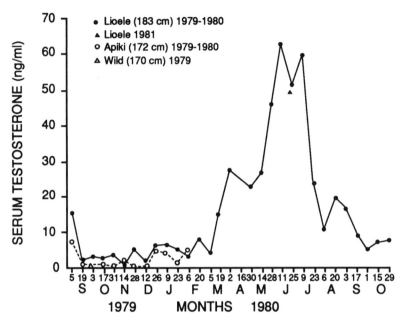

Figure 78. Serum testosterone concentrations for three male Hawaiian spinner dolphins.

1 ng/mL to over 60ng/mL for the male Lioele (fig. 78). The increasing titre began to appear in March 1980 and peaked in June and July. A single sample from June 1981 revealed approximately the same level as in the previous year.

The Female Spinner

Estradiol concentrations for both females were mostly low or intermediate (fig. 79). However, elevated levels were noted for both females, Kahe in June 1991 and Kehaulani in October 1979 and October 1980.

Plasma progesterone concentrations for both females showed very strong peaks preceded and followed by periods when very low levels were measured (fig. 80). The elevated progesterone levels for Kehaulani for September–October 1980 suggested ovulation. The October surge was preceded by elevated estradiol and low progesterone, which might indicate the follicular and luteal phases of the cycle. Kahe's record is more difficult to understand. In both 1980 and 1981, the progesterone level rose in June or July and remained elevated (fig. 80), usually a sign of pregnancy, although she did not abort a fetus or give birth.

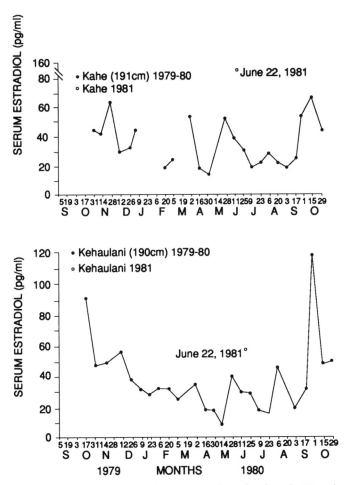

Figure 79. Serum estradiol concentrations for female Hawaiian spinner dolphins Kahe and Kehaulani.

Behavioral Correlates

Paired swimming with a member of the opposite sex, a quiet pattern without contact, did not appear to be correlated with hormone level. This is perhaps not surprising in a schooling species. The total time spent in heterosexual pairings did not vary from chance levels. There did seem to be preferences, however; Lioele spent more time with Kahe than with Kehaulani. These patterns can perhaps be expected considering the small size of the school and the relative lack of choices available to the dolphins.

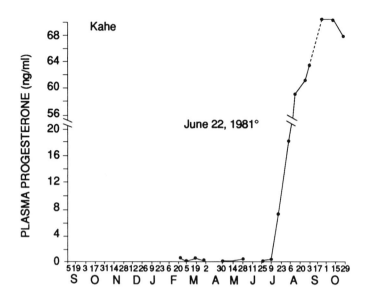

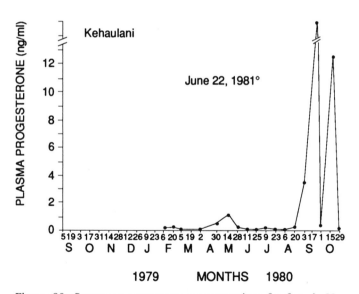

Figure 80. Serum progesterone concentrations for female Hawaiian spinner dolphins Kahe and Kehaulani.

More interesting were the sexually related behavior patterns exhibited by the dolphins, which in some cases did vary seasonally and with the sex of partners (table 7). Six such patterns were examined: genital-to-genital contact, beak-to-genital propulsion, other genital contact, nongenital contact, mutual ventral presentations, and chases.

Genital-to-genital contact occurred only between male and female dolphins and only when the male's testosterone titre was high (more than 48 ng/mL) (Kolmogorov–Smirnov $D = 0.74$, $n = 3$, $p < 0.05$). On one occasion, intromission was seen. No clear relationship to female hormone level could be ascertained; both baseline and high levels of estradiol and progesterone were involved. It should be noted that the adult male had sustained a back injury from leaping out of the tank and that he swam in a decidedly abnormal fashion, tilted slightly to one side. This may have inhibited his performance of normal mating behavior.

The occurrence of beak-to-genital propulsion was most closely related to the female hormonal cycle. The male rarely propelled the females, and there was no clear relationship to his testosterone levels. However, the highest frequency of beak-to-genital propulsion occurred during the follicular phase of the only clearly defined ovulatory cycle of the study. Kehaulani (the female) propelled Lioele (the male) seven times on September 30 when her estradiol titre was the highest of the study and when her progesterone level was low ($D = 0.82$, $n = 8$, $p < 0.01$). The speculation that the propelling animal may ensonify the recipient with click trains becomes more interesting in view of this relationship (see chap. 14). It could involve a receptive animal stimulating a partner to respond, or it could involve chemosensory cues, or both.

Mutual ventral presentations in which male and female tilted bellies toward one another were related to Lioele's testosterone levels ($D = 0.44$, $n = 10$, $p < 0.05$) and to Kahe's estradiol concentrations ($D = 0.67$, $n = 5$, $p < 0.05$). This was not consistent for both females, suggesting that the significance of Kahe's relationship may be due to the coincidence that her highest estradiol value and Lioele's high testosterone concentrations occurred together on June 22, 1981, when a sample was taken. In a larger dolphin school, the chance of such concordance between some members is much greater than for our tiny captive school. With more dolphins present, the chances of concordance of reproductive patterns between male and female might be greatly improved and mutual presentation could serve as a signal of this state.

The other variables such as other genital contacts, nongenital contacts one-way ventral presentations, and chases did not produce significant correlations with hormonal state.

In summary, even though sample sizes were small, three behavior patterns did relate at a statistically significant level to reproductive

Heterosexual behavior patterns: male relations to females (no. occurrences/hr) relative to serum testosterone conc. of the male

Female behavior patterns relative to male (Lioele)	Lioele		
	Baseline (0–8 ng/ml) $n = 4$ days: 26.06 hours $\overline{X}\pm$s.d.	Intermediate (8–30 ng/ml) $n = 9$ days: 62.69 hours $\overline{X}\pm$s.d.	High (> 30 ng/ml) $n = 4$ days: 30.77 hours $\overline{X}\pm$s.d.
Genital-genital contact			
Mutual	0	0	0.13±0.135
Beak-to-genital propulsion			
Give	0	0.02±0.053	0
Receive	0.24±0.480	0	0.03±0.065
Other genital contact			
Give	0	0.02±0.050	0.07±0.130
Receive	0.11±0.210	0.01±0.043	0.03±0.065
Non-genital contact			
Give	0	0.03±0.062	0
Receive	0.08±0.165	0.13±0.290	0.17±0.127
Mutual	0.25±0.289	0.05±0.076	0.17±0.168
Ventral presentation			
Give	0.04±0.080	0.03±0.062	0.17±0.168
Receive	0.29±0.300	0.12±0.149	0.03±0.065
Mutual	0.08±0.087	0.02±0.050	0.23±0.167
Chases			
Give	0	0.03±0.066	0
Receive	0	0	0

Heterosexual behavior patterns: female relations to male (no. occurrences/hr) relative to estrodiol of the females

Behavior patterns relative to female	Kahe		
	Baseline (0–50 pg/ml) $n = 11$ days: 76.31 hours $\overline{X}\pm$s.d.	Intermediate (50–90 ng/ml) $n = 4$ days: 29.07 hours $\overline{X}\pm$s.d.	High (> 90 pg/ml) $n = 1$ day: 7.53 hours $\overline{X}\pm$s.d.
Genital-genital contact			
Mutual	0	0	0.26 –
Beak-to-genital propulsion			
Give	0	0	0
Receive	0.02±0.048	0	0
Other genital contact			
Give	0	0.03±0.065	0
Receive	0	0.11±0.210	0
Non-genital contact			
Give	0.06±0.111	0	0.13 –
Receive	0	0	0
Mutual	0.05±0.074	0.17±0.135	0.27 –
Ventral presentation			
Give	0.04±0.067	0.14±0.198	0
Receive	0.04±0.069	0.13±0.133	0.40 –
Mutual	0	0.07±0.078	0.40 –
Chases			
Give	0	0	0
Receive	0.01±0.045	0	0

SOURCE: Wells 1984

hormone levels. Genital-to-genital contact and mutual ventral presentations occurred when the males' testosterone levels were high, while beak-to-genital propulsion appeared to be related to ovulatory or associated events in the female. The other patterns that were checked seemed more related to social than to purely sexual contexts (table 6). The genital-to-genital contact is known in other delphinids to be a precursor to copulation, for example, in *Delphinus* (Essapian 1962).

Homosexually oriented behavior, such as beak-to-genital propulsion between females or copulations between males (Bateson 1974), can also be found between heterosexual pairs and thus can be viewed as the co-option of a sexual pattern into social concourse between dolphins. Attempted mating by a male calf with its mother (Caldwell and Caldwell 1972*a*) might have a similar connotation.

Wells (1984, p. 471) has commented on the difficulty of sorting out the context of sexually oriented behavior patterns among heterosexual pairs unless the context of behavior is known:

> Puente and Dewsbury (1976) found that *Tursiops* behavior patterns they termed "courtship" occurred more frequently on days during which copulations occurred than on days without copulations. However, all their observations were made during the presumed breeding season . . . , and no data are available for comparison with frequencies outside the breeding season. It would be expected that behavior patterns most closely linked to reproduction would increase in frequency during the reproductive season. Similar behavior patterns observed at other times of the year or between inappropriate participants could be assumed to occur as part of the ordering of dolphin schools, or developing or maintaining relationships between individuals. Changes in the frequency of mating activity and durations of heterosexual pairings on a seasonal basis have been reported for *Tursiops* in several locations. . . . For example, males are reported to spend more time with females during the spring, and more copulations occur then. In one case it was noted that female–female social interactions were curtailed in the spring. . . . These changes were presumed to coincide with the breeding season, but no precise measures of reproductive condition of the involved dolphins was made.

Wells (1984, p. 471) goes on to note:

> Spinner dolphins often swim in large schools of mixed sex, whereas coastal bottlenose dolphins often swim in single-sex schools that occasionally meet and mix for periods of variable length with schools of the other sex. . . . The constant availability of members of the opposite sex may preclude the need to change typical swimming association patterns.

A hormonal definition of a "seasonal rut" is simpler for males than for females because elevated testosterone concentrations may be rather prolonged as compared to the short-lived hormonal events associated with ovulation. Wells (1984, p. 47) states:

> If spinner dolphins follow the typical ovulatory cycle described by Cupps [et al.] (1969), then elevated estradiol levels in conjunction with low progesterone concentrations, followed by reduced estradiol levels and a progesterone surge, should define an ovulation and presumably should define the time during which behavioral estrus should occur. Measurements of estradiol surges alone are not enough to indicate imminent ovulations.

The observations reported here probably suffered from the biweekly schedule of observation. Only one clear ovulatory event was described in 8.5 months of observation. Because such events are short in dolphins, others may have occurred when we were absent. Measures of the course of change in reproductive hormones are rare for dolphins, but can clearly be used to define the course or reproductive events if repeated frequently enough through the season.

When one considers the very different courses of sexual readiness in male and female spinner dolphins it appears likely that a spreading of the sexual readiness of female spinners is what determines the diffuse bimodal appearance of newborns in the population.

Our description of the life of Hawaiian spinner dolphins now moves on to describe some aspects of their locomotion that were observed through the windows of our viewing vehicle, *Maka Ala.*

TEN

Locomotion

Kenneth S. Norris and Christine M. Johnson

Most dolphins go about the majority of their daily activities at very modest swimming speeds, and Hawaiian spinner dolphins are no exception. Spinner schools typically move slowly. Most of the time in the bay, they could be followed by our slow-moving viewing vessel, which means that they seldom if ever exceeded 5–6 knots. Only during the last dash to sea in the evening did they speed up and begin to leap. Nighttime schools move rather slowly along the coast. But one must remember that these animals were also diving, sometimes to considerable depths as the school moved along, and our impressions of school speed were obtained from triangulation on the school at the surface.

Under duress, spinners can swim much faster than they normally do. The closely similar Hawaiian spotted dolphin (*Stenella attenuata*) has reached what from hydrodynamic considerations must be close to a top speed of 11.03 m/sec (21.4 knots). This speed was achieved under experimental conditions following a lure towed at a measured rate (Lang and Pryor 1966). The larger and chunkier Pacific bottlenose dolphin trained to swim at top speed down a measured channel reached only 8.7 m/sec (16.1 knots), even though it also swam near what was apparently top speed (Lang and Norris 1966). The dolphin swam down the 61-m course entirely on its side without breathing, a point to which we shall return.

Arcuate leaps are associated with high speed swimming. Unless a dolphin school is disturbed, such leaps are seen only during certain activities when rapid locomotion is the norm. During rest, the dolphins continue to swim, but no leaps at all are seen and the animals surface almost surreptitiously to breathe. Later, as the school begins to move toward the offshore feeding grounds, the subgroups of a school may begin to cascade from the surface in the first arcuate leaps of the afternoon.

CROSSOVER SPEED

The various percussive leaps we described in chapter 4 are frequently seen in slowly moving schools, while arcuate leaps occur in rapidly moving schools, suggesting a clear separation in their function. The percussive leaps seem clearly to be associated with sound production, while the smooth leaps are performed with a quiet reentry and begin to occur as dolphins approach what has been called the *crossover speed* by Au and Weihs (1980) and Fish and Hui (1991).

These authors define crossover speed as the point at which it becomes more efficient to leap than to swim wholly beneath the surface. In effect, the leaping dolphin is able to provide a few tail beats of propulsive thrust with the flukes still in water while more and more of its body is subjected to the lesser drag of air, and in the process, it is able to plunge forward about two body lengths before reentry. Au and Weihs (1980) calculated this speed at about 5 m/sec (10 knots) for *Stenella*. In our records, such leaps often began to appear well before predicted crossover speeds are achieved as the dolphins break the water surface more and more. Then as the school picks up additional speed, the animals plunge along until, near the predicted crossover speed, the entire school cascades from the water in a series of leaping subgroups.

Small animals have lower crossover speeds than large ones because their shorter body lengths translate directly to higher drag at a given speed (Lang 1966, Hertel 1969). And as the surface sea becomes turbulent with waves and as cascades of compliant bubbles are dashed into it by the tumbling peaks of swells, it becomes a less and less efficient place for dolphins to swim. The dolphins themselves clearly pick and choose where to leap. We have not noticed them exiting through turbulent water but only through the smooth lee slopes of swells. From below, one can watch them exit through the glassy lee slopes and swim below the water that is opaque with bubbles from the whitecaps above.

PROPULSIVE PATTERN

From observations below the surface and later from our films, we were able to reflect on another long-standing problem in the literature of cetacean locomotion. Based upon consideration of the anatomy of propulsive muscles of dolphins, it has been suggested that the power stroke of a swimming dolphin is delivered on the upstroke (Purves and Pilleri (1978). This conclusion was reached because the epaxial muscle mass of some dolphins, located above the lateral spines of the vertebrae, appears larger than the hypaxial mass. Consequently, the downstroke of the flukes was assumed to be largely involved with returning the flukes for a new upstroke.

When one watches a swimming dolphin in the ocean, as we have done in our underwater films and from our viewing vessel, no such asymmetry of the power strokes is apparent. Instead, both excursions of the flukes appear equal both in amplitude and in approximate time of excursion. In other words, both the upward and downward excursions of the flukes appear to be power strokes. Pabst (1990) seems to have solved the dilemma when she reported on the function of the subdermal connective tissue sheaths (SDS) of propulsive muscles in the bottlenose dolphin (*Tursiops truncatus*). She found that these crossed helical fiber arrays that encase propulsive muscle masses were able to confine and redirect the power of a muscle mass as much as 90° from the general axis of the gross hypaxial or epaxial muscle masses themselves. This work shows that simple cross-sectional calculations of hypaxial–epaxial muscle mass distribution do not relate directly to the distribution of the power stroke and that there seems to be no contradiction with anatomy in the production of equal power strokes in both directions.

Dolphins, of course, frequent the sea surface since they breathe there. This means that much of their locomotion is done at shallow depths where the power of the up and down strokes of the flukes will not have equal propulsive results. Because on the upstroke the flukes often press upward against the compliant air–water interface, propulsive power is lost in the production of what field workers call "dolphin or whale footprints." These are subcircular boils of water that mark each upstroke and are especially obvious if one follows a large whale. One young male bottlenose dolphin trained to swim at near top speed (Lang and Norris 1966) solved this problem of power loss by turning on his side and hence beating his flukes laterally against a symmetrical pressure field. Bottlenose dolphins, chasing after fish at the surface, often do the same thing in nature.

AXIS OF LOCOMOTION

Another feature of swimming mechanics that minimizes the problem of power loss at the surface became obvious to us as we examined our underwater films of dolphin locomotion. It is simply that the power stroke of spinner dolphins oscillates up and down around the plane of the dolphin's belly rather than the plane of its midline. When a spinner dolphin glides, its tail trails behind it at the level of its ventral surface rather than at midbody, as one might expect. The top of the stroke is reached when the flukes rise to about the highest level of the back. The bottom of the stroke extends downward an equal distance below the resting point, or to about the tips of the extended pectoral fins (fig. 81). Frame-by-frame analysis indicates an equal time expended in the up and down strokes,

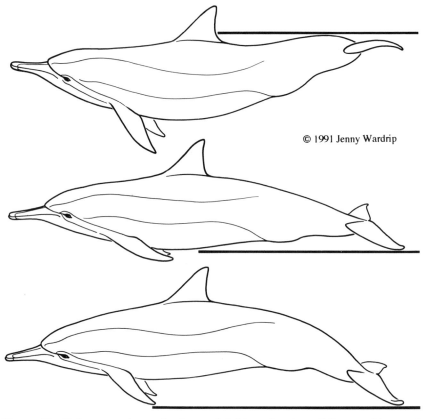

© 1991 Jenny Wardrip

Figure 81. Fluke excursions of spinner dolphins, showing gliding position and maximum excursions up and down.

although the method we used is not very precise due to the rapidity of fluke movement and the relatively slow frame rate of our camera (18 frames per sec).

The effect of this upper limit of fluke excursion is to keep the flukes well below the water surface, even on powerful upstrokes. We came to think that the dorsal fin tip may define the surface of the water for a swimming animal, providing it could sense the sea surface with this appendage, allowing the dolphin to breathe while still keeping its flukes below the water. Knowing where its dorsal fin and pectoral fin tips are may also allow the animal to gauge the position of its flukes in other contexts, such as group caressing. As mentioned in chapter 7, a dolphin may actually be able to see these fin tips as it swims along.

Thus, the arrangement of muscles, the position of the flukes relative to the fin and pectoral tips, and the mechanics of the power stroke all combine to allow dolphins an efficient means of providing propulsive power in the near-surface sea. At the same time, these provide indicators of fluke position so that interference of one swimming animal by another is usually minimized or avoided, even during close formation swimming.

ELEVEN

Breathing at Sea

Kenneth S. Norris and Christine M. Johnson

When the odontocete ancestor first became aquatic, it acquired a complex of difficulties relating to the handling of respiratory air. Over the many millions of years that followed, an exquisite series of adaptations has freed these mammals to swim freely, to dive and feed, and to phonate below the sea surface, often during long breath-holding sequences. This history of adaptation involved the alteration of many structural and physiological features of these animals. As a prelude to reporting on some simple but revealing observations about how spinner dolphins breathe at sea, we first review some of the major structural and functional developments of their respiratory–phonatory complex.

BREATHING WHILE SWIMMING

At first, the apical nose of land mammals must have caused the early cetacean ancestor to arch its neck upward to take a breath, thus greatly limiting the efficiency of its locomotion. Simple or not, this problem was only slowly solved. Through the evolutionary history of cetaceans, the nostrils gradually moved backward atop the head, stopping only when the superior nares reached the anterior face of the skull. This new nostril position allowed the animal to maintain clean sinusoidal swimming during breathing (Howell 1930).

BREATH HOLDING

The regular breathing rhythm of a terrestrial mammal is inappropriate at sea, and a complex of adaptations eventually adjusted cetaceans to breath holding. An apneustic pattern developed in which cetaceans were able to hold their breath for considerable periods—from a few minutes in most dolphins to perhaps as long as 2 hr or more in modern sperm whales (Watkins and Moore 1982). Once a sperm whale surfaces from such a dive, it breathes many times before it dives again. During this time, it seems to be paying off a lactic acid debt produced by anaerobic metabolism. The aerobic–anaerobic contribution to diving performance has been the subject of considerable controversy. Recent data suggest that elephants seals, for example, may carry out their 20–25 min dives without going into anaerobiasis. A considerable anaerobic component of respiration in the longer dives of sperm whales seems unavoidable (see also Kooyman et al. 1980).

These adaptations allow aquatic mammals great freedom to dive. But early in their evolution, swimming for long periods at depth must have posed additional physiological and anatomical problems related to oxygen storage, heat conservation, and microbubble release (the "bends") resulting from the supersaturation of tissues with nitrogen (McGinnis et al. 1972, Hempleman and Lockwood 1978, Mackay 1982, Huntley et al. 1987). An exquisite series of physiological and anatomic adjustments have extended the time period of the breath-holding until very long excursions underwater are possible in a number of modern cetacean species, including the sperm whale.

SEPARATING FOOD AND AIR

The putative terrestrial cetacean ancestor, like most terrestrial mammals, almost certainly utilized a common breathing and eating canal, the nasopharynx. But such a dual-purpose passage could have been a severe handicap to a mammal attempting to hold its breath, capture food, and phonate underwater at the same time, especially if a stream of air is used to actuate the phonation mechanism. An adaptation of all modern odontocetes was a by-pass system in which inhaled air and food were kept separate. The larynx of odontocetes (but not mysticetes) has an elongate extension that crosses the esophagus and is held tightly against the inferior bony nares by a nasopalatine sphincter (Slijper 1936, Lawrence and Schevill 1956).

The odontocete–mysticete split apparently occurred in earliest Oligocene times (Fordyce 1980), perhaps allowing this separation of breathing and alimentation, after which the odontocetes became the sophisticated echolocators that they are. A key adaptation allowing this capability to emerge must have been the dolphins' ability to capture and recycle respiratory air for use in underwater sound production in the upper nasal passages (Cranford 1988, Heyning 1989, Lawrence and Schevill 1956, Mead 1975, Norris 1968). The separation of the odontocete air and food canals was probably a crucial component in this development (Fraser and Purves 1960).

PHONATION

A number of observations and experiments pinpoint the site of odontocete sound generation at a locus in the nasal passages of the forehead superior to the bony nares (see, e.g., Norris et al. 1971, Dormer 1979, Ridgway et al. 1980, Mackay 1988, Cranford 1988). According to these workers, an echolocation click train is generated by the movement of an internal air stream past a generation point in the nasal passages superior to the skull. Cranford (1988) places click generation at paired structures on each side of the spiracular cavity, which he terms the *monkey lip–dorsal bursae complex*, located just below the vestibular sac floor. The air after being forced past these structures is finally captured internally in the dolphin's vestibular sac and held for recycling. A minority view based primarily on anatomical considerations has proposed phonation at the larynx (Purves and Pilleri 1983), but this has not been supported by experimental results that show the larynx to be passive during click and whistle emission (Ridgway et al. 1980).

Dolphin sound emissions typically occur in bursts of about 2 sec or less, and such events are correlated with internal opening of the passages of the nares while the blowhole remains closed. Conserved air returns inward through the bony nares to the basicranial space where the air is repressurized for the next sequence of sound emission (Norris et al. 1971, Dormer 1979, Ridgway et al. 1980).

PRESSURE EFFECTS

As cetaceans dive, the pressure of the surrounding sea is transmitted to all internal air spaces, even if surrounded by rigid bone, through the elastic vascular system. This allows pressure compensation to occur in which the vascularized retial linings of air spaces, including the middle ear, engorge and exactly fill the air space in relation to the water pressure upon the animal (Fraser and Purves 1960).

LUNG COLLAPSE

None of this complex of respiratory adaptations would have worked had the ancestral cetacean not also developed a means of collapsing delicate lungs during each expiration, without retention of internal air pockets. Trapped air could force bubbles through the lung walls into the blood stream, resulting in air embolism and death (Kooyman and Andersen 1969).

In terrestrial mammals, the rib cage is typically rigid, and the viscera and the fine structure of the lungs themselves are so arranged that during a dive not all air is squeezed evenly from the lungs into nonrespiratory spaces (Kooyman et al. 1981). So, along with the other adaptations previously mentioned, the posterior rib cage of cetaceans has been loosened. When the viscera are pressed upon by ambient water pressure, they move under and up against the air space of the lungs. In the odontocete lung, a sequence of tiny cartilage rings of graded stiffness is present in the bronchial and trachial passages. As the cetacean dives, air is automatically squeezed seriatim from each part of the lungs, leaving them totally bubble-free. Within the abdominal and thoracic organs, all that remains after the lungs are compressed is essentially "incompressible" tissue which produces little or no discontinuity in the pressure field imposed upon the diving animal (Kooyman 1973, Ridgway and Howard 1979).

We have written mostly of structural adaptations, but the reader should understand that these are supported by a profound series of physiological shifts of the circulatory system, blood character, and biochemical process (see Huntley et al. 1987) which we do not discuss here.

BREATHING AT SEA

Still, these adaptations solved only part of the problems a marine mammal ancestor faced in going to sea. Unless it leapt from the water with each breath, the marine mammal rising for a breath of air could never raise its nostrils more than a few centimeters above the sea surface. It had to inhale what it found there, and it had to exhale and inhale in the very short period of time of a surfacing event. If the sea was truly rough, this could mean breathing in a cascading field of half-water half-air. We wondered, was such inhaled water simply "spouted out" on the next breath or did the dolphin somehow find pure air at the air–sea interface, even in a storm?

A secondary but related problem relates to the speed of the respiratory event. When dolphins travel rapidly, their heads are exposed to the air for only a second or so during surfacings, before they plunge back into the sea. How could both exhalation and inhalation be completed within such a short time? After all, measurements indicate that a dolphin replaces most of the air in its lungs with each breath, not just a fraction of it as humans do (Kooyman et al. 1981).

The opportunity to observe these phenomena came when we noticed that we could observe entire underwater sequences of surfacing and breathing from the underwater observation vessel. So, we began to accumulate motion picture segments of such events. These were then analyzed frame by frame using a film editor. Knowing the frame rate of the camera (18 to 24 frames/sec), we could move from frame to frame observing and timing very rapid events with some precision.

Our films reveal a behavioral sequence in which the briskly swimming dolphin produces a rapidly expanding bubble over its anterior head as it rises, blows water away from its emerging blowhole, and breathes inside the cleared space, all before the water collapses over it again. When swimming slowly, dolphins may reduce the velocity of exhalation until air is swept backward off their heads by the passing water stream.

Exhalation

The first suggestion in our films of an impending breath was often found when a swimming dolphin bent its anterior body toward the surface from the horizontal swimming trajectory. Typically, it glided toward the surface at an angle that brought its snout tip and dorsal fin tip up against the surface simultaneously, or nearly so. As the snout tip touched, and while the blowhole was still 2–3 cm below the surface, expiration usually began. In our films, we often saw the start of a breath as the extrusion of a silvery bleb of air over the blowhole. In a briskly moving dolphin, this initial bleb was typically directed in an anterior direction close over the surface of the dolphin's head. In a slowly moving dolphin, it usually slid backward over the head. At this point, the blowhole opening was still a crescentic slit (fig. 82).

The convoluted internal air path taken when the blowhole is slightly open guides the about to be expired air along a curved path. The air stream, if forcibly exhaled, moves under the hard posterior lip of the blowhole and then blasts forward, essentially in the longitudinal plane of the animal's body. This directional component causes the air stream to exit forward, right over the surface of the head. This causes the dolphin to swim into its own expired air at the surface. As is known from the work of Kooyman and Cornell (1981), the velocity of such an exhaled sheet in air can be very high—as much as about 65 m/sec (213 ft/sec) for the bottlenose dolphin, estimated for the volume after passing the blowhole. (Because of the Bernoulli effect, the velocity at the restriction of the blowhole itself would be higher; G. Kooyman, pers. comm.) In the much larger gray whale, the paired nostrils are pursed during exhalation producing a blast at about 176–220 m/sec (577–721 ft/sec). Inhalation, which occurs with the nostrils wide open, is much slower—44 m/sec (144 ft/sec). When comparing this performance to that of the dolphin, remember that the velocity of the whale breath is strong enough to produce the obvious spout or blow of a great whale (Kooyman et al. 1975). Dolphins do spout (fig. 83), but the spouts are much smaller than those produced by whales.

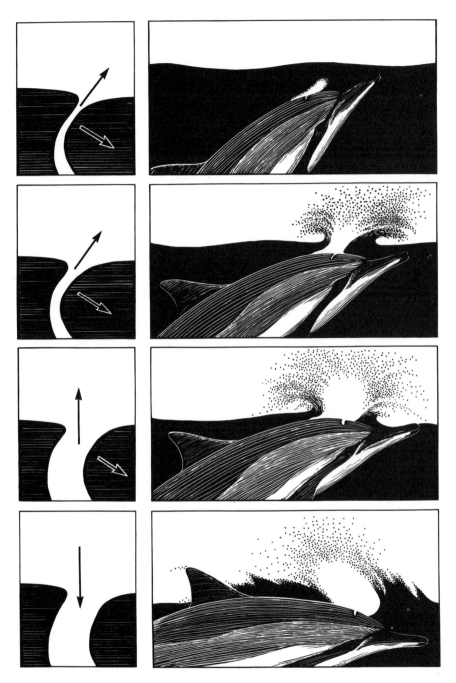

Figure 82. Respiratory cycle of a surfacing spinner dolphin, as determined from frame-by-frame analysis of moving picture films.

Figure 83. Spinner dolphin spouts at Kealake'akua Bay.

The dolphin's high velocity exhalation imparts momentum to the water around the escaping bubble just as the dolphin's head breaks the surface. This creates a radial spray of water droplets blown outward ahead of the rising dolphin, which scours seawater from the air above the rising animal's blowhole with a spray of watery "bullets." In the proper lighting, we have sometimes seen this outward blast of droplets, which appeared much like the spray from a radial lawn sprinkler or as a coherent, more vertical spout (fig. 82).

At first emergence, this plume of fast-moving droplets is directed mostly forward into the line of the dolphin's ascent because the blowhole is still incompletely opened and the air is still being deflected forward by the rigid posterior blowhole lip and by the partially open blowhole valve.

As the animal ascends farther, the blowhole opens more widely and the outrushing air column becomes more and more vertical with respect to the dolphin's head. The excurrent blast of air thus moves from the horizontal to the vertical direction as the animal rises into its own exhalate. The water over the blowhole sprays outward in all directions as the animal begins to inhale in the highly transient vacuity thus cleared by its own exhalate. The roll of the dolphin then begins to take it downward beneath the sea again.

Inhalation

At this point, the dolphin may hit the surface of the water with its anterior head, creating a transient, air-filled cavity in the surface of the sea into which it arcs downward in a dive. Inhalation is completed inside this highly transitory space, which probably allows the respiratory event to be 0.1–0.2 sec longer than it otherwise would be and allows complete inspiration in water-free air. The Dall's porpoise (*Phocoenoides dalli*) creates the most obvious of such breathing spaces in the water surface, but these are also produced by spinner dolphins during the last stage of their breathing sequence.

We have timed a series of complete breathing events from our films (fig. 84). The total respiratory durations are smaller than those recorded from beached captive bottlenose dolphins in which total breath sequences typically ranged from 0.7 to 1.0 sec in duration (Kooyman and Cornell 1981). The spinner dolphins we photographed at sea sometimes breathed, out and in, in as little as 0.5–0.6 sec total elapsed time, although the average total time was 0.77 sec.

About twice as long is taken during the inhalation phase as during exhalation. We wondered if this was purely a function of the need for high velocity air on exhalation to clear a space in which to breathe? Perhaps it might also be involved in a reduction in air velocity within the respiratory tree when air is being inspired at the end of a breath sequence. This might reduce any impact or "pressure transient" effects that could be produced by a high velocity air column entering the respiratory tree many hundreds of times a day. In fact, the structure of the toothed whale airway, from blowhole to alveolus of the lung, seems designed to reduce the velocity of air as it is inhaled and to speed it during exhalation. (This occurs mostly by a Bernoulli effect at the narrowed opening of the blowhole during early exhalation, but also by diameter relationships throughout the collective airway; see Belanger 1940 and Denison and Kooyman 1973 for structural arrangements within the lung.)

Locating a Breathing Place

Does this adaptation of clearing a space in which to breathe avoid the necessity of choosing the best breathing place? It seems not. We have watched surfacing dolphins from below the water in moderately rough

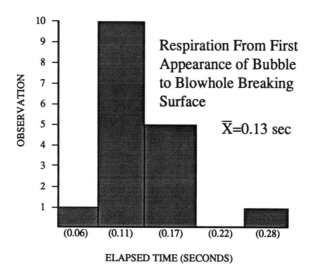

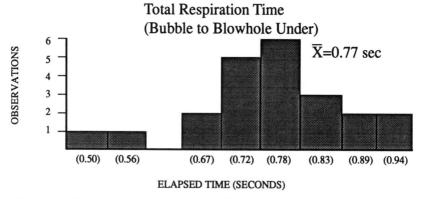

Figure 84. Respiratory sequences timed from frame-by-frame analysis of moving pictures.

weather—when there were cascading breakers from white-capped seas. In such a case, the animals were not tossed about underwater as one might expect, but instead moved gracefully in the orbital motion of the swells, keeping excellent station with one another. Then, when one of them rose to breathe, it surfaced into the glassy water between turbulent areas, places that must be as wholly evident to them as they were to us— underwater mirrors in the water surface, existing amidst cascades of bubbles.

One of us (B. Würsig) has watched dolphin movements during a se-vere storm. The dolphins leapt above the frothy water surface, presum-ably to breathe. So it is probable that surface respiration by dolphins in-volves three stages dependent upon sea state: leisurely breathing in calm water without an exhalant blast, exhalant clearing of a breathing space at moderate speeds or in moderately turbulent seas, and leaping to breathe during swift locomotion or storms.

In these ways, it became clear that the breathing events we take for granted on land are done very differently at sea. The ease with which modern dolphins take their surface breaths is clearly the end-product of a long evolutionary process involving many parts of their bodies, their physiology, and their behavior.

TWELVE

Food and Feeding

Bernd Würsig, Randall S. Wells, and Kenneth S. Norris

Spinner dolphins are primarily feeders on small mesopelagic prey that they obtain by diving. Many of these prey species are vertical migrants coming toward the surface at night. Spinners routinely take several species that never reach the surface. Populations that frequent island shores or shallow embayments, such as the Hawaiian spinners discussed here or the Gulf of Thailand spinner (Perrin et al. 1987), also take bottom-dwelling species, and a small increment of surface-dwelling forms also enters the diet.

This chapter first describes the mechanics of food capture and manipulation and then reviews and adds to the few published accounts detailing their diet. The few observations we have made that relate to the feeding behavior of schools are described.

Spinner dolphins are among the longest beaked of all delphinids. They possess long homodont tooth rows of slim, curved conical teeth splayed outward on the margins of each ramus of the jaw and interdigitating with the teeth on the opposing jaw when the mouth is closed. These slim little teeth, numbering 48–62 per row (Perrin 1990), reflect the small prey of the species, a feature well borne out by examination of stomach contents. Although some prey species are taken that grow to large size, in these cases the spinner dolphin specializes in the juvenile stages.

MECHANICS OF FOOD CAPTURE AND HANDLING

Norris and Møhl (1983) outlined some aspects of the evolution of food getting mechanisms in odontocetes, including the appearance of the

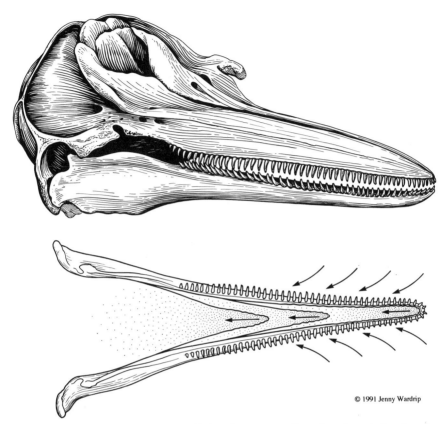

© 1991 Jenny Wardrip

Figure 85. Diagram of spinner skull showing outwardly splayed teeth (top) and the action of the piston tongue (bottom).

cage jaw early in odontocete evolution and the *piston tongue*, which is present in all modern delphinids that have been inspected, including the spinner dolphin (fig. 85). These adaptations relate to two problems that dolphins face when feeding.

The Cage Jaw

The first problem occurs when an aquatic predator closes its jaws upon a prey organism and water tends to be forced out of the jaws, carrying the prey away in the excurrent stream. The cage jaw consists of long narrowed jaws and splayed-out, slim conical teeth. These teeth allow food to be trapped by their interlocking tips before the excurrent stream forces it away. This adaptation has appeared in many aquatic taxa, from gars and needle fish to the earliest secondarily aquatic reptiles, the mesosaurs of Pennsylvanian and lower Permian time (Oelofsen and Araujo 1983), and many cetaceans, especially early forms. The cage jaw is widespread in long-jawed delphinids, including the spinner dolphin.

The Piston Tongue

The piston tongue is an arrangement of the tongue and its supporting bones and muscles that allows the tongue's rapid posterior retraction. Thus, water and prey are pulled *into* the mouth as the jaws close. The delphinid tongue, including that of the spinner dolphin, possesses this arrangement. The spinner dolphin's tongue extends very far anteriorly in the gape, to just behind the narrow mental symphysis of the lower jaw, or within about 1 cm (0.4 in.) of the jaw tip. The tongue is equipped with lappets that insert between the teeth and effectively extend it to the back of the tooth row. This arrangement is especially obvious in suckling young and is probably involved in ensuring that milk is not lost to the surrounding water or mixed with salt water during nursing.

Ingestion

When we observed captive spinners ingesting food, they typically took the prey item (in this case, a frozen Columbia River smelt, or the eulachon, *Thaleichthys pacificus*) about one-third of the way into the gape from the tip of the beak. They then swam forward, drifting the immobile food farther and farther into the gape until it was engulfed. As they did so, their jaws typically "chattered" over the fish, holding, releasing, holding again, and so on.

Although these dolphins were ingesting rather large dead prey, they emitted soft echolocation trains until the very moment the fish was closed upon by the jaws. This often caused the upper jaw to be raised so that the plane of the upper jaw was directed a few degrees above the drifting fish. We concluded that some portion of the dolphin's sound beam emanated directly at the fish, perhaps from the upper jaw tip where click trains are known to be emitted from the mesorostral cartilage, which extends slightly beyond the bony jaw tip, just under the skin (see Diercks et al. 1971). However this sound field might have been emitted, it seemed sufficiently broad that it could ensonify the fish right into the animal's gape. Since the spinner dolphin feeds in black nighttime water, the prey must be followed by using echolocation or bioluminescence. Whichever the case, the spinner echolocation system seems able to track prey very close to the jaw tip. Such an adaptation seems of obvious significance, even if it is not wholly clear how it works. It involves the use of sound in the difficult-to-typify near field. This anatomical–acoustic arrangement is perhaps ubiquitous in delphinids.

The long, narrow spinner dolphin jaw is presumably also an adaptation that allows the dolphin to grasp living prey well ahead of its large body and thereby to cause a minimum of escape reaction from prey during attack sequences. When we observed the ingestion of dead fish by captive spinners, we did not see the action of the piston tongue come into play, perhaps because the food was many times larger than the usual natural food and because the dolphin had learned that it would make no attempt to escape.

OWNERSHIP OF PREY

As has been noted with other odontocetes (Norris and Prescott 1961), it is possible that the "ownership" of food may be respected among schoolmates in spinner schools. We did not see struggles for a prey item with other dolphins in the tank, once a given spinner dolphin took it. On the contrary, there is evidence of prey sharing in other cetacean species, at least among taxa that feed on large food that takes some time to process and swallow. For example, the false killer whale (*Pseudorca crassidens*) has been observed to take a large fish crossways in its jaws (*Coryphaena hippurus*) and allow other school members to take pieces from it (Brower and Curtsinger 1979, Norris and Schilt 1988).

Norris and Prescott (1961) report what looked like "ownership" of captured prey fish (large yellowtail, *Seriola dorsalis*) by individual captive bottlenose dolphins. Dusky dolphins, and other dolphins that herd schooling fish into tight balls, seem to take turns feeding (Würsig and Würsig 1980). However, it is not known whether such chasing is totally equitable and what complexities to dominance-related access to food might be present in such societies.

DIET AND FEEDING DEPTH

Probably the first study of spinner dolphin stomach contents was that of Cadenat and Doutre (1959), who noted that spinner dolphins taken off the African coast had eaten mostly mesopelagic fish of the family Scopelidae [= Myctophidae].

Fitch and Brownell (1968) provided the first direct evidence of the spinner dolphin diet in the Pacific Ocean. Their work, which dealt with dolphins from the eastern Pacific tuna grounds, concentrated on fish otoliths found in stomach contents; thus, other parts of the food intake such as squids and shrimps were not enumerated. They collected otoliths from five spinner dolphins and one spotted dolphin that had been netted in the eastern tropical Pacific during daytime by tuna fishermen, between about 12°N and 20°N latitude. Since these dolphin species occur together in that part of the ocean, and since there are questions both about what bonds the species together and whether one (*Stenella longirostris*) might be nocturnal and the other (*S. attenuata*) diurnal, we quote Fitch and Brownell's (1968, p. 2566) paper in some detail. The authors state:

> Although the five *S. longirostris* had fed upon 6–15 kinds of fish, two kinds of myctophids, *Benthosema panamense* and *Lampanyctus parvicauda*, comprised over 50% of their piscivorous diet, both individually and as a group. *Benthosema panamense* was also important to *S. graffmani* [= *S. attenuata;* authors], being the only species noted in one of the three stomachs and contributing all but 3 of the 1238 otoliths found in a second. This tropical lanternfish normally undertakes a diurnal migration, but sometimes dense "balls" (schools) of *Benthosema* remain at the surface during daylight hours, where they have been observed under attack by birds, fish and dolphins. Even though both dolphins had fed heavily on *Benthosema*, we believe that *S. longirostris* had captured its prey several hundred feet down, whereas *S. graffmani* [*S. attenuata*] had fed at or near the surface.

They go on to describe (p. 2566–2568) the various families and genera of fish found in the stomachs of oceanic spinner dolphins:

> Bathylagidae—The *Bathylagus* otoliths in the stomachs of *Stenella longirostris* may have been those of *B. pacificus*, a species that lives more than 650 ft (200 m) beneath the surface and does not undertake a diurnal migration. They were not from *B. stilbius*, one of the commonest blacksmelt in that area, and a known vertical migrator. None of the 15 otoliths in the three stomachs was from a large fish, perhaps 4 or 5 inches (100–125 mm) being about maximum.
>
> Bregmacerotidae—Only the stomachs of *S. longirostris* contained otoliths of *Bregmaceros bathymaster*, even though the species is known to migrate from the depths to surface waters at night. A full-grown *B. bathymaster* would hardly exceed 3 inches (75 mm) and 1 g.
>
> Centrolophidae—The most abundant otoliths in the stomach of *S. graffmani* . . . were those of an unknown stromateoid. . . . We believe they represent a species of centrolophid, but to our knowledge no member of the Centrolophidae is known from waters off tropical west Mexico. In any event, stromateoids (including Centrolophidae) are noted for their association as juveniles with floating debris, jellyfishes, and similar items, so the fishes these came from probably were eaten at or near the surface. Three of the *S. longirostris* stomachs we examined also contained otoliths of these stromateoids, but in much smaller numbers.
>
> Exocetidae—Twenty-nine otoliths in the stomach of *S. graffmani* . . . were from the shortwing flying fish, *Oxyporhamphus micropterus*. This tropical species is strictly an inhabitant of surface waters. A really large individual might attain a length of 8 inches (200 mm) and a weight of 2 ounces (55 g).

Gonostomatidae—Otoliths of *Vinciguerria lucetia* were relatively abundant in four of the five stomachs of *Stenella longirostris*, but only three were present in the *S. graffmani* stomachs. *Vinciguerria lucetia* inhabits depths of 650 to 1000 ft (200–300 m) during daytime, but migrates toward the surface at night and often can be dipnetted from under a bright light suspended above the surface. At maximum size they are just over 2 inches (50 mm) long and weigh but a few grams. Seven otoliths in one *S. longirostris* stomach . . . appear to be those of a gonostomatid but our comparative collection of gonostomatid otoliths was small, so we were unable to make a positive identification.

Myctophidae—At least eight species of lanternfish had been fed upon by *S. longirostris* and three by *S. graffmani*. Six of these, *Symbolophorus evermanni*, *Myctophum aurolaternatum*, *Diogenichthys laternatus*, *Benthosema panamense*, *Diaphus*, and *Hygophum*, are found in surface waters at night, but the two species of *Lampanyctus* identified from *S. longirostris* stomachs are deep (~400 m) during daylight hours and usually remain fairly deep (~200 m) at night. Three to 5 inches (75–125 mm) would be about a maximum length for all of the myctophids except *Diogenichthys*, which seldom attains 2 inches when full grown.

Paralepididae—The 30 barracudina otoliths in the stomach of *S. longirostris* . . . are difficult to assign to a genus because of a shortage of comparative material. They are not from *Paralepis*, *Notolepis*, *Sudis*, or *Lestidium*, and several other genera are unknown from the eastern Pacific. Barracudinas are most abundant at depths of 1000–2500 ft (300–800 m), but range both above and below these depths. Adults of *Stemonosudis* are unknown in collections, but otoliths in this dolphin stomach unquestionably were from adults. Based upon otoliths from other family members, these fish could have been 10–12 inches long (25–30 mm).

Scopelarchidae—One otolith (from *S. longirostris*) was from a 4- to 5-inch (100–125-mm) *Benthabella*, a fish that usually lives 800–1000 ft (250–300 m) down during daylight hours. Young individuals migrate into surface waters at night, but habits of adults are unknown.

Fitch and Brownell (1968, p. 2572) concluded that three of the cetaceans they investigated (*Kogia simus*, *Stenella longirostris*, and *Lissodelphis borealis*) had been feeding 250 m (800 ft) or more beneath the surface. However, *S. graffmani* and *Phocoena sinus* appeared to have fed within 30 m (100 ft) of the surface and probably even shallower. They also noted that at least 8 of the 16 species fed upon by *S. longirostris* are known to ascend into (or nearly into) surface waters at night, but they doubted that the spinner dolphin did much foraging for food in these upper water layers.

It seems fair to conclude that the nocturnal spinner dolphin dives deeply for at least some of its food, although other species in its diet could have been obtained in the dark nearer the surface. The mix of depth preferences in the spinner dolphin diet could be associated with dive patterns. Toward dusk, the dolphins off Hawaii begin to dive to meet the rising scattering layer. In the dark, their dives might be shallower and the diet composed of both shallower water species and those rising from deep layers.

RESOURCE PARTITIONING

Another study, which included both spotted and spinner dolphins and tuna, was carried out by Perrin et al. (1973) using a fairly large sample of each taxon (79 tuna, 140 spotted dolphins, and 46 spinner dolphins, all from the same series of six net sets). They found that the tuna and the spotted dolphins had overlapping feeding patterns; both took the squid (*Dosidicus gigas*) and the bullet mackerel (*Auxus* spp.) in considerable numbers. However, the spotted dolphin took considerable numbers of the epipelagic flying fish (*Oxyrhamphus*), while this species was a minor component of tuna stomach contents. Also, spotted dolphins and tuna each specialized in some food items that the other did not take. For example, the portunid swimming crab (*Euphylax dovii*) was a very important item of diet for tuna, but wholly absent from dolphin stomachs.

Spinner dolphins contained a very different, though not totally separate, collection of species. Spinners fed primarily by diving into the scattering layer below the thermocline, where they ate a variety of small scattering-layer fishes, some of which are not vertical migrators. They also ate a few fish that either are restricted tò the epipelagic zone (one *Oxyrhampus* was found in one spinner stomach, for example) or are vertical migrators and could have been taken near the surface.

Perhaps the most interesting finding of Perrin et al. (1973) was that when spinner porpoises were sampled (three hauls), the rate of occurrence of empty stomachs and the state of digestion of stomach contents indicated that they had not fed at the same time as the tuna or the spotted dolphins. They found that of 49 spinners examined from the eastern tropical Pacific, 65.3% had empty stomachs. Norris and Dohl (1980*b*) have suggested that this fits well with their observations of Hawaiian spinners that rested during the day. At the time of entry into the rest coves, they often showed a high incidence of defecation, as if the digestive process for the night's food was being completed. Digestive passage time in *Stenella* has yet to be measured.

Scott (1991) has further studied *Stenella* diets from the eastern tropical Pacific and has shown that the picture of a diurnal spotted dolphin and a noctural spinner dolphin is something of an oversimplification. He states:

> Information on the feeding habits of spotted dolphins collected by the IATTC and NMFS researchers indicates that none of the stomachs collected from dolphins captured between 1200 and 1600 h had recent prey

items in them—precisely during the period of day when the largest average herd sizes were observed. . . . Virtually undigested fish or squid were found only in the stomachs of dolphins caught in the early morning (0700–0930 h). These data, then, do not support the hypothesis that spotted dolphins aggregate in response to a diurnal aggregation of their prey. A similar feeding hiatus may be present in yellowfin tuna as well.

These observations seem to show that, while the spotted dolphin does feed during the early and middle hours of the day, it seems not to feed in the afternoon.

THE TUNA–DOLPHIN BOND

Yellowfin tuna (*Thunnus albacares*) and spinner and spotted dolphins are frequently netted together by tuna seiners in the eastern tropical Pacific. In fact, the strength of this bond is the basis of the dolphin-based seine fishery. Fishermen can encircle whole schools of dolphins, knowing that under them may swim large tonnages of yellowfin tuna.

In the broadest view, the association seems to be partially food based. Clark and Mangel (1984), using game theory, have shown that when food is patchy, as it is in the eastern tropical Pacific, group foraging is an evolutionarily stable strategy. The circumstances of open sea foraging, then, may drive various participants into a group process, and it can become one of considerable ecological complexity, also involving the strategies the partners may use to reduce predation. It does not seem to be an "either–or" situation, that is, either food or predation based, but rather a more total response of the individual circumstances of life for the various participants. We will say more about these tendencies and the complexity involved when we discuss multispecies aggregations in chapter 15.

The relationships between dolphins and tuna, and among the different species of dolphins, seem in no case to be symmetrical. That is, the participants have different roles relative to each other in the aggregation phenomenon. Reasonable evidence suggests that the fish follow the dolphins (see Scott 1991 for a review), although there are proponents of the opposite relationship (Au and Pitman 1986). The bond is predictable for the spotted dolphin and even large tuna. It is much less reliable for spinner dolphins and smaller tuna. Scott (1991) notes that the percentage of sightings of dolphin schools that lead to net sets is only about one-third as great for spinner and common dolphins as it is for spotted dolphins. The spotted dolphins are the primary species, as the fishermen say, "that carry tuna."

As for the daytime bond between dolphin species, spotted dolphins are frequently found during the day in single-species schools, while spinner dolphins are much less often netted alone and are usually found in association with spotted dolphins. Scott (1991) notes that spinner dolphins are primarily nocturnal feeders and says, "It would seem unlikely that dolphins that feed at night would aggregate during the day because of any feeding advantage." The most parsimonious explanation seems to be that the spinner dolphins are seeking an association with the spotted dolphins that is not food based.

Fishermen report that spinners "carry smaller tuna" (Captain Harold Medina, pers. comm.). From the considerations just described and the following observations, we can summarize what seems to us to be a plausible explanation for the associations of dolphins and yellowfin tuna in the eastern tropical Pacific.

We suggest, as has W. F. Perrin (pers. comm.), that the spotted dolphin (*S. attenuata*) is nuclear to the association. We suggest that the larger tuna follow this dolphin species because, by use of its echolocation, it is able to lead the fish to food patches of the species and sizes they prey upon. We further suggest that the spinner dolphins may associate during the day with spotted dolphins because resting spinners, who (judging from Hawaiian spinners) are silent and not echolocating at this time, achieve a protective association that may leave them in the vicinity of food as the day ends.

We base these ideas on several observations. Spinner and spotted dolphins are clearly not equal partners in their association with tuna. Spotted dolphins more reliably "carry tuna" than do spinners. In a tuna net, spinner dolphins were observed by Norris et al. 1978) to be peripheral to the spotted dolphins; they tend to swim deeper than the spotted dolphins and mostly outside the main mass of the netted school. Hawaiian spinner dolphins feed at night and rest during the day, while the spotted dolphins appear to be heavily crepuscular or daytime feeders, as are the tuna (Scott 1991), although they may engage in afternoon "rest" periods when feeding does not appear to occur.

Spotted and spinner dolphins caught in tuna seines during the day show different digestive stages (Perrin et al. 1973). Whole food is found in the stomachs of spotted dolphins taken during the day, and either empty stomachs or just otoliths and beaks tend to be found in the stomachs of spinners caught after about noon. Yellowfin tuna and spotted dolphins show clear overlap in major items in their diet, especially the squids (Perrin et al. 1973).

Dolphins possess greatly superior long distance food-finding capability compared to tunas because they possess echolocation, a teleportive sense, or one that very rapidly provides environmental information from a distance farther than vision will serve (Evans and Norris 1988). The long distance sensory capacity in which tunas may be superior is

chemoreception, and even this may not be correct given the highly developed taste buds of dolphins (Caldwell and Caldwell 1972*b*, Ridgway
1990). Such a chemical sense requires tracking down an odor trail and
does not provide the immediate information of the whereabouts of prey
that sound can.

These points taken together suggest that the spotted dolphins and
tuna swim together because the major items of their respective diets
overlap. The dolphins are thought to be nuclear because the tunas re-
accumulate under them if stripped away by fishing and because we have
seen tunas follow dolphins in the net (Norris et al. 1978). A similar conclusion was reached by tuna fishermen who worked before the days of
seining. At that time, tuna were held alongside bait boats by throwing
small schooling fish into the water (chumming). The tuna were reported
to remain near the bait boat as long as the associated dolphins milled
nearby. If the dolphins left, so did the tuna (Cosgrove 1991).

In many parts of the world, tuna swim deep in the general depth of
a deep thermocline, which is presumed to concentrate their food.
Whether there is an association of tuna with dolphins at such depth is
unknown. But in the eastern tropical Pacific tuna grounds, where the
thermocline is shallow, tuna cluster under dolphins, even though the
mammals must return frequently to the surface to breathe. It seems
likely that the tuna benefit because the spotted dolphins can seek out
prey patches better than they can, using echolocation, and that the disruption of air breathing is inconsequential enough that the bond between fish and mammal can persist.

THE FOOD OF HAWAIIAN SPINNER DOLPHINS

Norris and Dohl (1980*a*) examined the stomach contents of four Hawaiian spinner dolphins caught early in the day and three additional specimens taken in the afternoon (table 7). The former contained remnants
of food items: squid beaks, chitinous shrimp remains, fish bones, and
otoliths. The latter all had empty stomachs. Because formaldehyde pres-
ervation had eroded the otoliths in the collection, the authors were only
able to report on the cephalopod and shrimp remains. Epipelagic squids,
though common in Hawaiian waters, were absent from the stomachs.
Such relatively deep water squids as *Abralia astrosticta* and *A. trigonura*
were common in the stomach samples, although the latter is rarely reported from Hawaiian waters. Dr. Richard Young of the University of
Hawaii made the taxonomic determinations of squids and reported that
A. trigonura is a vertical migrant that moves from a depth of about 500 m
(1640 ft) during the day to the upper 100 m (330 ft) at night, while *A.
astrosticta* is known from only a few captures in bottom trawls, with some

TABLE 7 Summary of Food Items Known from *Stenella longirostris**

Family/Genus/Common Name	Hawaiian Waters						Eastern Tropical Pacific					
	This is Study KSN-80-4			Norris and Dohl (1980)			Perrin et al. 1973			Fitch and Brownell 1968		
	Total	%	Number/%	Total	%	Number/%	Total	%	Number/%	Total	%	Number/%
Fish	500	100	1/100	Otoliths not available for identification			5668	100	6/100	13839	100	5/100
Myctophidae: lanternfishes	263	53	1/100				4189	74	6/100	12200	88	5/100
Diogenichthys	0	0	0				3411	60	6/100	1423	10	5/100
Lampanyctus	5	1	1/100				388	7	5/83	2746	20	5/100
Benthosema	0	0	0				281	5	6/100	7606	55	5/100
Myctophum	4	0.8	1/100				29	0.5	4/67	222	1.6	4/80
Hygophum	27	5.4	1/100				5	0.1	4/67	41	0.3	4/80
Diaphus	21	4.2	1/100				2	—	1/17	1	—	1/20
Notoscopelus	6	1.2	1/100				0	0	0	0	0	0
Ceratoscopelus	3	0.6	1/100				0	0	0	0	0	0
Symbolophorus	0	0	0				5	0.1	3/50	16	0.1	2/40
Unid. Myctophid	201	40	1/100				68	1	5/83	145	1	3/60
Photichthyidae: lightfishes												
Vinciguerria	0	0	0				192	3	5/83	1317	10	5/100

Taxon									
Bregmacerotidae: deep sea cod									
Bregmaceros	4	0.8	1/100	842	15	4/67	243	1.8	5/100
Melamphidae: bigscales				424	7	2/33	0	0	0
Melamphaes	0	0		11	0.2	1/17	0	0	0
Scopelogadus	0	0		413	7	2/33	0	0	0
Bathylagidae: blacksmelts									
Bathylagus	0	0		11	0.2	1/17	0	0	0
Scopelarchidae: pearleyes	3	0.6	1/100	5	0.1	1/17	0	0	0
Stromateidae: butterfishes	0	0		2	—	1/17	0	0	0
Paralepididae: barracudinas	0	0		1	—	1/17	30	0.2	1/20
Unidentified	0	0		2	—	3/33	0	0	0
Postlarval fish				0	0	0	0	0	0
Apogonidae: cardinalfish	13	2.6	1/100	0	0	0	0	0	0
Carapidae: pearlfishes	3	0.6	1/100	0	0	0	0	0	0
Holocerdae: squirrelfishes									
Myripristis–Uu	1	0.2	1/100	0	0	0	0	0	0
Bythitidae: livebearing brotulas									
Brotula			1/100	0	0	0	0	0	0
Unknown larvae	10	2	1/100	0	0	0	0	0	0

TABLE 7 (continued)

Family/Genus/ Common Name	Hawaiian Waters						Eastern Tropical Pacific					
	This is Study KSN-80-4			Norris and Dohl (1980)			Perrin et al. 1973			Fitch and Brownell 1968		
	Total	%	Number/ %	Total	%	Number/ %	Total	%	Number/ %	Total	%	Number/ %
Pempheris-like sweepers	9	1.8	1/100				0	0	0	0	0	0
Cirrhitid-like hawkfish	2	0.4	1/100				0	0	0	0	0	0
Blenny-like blennies	1	0.2	1/100				0	0	0	0	0	0
Miscellaneous	187	37	1/100				0	0	0	0	0	0
Cephalopods	Present but not identified									Present but not identified		
Onychoteuthidae Onykia							27	4/33				
Ommastrephidae Dosidicus							45	3/25				
Symplecoteuthis							14	6/50				
Unidentified							23	1/8.3				
Enoplateuthidae Abraliopsis							152	5/42				

Cheuthidae				
Abralia	46	4/100	2	1.8.3
Histioteuthis	1	1/25		
Crustaceans				
Gammarid amphipod			1/8.3	
Shrimp				
Pasiphaeids	14	2/50		
Sergia	35	2/50		
Acatntephyra	1	1/25		

*For fish, we show the total number of otoliths collected, the percentage of the total that is comprised by each taxon, the number of dolphins in which each kind of otolith was found, and the percentage of the total number of dolphins examined in each study in which the otoliths were found. The presence of the various invertebrate taxa is indicated.

small individuals having been taken in mid-water. The Norris–Dohl
squid specimens were adults. The shrimps were of a single species,
Sergia fulgens. Although this species migrates vertically, it stays below
about 150 m (500 ft) depth during the day. Thus, these ecological correlates of food items match the general patterns deduced by Fitch and
Brownell (1968) for open ocean spinner dolphins.

The stomach contents of a spinner dolphin that died during capture
for radiotagging was analyzed for fish remains. The dolphin, taken on
April 22, 1980, at 1:30 P.M., had no fleshy remains in its stomach but
contained a collection of otoliths that were preserved in ethanol and
identified by John Fitch of the California Department of Fish and Game.

Most myctophid otoliths from this specimen were not identified below
the family level, but *Hygophum* and *Diaphus* were noted to be abundant.
These two genera comprised a small percentage of the total number of
otoliths in the other studies. Both *Lampanyctus* and *Myctophum* were also
noted but not enumerated. The adults of these lantern fishes are on the
order of 10–15 cm (4–6 in.) long. Fitch and Brownell (1968) noted that
Lampanyctus does not move above about 200 m (660 ft) depth, even at
night, while the other genera are vertical migrators that may reach the
surface.

Considering all the spinner dolphin stomach samples that have been
studied, we can say that all prey items are less than about 200 cm (8 in.)
long. While the adults of some of the taxa found in these stomachs, including reef species, may reach lengths of 80 cm (31 in.) or more, only
postlarval forms were taken by the dolphins.

DIVING PATTERNS

The Hawaiian radiotracks and stomach contents both support the view
that spinners feed primarily by diving and that they may reach depths of
250 m (820 ft) or more. Some fishes in the diet were probably taken near
or at the surface (flying fish, for example). The dolphins apparently
reached the bottom from time to time, judging from the presence of
squid species otherwise known only from bottom collections.

Norris and Dohl (1980*b*:836) observed the behavior of spinner dolphin schools during what were thought to be the first feeding dives of
evening. The dolphins had left their daytime retreat in Kealake'akua
Bay and had moved offshore into relatively deep water. They say:

> Feeding schools were observed on three occasions at dusk. Each was com-
> posed of widely scattered groups, covering as much as 3 km in widest di-
> mension, moving together. Diving was subsynchronous. Before a dive oc-
> curred, groups were evident and there was much aerial behavior across the

entire width of the school. Then groups of the school dove individually, all following within approximately a minute or two. Dives were long, averaging 3.5 min according to our records. Surfacing was approximately as coordinated as diving; that is, the various groups straggled to the surface over a minute or two.

It was striking to see these very broad diffuse schools reverse their course in relative synchrony (within a minute or two) even at dusk, indicating a communication mechanism that could pass information rather quickly across the school in dark water.

The radiotracked dolphins that were followed in this study produced a pattern of dives and surfacings that continued throughout the time the dolphins patrolled along the island slopes in the dark.

Our discussion now moves on to a consideration of schooling and social behavior, much of it observed from our underwater viewing vehicle, the *Maka Ala*.

THIRTEEN

Schools and Schooling

Kenneth S. Norris and Christine M. Johnson

A spinner dolphin never leaves the protective confines of its school. Although at first sight a dolphin school seems to be just a cluster of animals traveling together, we are beginning to realize that it is far more than that. Expressed in a school are patterns, signals, and processes that are relevant at various levels of organization above the individual. An entire lineage of toothed whales may be identifiable by a vocal dialect held in common by all members, as in the case of killer whales (Ford and Fisher 1983). Resting schools of spinners may merge into larger feeding aggregations at night, and underlying this great daily change in group size and composition are mechanisms that allow the two sorts of schools to retain their structural integrity and their protective functions. We describe these in our discussion of *bout behavior* (chap. 14). At any given time any individual school of dolphins is much more than the sum of its parts. The sensory capacities and what is detected by individual dolphins appears to be integrated into a supraindividual system that makes such a school into a protective unit far more capable of dealing with predators than any individual.

The means of communication that mediate these various classes of supraindividual behavior are specific to their various levels. Dialects function between lineages of animals and, in the case of this study, it is no wonder that we did not find such signals. We worked entirely within a single fission-fusion society of dolphins that occupied the entire island of Hawaii. Had we worked both on Maui and Hawaii, for instance, we might have detected such signals.

The behavior of individual schools of spinner dolphins is mediated by rhythmic patterning or oscillation of behavior within the school that regulates the passage of the entire group of animals through its daily round.

We will return to this matter of rhythmic patterning later in the book, but here we will approach the question of schooling mostly from the other direction, from looking primarily at individual interactions between dolphins, the signaling that underlies them, and how the behavior of the school as a whole comes to result from such interactions.

Even a casual glance will show that the usual spinner school is composed of a series of small clusters of animals. These subgroups, which vary from about four animals to as many as twelve, may represent the smallest reactive units involved in the school protective system. Only once did we see the minimum school of four swimming as an isolated school, and when our swimmer approached it, he was threatened, which did not occur on similar approaches to larger schools. Never have we seen a single spinner dolphin swimming by itself.

When a few of these units cluster together they form the more typical spinner dolphin schools; in Kealake'akua Bay such schools usually number about 20–45 animals. We propose that such a school is capable of providing the matrix for communication that knits all the member dolphins together. It provides the substrate for a sea-going culture; assists in food location and perhaps capture; provides places for nurture, teaching, and reproduction; and perhaps helps reduce the energy costs of locomotion (Norris and Dohl 1980b, Norris and Schilt 1988). At numbers higher than about 45 animals, dolphin spinner schools begin to show a tendency to subdivide and sometimes even to move independently, although frequently associated. By the time 80–90 dolphins are assembled, this tendency toward fission becomes fairly obvious.

Only during deep rest is the tendency to clump in subgroups of modest size suppressed to the point that the larger school comes to dominate the geometry.

FUNCTIONING OF SCHOOLS

Most observations and experimental work on schooling have been performed on fish (see Pitcher 1986, for a summary). Other studies include bird flocks (Davis 1980), groups of echinoderms (Pearse and Arch 1969), crustaceans (Hamner 1984), and molluscs (see Norris and Schilt 1988, for a review). Remarkably, the major rules that regulate these groupings seem to have far more commonalities than differences.

When dolphins are threatened, or when they rest, their schools seem to operate in much the same way as polarized open water fish schools, with at least one exception. Fish schools seem much more capable of close maneuvering than do oceanic dolphins. Tunas, though sometimes about the size of dolphins, will escape through holes in tuna seines, while dolphins crowded in the same seine may sink into a group catatonia (Coe 1980, Cosgrove 1991). The difference appears to lie in the fishes' simultaneous use of vision, hearing, and mechanoreceptors. The latter sense operates at very close range. Dolphins, on the other hand, appear to rely on vision, hearing, and echolocation, all of which require considerable space between animals to be effective.

We only mention in passing the very large assemblages of dolphins, or other animals (Clark and Mangel 1984) sometimes numbering in the thousands, that may gather around a food source (Scott and Chivers 1990, Scott 1991).

Nonetheless, we do not imply that the larger schools are without additional functions for member dolphins, only that they are mostly extensions of the smaller schools. Large groups of dolphins can clearly provide cover for individual dolphins by their sheer numbers (Williams 1966). When a dolphin is surrounded by other calmly swimming dolphins, it is unlikely that an attack is underway, and the predator cannot see very far through such a mass of animals to mount an attack upon an animal in the "shadow" of another.

SCHOOL FORMATIONS

Echelon Formation

As spinner schools passed by the viewing window of the *Maka Ala,* one of the first things we noted (see also Norris and Dohl 1980*a,b*) was that the dolphin schools are clusters of animals arranged in a recognizable but fluid geometry of members. They do not tend to form clean ranks of animals swimming precisely side by side, nor do they string out cleanly head to tail, although they may temporarily approach both of these formations at times. Most commonly one animal is stationed at about the level of its neighbor's pectoral fin. We call this the *echelon formation.* As the school moves this way and that, the distance between subgroups is extended and shortened. For example, when the school turns, the outside subgroups must race to maintain position while the inner ones may slow down.

We concluded that the staggered arrangement of individuals in subgroups could take advantage of the flow fields of neighbors in echelon formation, just as has been found to be true of the flying formations of birds (Lissaman and Shollenberger 1970). The dolphins' positions might well reduce drag by recapturing some of the energy lost to turbulence created by swimming neighbors, although we can bring no direct evidence to bear upon the question (see chapter 9).

Sensory Windows

Concurrent with any possible hydrodynamic arrangement, this same three-dimensional echelon formation leaves a wide *sensory window* available for each dolphin of the subgroup. By this we mean that each dol-

phin seems positioned to "see out" into a fairly wide sector of the surrounding sea. To be sure, when schools are large and very active,
especially when the dense caressing groups are seen, some inner animals
must depend upon outer animals for environmental information. But
such tight groups never, in our observations, comprise the whole school.
This proposition squares with the highly directional nature of dolphin
echolocation emission, which allows each exterior animal to scan a sector
of sea ahead of the moving school.

SENSORY INTEGRATION SYSTEM

We came to suspect that the spatial disposition of animals within dolphin
schools, in particular the interdolphin distance and the fluid geometry
of echelons, was in part an expression of a crucial signaling system that
allows organization of information transmission and response throughout the school. Norris and Dohl (1980a,b) and Norris and Schilt (1988)
proposed a school organization to describe this system and called it a *sensory integration system* (or *SIS*) (see chapter 8).

Sensory integration means that the individuals of a school swim as
parts of a supraindividual signaling system that allows sending and re-
ceipt of information and the passage and amplification of environmental information from the collective sensory windows. Very faint environmental information can be amplified so that all schoolmembers can
receive it if a schoolmember receives a signal from the environment and
then initiates its own signal through the school. The SIS can provide
early warning of predators, and food or other environmental features
can be localized. Mimicry among schoolmembers can "initialize" a communication between specified individuals and thereby indicate such
things as relationship and hierarchy.

Because the dolphins can be expected to react not only to signals from
their immediate neighbors but also from more distant members, information within the school should be able to travel very rapidly across a
subgroup, much more rapidly than if it were required to pass, seriatim,
from individual to individual across the school, a crucial capability of the
school in predator avoidance.

Sensory Summation

An unexplored but probably important aspect of the SIS as a protective
system is that it can provide sensory summation effects to its members.
That is, because the SIS can integrate the sensory inputs of many animals sensing in various directions, both the precision and sensitivity of
the school as a whole to environmental stimuli may be increased. An example of how this works can be shown from studies of the physiology of
receptor systems. The precision of response of a single temperature re-

ceptor in the skin may only be 0.1°C, for example, but the organism as a whole may be able to respond to a 0.01°C change because the responses of large numbers of receptors are involved. In an analogous way, the individual dolphins of a school, each acting as a sensory receiver, should in aggregate be able to contribute to a similar increase in sensitivity of the school by the process of areal or spatial summation (Kandel and Schwartz 1985).

For sensory integration to work optimally, individual animals need to sense outside the school envelope (through their sensory windows) and to sense the various dolphins in a school. The dolphins must be able to sense an approaching shark or a predatory cetacean and at the same time be able to transmit such information throughout the school.

Sensory Involvement

Except during rest when vision holds sway, information transmission within a school during daytime seems to be multimodal involving multiple senses. This should allow what Owings et al. (1986) described as *managerial communication,* which depends on both visual and acoustic signs and signals being used simultaneously. Such a managerial system uses multiple pathways for information that may involve different messages to various recipients. For example, a mother and young bottlenose dolphin can mimic each others' whistles (Caldwell et al. 1990), allowing them to identify each other as being in the same family unit, while other animals can listen in (Tyack 1991) and may draw different action imperatives from the signal.

At night or in murky water, the *phatic signaling system* (as outlined in chap. 8) seems to describe how the essential integrity of the protective school may possibly be maintained by sound alone. But during daytime rest, school coordination seems to be visually mediated to an important degree. From the *Maka Ala's* transparent bow, we were able to see that dolphins entering Kealake'akua Bay and about to descend into rest would not remain in the bay if the water was murky. They came in, explored their habitual resting grounds for a short time, and quickly left, still in the alert state.

Our tentative explanation was that when vision was impeded, either early warning about predators would be ineffective or the effectiveness of the visual mediation of school organization could be reduced. Therefore, the dolphins did not descend into rest and would not remain in the area.

SCHOOL PATTERNS

The protective strategies and spatial arrangements of acoustically mediated dolphin schools are strikingly different from those mediated wholly by vision. These differences seem to be related to two features of acoustic versus visual communication systems in the marine environment.

First, the directionality of acoustic fields seems to be involved in this difference. When a dolphin school comes toward a hydrophone, typically one first hears cascades of clicks and other signals as the dolphins approach, and as they pass abeam, the sounds reduce until only sporadic sounds are heard. It is possible that this sensory asymmetry promotes the back and forth courses of foraging dolphins, as if they keep general track of predators by means of this early warning system and then retrace their paths through recently scanned water.

Second, because sound typically travels much farther than light at sea and because its transmission is independent of the day–night cycle or of depth, it is the most effective mode for long distance predator detection and is expected to be the sole means of accurate long distance detection whenever the water is dark. Thus, feeding schools swimming at night or diving deeply may be widely spread, especially in comparison to the observed tight formations of visually mediated resting schools (Hunter 1968a). Because such dolphins are protected by an early warning system that gives them considerable time to react, we assume that they can afford to spread out into very loose schools. The spreading may, however, relate to reducing the energetic cost of food finding.

THE QUESTION OF SCHOOL LEADERSHIP

All of these considerations leave unanswered the question of how a dolphin school is guided. On casual inspection, leadership in dolphin schools seems subtle at best. As we have observed, the animals that swim in front of a school at one time may end up in back when the school reverses course. Where, then, does school guidance come from? After all, schools do arrive at specific coves to rest and do use very specific areas along the island's coast, while there are many other areas that they never seem to frequent. In other words, there is certainly some way in which these fluid schools *are* directed.

These questions led us to look carefully at what happens to individual dolphins when a school makes a course change and how they are deployed within the school envelope. The term *school envelope* refers to the labile outer three-dimensional boundary of a school.

Turning maneuvers were studied in two ways. First, they were examined by frame-by-frame analysis of underwater moving pictures taken from the *Maka Ala,* and second, they were observed directly in the captive school. In the captive group, one could watch the shadows cast on the tank bottom by dolphins swimming above and could often perceive which animal had turned first in the little school.

We watched incipient turns in calmly moving schools, both visually and on films. The turns we observed occurred when a single animal bent its anterior body away from the general line of school travel and moved at a shallow angle toward neighbors. At a given point, adjacent animals on both sides and above and below responded by reestablishing the old alignment with the turning animal, until in series, the whole school had changed direction. In the captive school, we sometimes heard a low intensity two-parted click given when the adult male initiated a turn.

We found that there seem to be two means of imparting direction to a dolphin school. First, in loose undisturbed schools, turns are initiated by certain animals, in our observations, of either sex, with the other animals changing course very shortly afterward. Second, in frightened schools, turns seemed to occur as a direct reaction of several members nearest the threat, falling off with increased distance from the source of threat without regard to age or sex.

Any experienced animal, therefore, might be able to regulate the movements of the school under normal circumstances. Using high speed films of wild schools and the captive school, we were able to locate and follow turn initiation by dolphins swimming on the margins of schools, in the middle of the schools, and even near the posterior margins of schools. This latter observation led to the conclusion that, unlike humans, dolphins have an essentially complete visual field and can see very well behind themselves (see chap. 6) so that geometrical position within a group may matter much less to them that it does to us.

RELATIONSHIP OF VISION TO DOLPHIN SCHOOLING

The dolphin optic nerve appears to cross completely at the optic chiasma (McCormick 1969) and therefore optic function is thought to be significantly independent for each eye (see chap. 7). This condition may promote hemispheric independence in the visual domain, allowing signals from one side or the other to be reacted to separately without uncertainty and, hence, without loss of time. Electroencephalograph patterns of bottlenose dolphins, for example, are either synchronized between the hemispheres or generated independently in either one (Mukhametov et al. 1977).

It is commonly observed by horse trainers that signals trained for reception by one eye of the horse may not be transferred automatically to the other eye without training directly for that eye. Thus, a horse trained to allow a rider to mount from one side may balk if the rider attempts to use the untrained side. The possible implication of this is that such a group-living species may be able to receive and deal independently and rapidly with signals coming from each of the two sides (Ridg-way 1990).

For example, a dolphin school or a horse herd under attack may receive a message about school or herd movement with one eye and the activities of a predator with the other. Such an arrangement might seem capable of producing confusion in a schooling animal, but perhaps this is just a superficial human viewpoint. The effect may be quite the opposite.

If the eyes acted together in binocular fashion, a "processing decision" may be needed to integrate the two inputs. However, if the eyes functioned separately, the stronger signal might prevail without loss of time. If such an anatomical arrangement is widespread among schooling, flocking, and herding animals, the scenario presented above would take on considerable generality and interest.

SCHOOL TIGHTENING AND SCHOOLING DISTANCE

We came to doubt that so-called school leadership by individuals continued to exist during times of stress. Threatened spinner dolphin schools appear to be highly responsive reaction systems whose members typically tighten ranks at the first hint of danger (Hobson 1978). Such tightening is proposed as a maneuver that brings schoolmembers close enough that the visual signals of neighbors come to subtend the maximum visual angle in the eyes of the adjacent viewer *without requiring eye movement.* This we postulate minimizes the reaction time of the school as a whole. We further propose as a model for testing that the filling of the visual field with the critical pattern components of neighbor, especially those of the pectoral fin region, without the necessity of scanning eye movements defines this critical distance.

Conversely, if two animals move farther apart in a school, subtle intention movements come to subtend smaller and smaller arcs in the visual field and thus to become increasingly ineffective in mediating rapid and precise school movements. This maximally effective distance between animals we call *the schooling distance,* and as far as we know, it is undefined in any species. Spinner schools will tighten to about pectoral tip touching distance.

When schools of spotted dolphins are artificially crowded closer than their schooling distance, as they are when tuna fishermen crowd them into a backdown channel prior to sluicing them out of the tuna net, a typical reaction of the dolphins is to sink passively and stack like cordwood upon one another in the net (Coe, 1980, Norris, 1991). This, we propose, represents the dolphin school being compressed more closely together than their schooling distance allows, with no room to compensate by moving away. The dolphins, bereft of means of coordination with their neighbors slide into the Delius stress syndrome (Delius 1970).

We have observed turning maneuvers by tightened schools in both fish and dolphins. I (Norris) have watched the obvious tightening of a dolphin school as it reacted to pursuit, which I regard as a sure sign that the general organizational features of polarized schooling were at work. In this case (observed long before the U.S. Marine Mammal Protection Act of 1972 was in place), the vessel's captain shot a rifle into the water behind a retreating school of bottlenose dolphins, with obvious tightening the result.

SCHOOL BEHAVIOR IN RELATION TO ATTACK

Because of the tightening relationships just described, predator attack should be a time of maximum synchrony of schoolmembers. Only once during this study did we see what may have been an attack on a wild spinner dolphin school, but brief as it was, the encounter was instructive. On May 2, 1970, we were cruising slowly in a relatively shallow water cove just south of Palemano Point, near Kealake'akua Bay, Hawaii. Suddenly an entire dolphin school (of at least 40 animals) leapt from the water *en masse* in a circular roseate a few dozen meters from our beam, as if it had swum over a source of disturbance and its members leapt away simultaneously in all directions from the threat. The dolphins made one or two additional leaps away from the epicenter before coalescing and slowing.

This encounter provides two interesting insights. First, each dolphin moved in relation to its own position relative to the source of disturbance, and the dolphins of the entire school did so nearly simultaneously. Second, the fact that the dolphins jumped *en masse* away from the source of disturbance in a roseate and did not flee as a polarized formation provides clear evidence that leadership had been suppressed in favor of more individual flight responses by each schoolmember in relation to the source of the stimulus. The response of these dolphins seems identical to what Pitcher (1986) called *skittering* in fish schools, in which fishes nearest an attacking predator leap radially away from it, often over ranks of other fish, and back into the school again.

This observation led to another question. When external disturbance or threats to dolphin or fish schools are absent, the animals may wander extensively within a school without inducing any turning movements in neighbors. What allowed such freedom of movement, while under threat, the school moved so cohesively?

We finally concluded that such freedom of movement occurs only when conditions allow schooling animals to move far enough apart so that high speed signaling within a school is not required. In these terms, school tightening is a visual indication that the school has been threatened and is in the process of adjustment toward its maximum speed of reaction. This also means that the alert school is in the process of reducing the individual freedom of movement of its members.

Tightening in a dolphin school can sometimes be seen as a "wave of influence" passing through a fish school. This visible wave of influence is one of a class of reaction waves that have been observed within schools. We (Norris and Schilt 1988) call them *Radakov waves* in recognition of the Russian biologist who first described them for fish schools (Radakov 1973). These waves, as he reported, can propagate at modest speeds when a school is in the process of tightening, or they may move very rapidly through an already tightened school, as fast as about 8 m/sec (26 ft/sec). A key to understanding these phenomena lies in the process of tightening and in the organizational system of a school in which it occurs.

ANTICIPATION

When schools move in relation to external stimuli, they do so with members watching more than just their immediate neighbors, as has been found to be the case in schooling fish (Partridge 1981, 1982). A school or flock is not composed simply of one animal reacting to the next, and so on, or even of such pairs reacting to one another. It is a group response system in which schoolmembers watch not only their nearest neighbors but also those in ranks farther away.

This phenomenon has been described in bird flocks and given the name *chorus line effect* (Potts 1984). This means that if schoolmembers watch not only their immediate neighbors but also two or more ranks away, a given animal can *anticipate* a school movement and can react as much as about 2.6 times its own individual maximum reaction speed. It can anticipate a wave of reaction coming and begin to react in advance of any movement by its immediate neighbor. This is because there is a length of time needed both for mutual mental preparation for the physical act and for the physical motions themselves. If the mental processing can be anticipated only the actual motor action remains to be done and hence the total time needed to react can be a fraction of the usual time needed. Such prediction in advance seems to be behind the almost miraculous ability of schools to "melt" in front of an onrushing predator. For the process to work there has to be a minimum number of animals present, about six to eight, close to our minimum observed school size for spinner dolphins.

An example of how effective such signaling can be was shown when eight schooling fish were allowed to form a small school. They proved essentially invulnerable to a single predatory fish, while one or two prey fish alone were easily caught by a single predator (Landeau and Terborgh 1986). It is presumed that when the school became invulnerable, there were enough prey fish present that they could anticipate the movements of the predator.

CONFUSION

A critical function of a school is the production of the *confusion effect,* which is dependent upon there being numbers of near-identical prey in a school (Partridge 1981, Pitcher 1986). When a predator attacks, such as a shark attacking a dolphin school, it must fixate on a given prey item and then mentally "calculate" and follow an attack vector. When all prey look alike and when they flash rapidly in front of the predator, attack seems to be foiled because the predator proves unable to fixate and follow a given prey animal. Instead, it flicks its eyes from identical animal to identical animal as they flash in front of it, losing precious reaction time in the process (Norris and Schilt 1988). This effect seems to be lodged in a feature of the cognitive functioning of the brain: the rough subdivision of events and objects in the environment into approximately seven sectors. This occurs when the short-term memory function of the brain is in use, a feature seen widely in highly organized animals, both vertebrates and invertebrates (Miller 1956). This means that under the extremely short time involved in an attack, the predator must approximate its world, making it impossible for it to discriminate individuals in any detail.

When these geometric relationships within schools are added to the information transfer and protective functions proposed earlier, it becomes easy to see why spinner dolphins may not survive at sea in the absence of their schoolmates and the subtle organization they seem to provide.

FOURTEEN

Social Behavior

Christine M. Johnson and Kenneth S. Norris

Our conclusions about social behavior presented in this chapter are based primarily on two sets of underwater observations: those made underwater in Kealake'akua Bay from the viewing vehicle *Maka Ala* and those made from the underwater viewing vault (fig. 86) of an oval 15-m (50-ft) diameter tank located at the Oceanic Institute on the island of Oahu, Hawaii, in which a small school of captive spinner dolphins was held (see chap. 3). The *Maka Ala* observations were made during several months of reconnaissance in which we attempted to understand as much as possible of the totality of underwater spinner dolphin life, as a foundation for more focused future efforts. This work followed a much shorter reconnaissance, also using an underwater vehicle (Norris and Dohl 1980a). Important in shaping our ideas of dolphin social behavior were earlier studies of captive dolphins (McBride 1940, McBride and Hebb 1948, Tavolga and Essapian 1957, Tavolga 1966, Bateson 1974) that told us to expect a society organized along mammalian lines, not basically different from mammalian societies ashore. Other work with cetaceans had documented high levels of cooperative behavior, apparently across the entire suborder Odontoceti (Caldwell and Caldwell 1966, Connor and Norris 1982, Johnson and Norris 1986). Therefore, we sought to understand how such cooperation would find expression in spinner dolphin society.

Because of the danger of operating our viewing vehicle, *Maka Ala*, in even modest seas, the Kealake'akua Bay observations were restricted to daily events that occurred deep in the bay: morning arrival from nighttime feeding, descent into rest, rest, awakening, zig-zag swimming, and departure to the feeding grounds (see chap. 4). What we saw were almost always short and fragmentary segments that had to be observed

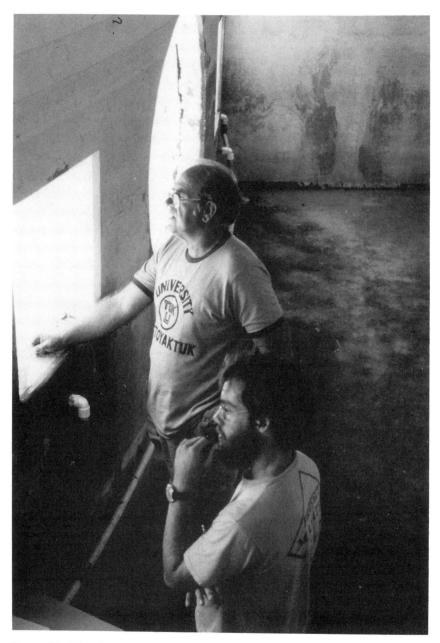

Figure 86. The viewing vault at Bateson's Bay where our underwater work with captive spinners was performed. K. Norris and R. Wells are shown.

repeatedly before we could have confidence in what we had seen. Our behavioral observations of all dusk and nighttime activities at sea were casual and from the surface.

Our viewing vehicle was designed to allow a prone observer lying on a mattress with his or her head in the plexiglas bow to see, videotape, photograph, and make concurrent acoustic recordings of wild dolphin schools. Voice communication between the pilot and observer was easily accomplished, and such teams soon learned how to approach and insinuate the vessel into bay schools. As we described in chapter 3, we have now constructed a new and better viewing vehicle for underwater behavioral observation (see Appendix B).

The captive school at the Oceanic Institute was very small; five at first, but due to the death of two dolphins, it was composed of just three dolphins during most of our observations. (See Appendix B for further information on these animals.) The remnants of this little school were released back to sea not long after our work with them was finished, so at this writing, we know of no spinner dolphins in captivity. Experience at Sea Life Park Oceanarium shows that, although the species can be maintained in captivity, it is clearly more difficult to maintain than some nearshore forms such as the bottlenose dolphin (*Tursiops truncatus*) and the belukha (*Delphinapterus leucas*).

We were able to compare some of our captive observations with tapes and records earlier accumulated by Gregory Bateson in a large show pool at Sea Life Park Oceanarium. When Bateson was actively working, he and his team of observers studied a mixed school of seven spinner and spotted dolphins that was undergoing show training in a large public exhibit pool (Bateson 1974). These observers concentrated on the hierarchical relationships within the school and on the shift between rest and active behavior.

SCHOOL WARINESS AND HABITUATION

One of the first things we noticed about spinner dolphins was that the wariness of their schools varies widely and, to some extent, inexplicably. Usually, but not always, during rest or during descent into rest, the dolphins tend to move slowly away from an approaching vessel. At other times, they are boisterous and often approach the viewing capsule, even looking in at the observer. In general, active schools were frequently the ones that sought out our vessels and rode the bow. We learned to test this attitude of the dolphins and tried not to impose ourselves on spinner schools that showed a reluctance to accept us, hoping to avoid negative conditioning to the vessel. Our feeling was that we succeeded in this attempt, at least to the extent that schools did not become increasingly wary of us. Because the level of wariness changed rapidly in a single school, we came to feel that this level was somehow connected to the cyclic behavior state of a school and not to its composition. A major difficulty with such habituation to our vessel was that we were not dealing with the same school day after day, but instead with subschools drawn from an ill-defined larger offshore entity. Therefore, the composition of the animal groups we sought to habituate to our presence varied from day to day, and only over a span of months were our attempts to enter schools repeated very often for a given dolphin or subgroup of dolphins.

EPISODIC OR BOUT BEHAVIOR

To a remarkable degree, events in the daily life of spinner dolphins appear to be diurnally cyclic as well as oscillatory among various behaviorial states within a given day. This can involve the entire school, which may oscillate between one behavior state and the next, and individuals as well, who shift from one behavior pattern to another, also regulated by oscillations between an old and a new state. Let's look more closely at these individual behavior states and their transitions.

After concerted observation of the group of nearly undisturbed captive spinners, it became obvious that, except for a daily rest period in which all dolphins participated simultaneously, the daily behavior of each dolphin was arranged in *episodes* or *bouts* during which a given behavior pattern (such as echolocation or aerial behavior) or class of patterns (social behavior) dominated its activity for a time before it cycled on to the next pattern. These bouts ranged from a few minutes to more than 30 min in length.

Not long after we first perceived the existence of these bouts, we began to see indications that they were real entities and not mere expressions of the statistics of observation. The dolphins engaging in them did so with obvious singleness of purpose, shunting aside other behavior in favor of what seemed almost like a "duty." Furthermore, we began to see that the bouts had obvious and stereotyped beginnings and endings that involved the oscillatory transition from one pattern to another. A bout began, oscillated upward in intensity, stayed relatively stable for a considerable period, and then oscillated to a lower and lower intensity. An experienced observer could eventually predict where in a bout a given dolphin was.

Terminations emerged as times of clear oscillation between the present pattern and the new pattern about to be performed. Each of the two animals in a bout had a clear and sometimes separate role. In low intensity caressing, for example, there was sometimes a caresser and a caressee throughout an entire bout.

To help visualize these relationships, we describe a composite example of a single bout, including its beginning and termination:

The two adult female spinner dolphins, Kehaulani and Kahe, have come together in a caressing bout. The bout begins as Kahe reaches out toward Kehaulani with her inside pectoral fin tip, running its anterior edge and dorsal surface over much of the latter's body, especially between the pectoral fins, along the abdomen, and onto the genital area. Kehaulani begins to solicit these caresses by swimming up next to Kahe or by turning under her so that her belly and sides are slowly run past the outstretched pectoral of the caressing animal, while they both swim forward at modest speed. Kahe's outside pectoral fin is held in normal swimming position so that the reaching of her inboard pectoral toward the caressee is made obvious.

As the bout proceeds, caressing becomes more and more constant and intense. Swimming speed increases. One can see that Kahe, who swims slightly above Kehaulani, extends and retracts her pectoral in synchrony with the vertical swimming undulations of Kehaulani; the extending and retracting allow precise contact. Kahe then pulls away as Kehaulani speeds up, sweeps around the more or less circular tank, and then circles in for further contact.

Contact now becomes intense and the partners swim in close juxtaposition much of the time. This state persists for a short time, with the two animals attending only to one another, circling constantly and maintaining close contact. Then, both the intensity of contact and its frequency begin to diminish.

Kehaulani breaks away, races to the bottom of the tank, vibrates her belly and genital area against the tank bottom, and then bursts from the water in a spin. She reenters and quickly returns to Kahe who caresses her. The degree of contact in the next few minutes clearly begins to oscillate slowly between close and intense caressing and more distant and less intense contact. With experience, the observer could predict that the bout was nearly at an end.

Kehaulani next drifted out of contact with Kahe and the pair circled, but she continued to swim quietly alongside her partner for a short time before drifting back into contact again, receiving body caresses for a brief time before drifting away again out of contact, and, for the first time, with a full body diameter between the two dolphins.

The dolphins drifted into contact again, this time for only seconds, and out again to pectoral touching distance very briefly, and then swim alongside one another with two full body diameters separating them. This position persists for a few seconds.

They come in contact again, this time for just moments before drifting into more distant formation swimming again. The synchronized pair approaches the lone male Lioele, and the partners go on opposite sides of him, and instead of coming together again, they now circle in opposite directions. Kahe assumes a position alongside the male, while Kehaulani races to the tank bottom, vibrates her belly intensely on the bottom, and then bursts through the water surface in a high intensity spin. She continues to spin alone while Kahe and Lioele swim together.

We later found that Purbrick (1977) had previously observed a similar cyclicity of behavior patterns in the larger captive spinner and spotted dolphin school then held at Sea Life Park Oceanarium adjacent to the observation facility we used, although she did not classify or describe them in detail.

The discovery of this temporal patterning led us to look for such sequences in dolphins in the wild. We first sought to determine if in nature a single animal repetitively performed one of the cyclic patterns we had seen in the captive school. This proved easy to observe for aerial behavior. We also found that bouts of social interaction were obvious though difficult to time because animals almost never stayed long enough around our observation vehicle for complete observations to be made.

The small size of the captive school (three dolphins) created an automatic organization of repetitive patterns of behavior. For example, if two animals caressed, the third was left alone to perform a single pattern, thereby imposing a cyclicity. This caused us to ask the question, "was the cyclicity we saw something that existed in nature?"

We soon saw rough and incomplete evidence of such cyclicity at sea. Although many animals were always available for interaction in wild schools, we still found single individuals repeating aerial patterns over and over, and we were sometimes able to describe fairly extended caressing sequences involving two or three individuals.

Though difficult to quantify, it was also clear that whenever we observed caressing and other intense social contact in a school from our viewing vehicle, the total number of animals involved was less than half the school. This could have been an expression of a division of such behavior into the three major bout types. In addition, whenever we were recording sounds from an active dolphin school at sea, a number of animals were always echolocating at any given time. The school goes about its business day and night in a constant cascade of click trains. This indicates that an important proportion of the animals in a school is always engaged in scanning using echolocation.

In captive dolphins, we found that individual or group bouts were overridden by the sedate formation swimming of the daily rest period during midday. This clearly also occurred in nature in the form of the rest period.

These cyclic patterns persisted continuously in the captive school throughout two continuous days and nights of observation, and sequences were noted on all other occasions when we looked for them. Bout behavior seemed clearly enough to be the normal course of events for the captive school. The conception became predictive of the general events within the tank.

These observations focused our attention on possible functions for such repetitive behavior and upon details of its occurrence in wild schools. It seemed possible that such bout behavior could represent the means by which schools of constantly changing membership and size could carry out the essential patterns that allowed dolphin societies to survive at sea. It could be a design for division of essential labor. In a society of constantly changing social makeup, aerial behavior could establish the phatic communication system (chap. 8), social interaction (including caressing and formation swimming) could reaffirm learned social bonds among kin and nonkin alike, and echolocation could allow location of both predators and prey.

So, roughly, we were able to confirm the simultaneous occurrence in nature of the three classes of behavior that we saw so clearly in captivity. The question of quantifying their cyclicity and timing in nature remains only partially answered.

Aerial Behavior Bouts

Aerial activity patterns and their relationship to the diurnal cycle were described in chapter 4. Here we deal with the organization of such aerial behavior into bouts.

Spins, observed both in nature and in captivity, are given in series by a single dolphin, sometimes for relatively extended periods. Usually a spin series by a single animal seen at sea starts with a high energy spin and is followed by a series of lesser spins. These usually decline in intensity as the series proceeds, until the last spin may not clear the water. On some occasions, we felt sure that it was the same animal performing multiple series of aerial patterns because of the dolphin's size, its location in a school, and special idiosyncratic features of its spins.

These idiosyncratic features provided the strongest evidence of such a repetitive series of aerial patterns. Spinner dolphins typically do not mix the various kinds of aerial patterns. Spin sequences do not contain tail-over-head leaps or head slaps. The details of aerial patterns proved to be idiosyncratic to a given individual. Some dolphins produce high arcing spins, and others vault lower over the water surface. Other "markers" of individuality can be noted upon close observation. In this way, we came to feel secure that we had followed single individuals performing aerial patterns over several minutes' time within wild schools. The numbers of leaps in a single spin sequence ranged from single spins to sequences of about 20 spins. A usual series was 3 or 4 spins per sequence.

Spin series at sea give the impression of being given with intensity and singleness of purpose, just as we had observed in captive dolphins. One example amplifies this point. We watched a young animal spin in an offshore school near dusk. It began its series about 100 m (330 ft) off our bow and each spin brought the animal closer to the boat. The final spin of a 14-spin sequence was so close to the boat that the dolphin reentered into our bow wave, became startled, and swam rapidly away, as if it had not previously noticed us.

One way of quantifying how many members of a school were in spinning bouts was to estimate the size of the school and then estimate the number of spinning animals. Although we did not obtain precise numbers, it is our opinion that in highly active schools this percentage is a modest fraction of the animals present, a third or less.

Caressing Bouts

In our observations of captive dolphins, caressing bouts consisted of a pair of dolphins singlemindedly stroking one another with any or all of their appendages for a considerable period of time to the exclusion of other activity. In captivity, such bouts tended to be fairly long; five bouts for example, ranged from 9 to 30 min in duration each. Our observations of such bouts at sea suggest that they are much shorter than in captivity and that multiple partners can be involved (fig. 87). Our impression was that because multiple partners were available at sea, a continual testing of the strength of interaction between caressing pairs occurred and thus partner switches were very frequent. It may be that the duration of an entire bout seen in captivity where no excess partners were available represents the same pattern that occurs with multiple partners seen at sea. It may also be that if, as we projected in chapter 3, caressing is important as a reaffirmation of relationships, especially between unrelated dolphins, its importance depends upon concourse with many animals and not just one during any given bout.

The higher the activity level of a wild school, the shorter the period of attention of one animal to another. In the highest energy schools, group caressing involving a dozen or more dolphins at once (our *wuzzles*) made specific partners difficult to discern, if they existed at all.[7] Such group caressing is very active behavior involving what seems to be an entire subgroup within a school. As the dolphins swim along, they move into close juxtaposition with one another and begin to weave alongside one another in frequent contact, often with more than one animal at a time. Such groups form tight fluid pods of dolphins in which much twisting

7. The term comes from W. E. Schevill of Woods Hole Oceanographic Institution who, when asked what the behavior was, replied without hesitation, "Why, it looks like a wuzzle to me."

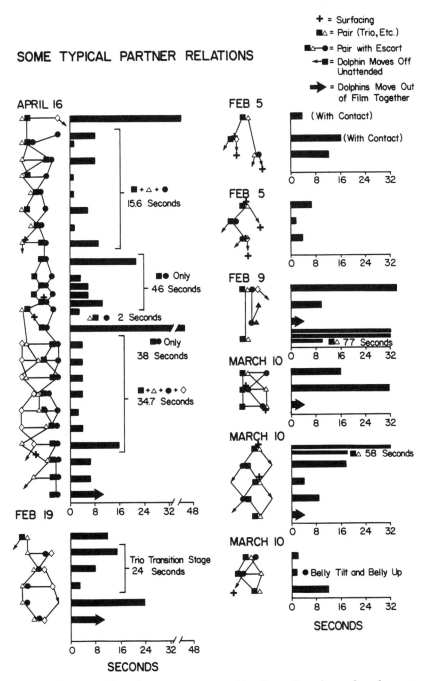

Figure 87. Examples of partner relationships from five days of underwater observations.

and turning is evident, even to a fairly distant observer, because a constant flashing of white bellies can be seen, often from three or four animals at once. Doubtless this flashing signal is as evident to the dolphins as it was to us and could serve to draw new animals into the caressing subgroup. We did see dolphins swimming toward and joining these groups.

Both in nature and in captivity, all available combinations of sex and age classes take part, from nondependent calves to adults of both sexes. Very small calves were sometimes seen in these melees, but they remained closely associated with their attending adults, leading us to believe that the adult had entered the group, not the young dolphin. Whether the young caress or not was not settled.

Caressing in these groups involves rubbing a partner with pectoral fins or flukes and sometimes involves bodily contact in the mating posture. During the summer period when many females came into estrus, more obvious mating patterns were observed in such groups and intromission became a common feature in our observations. Similarly, Wells (1984) found a significant correlation between high reproductive hormone levels and high frequency of genital–genital contact and copulation in captive spinners (see chap. 9).

Caressing seems to occupy a considerable part of a dolphin's day. While caressing did not occur during rest, in active schools in daylight it was usual to see 30% or more of a school engaged in caressing patterns at any given time. Judging from observations of captives, we believe these patterns may continue throughout the night.

Beak–genital propulsion was the most unusual and slowest caressing pattern we saw (fig. 88). As noted in chapter 9, this pattern is associated with the peak of the female estrus cycle. It was first described by Bateson (1974) and illustrated by Norris and Dohl (1980a). It is a low energy activity pattern in which partners apparently do not switch during a given bout.

In this pattern, an animal "invites" a second animal to come up from below it and place the tip of its rostrum in its genital slit. The lower animal often turns partially on its side, sometimes supporting the horizontal flukes of the upper animal from its head to its outstretched pectoral flippers. In this position, the lower animal, its tail bent down slightly from the horizontal, propels the upper animal forward, only breaking from the pattern to rise for breaths of air. Sometimes it swims belly-up or dorsum-up during such propulsion.

Because of the unusual postures of the animals and because clicked sounds are known to be projected at times from the upper beak of dolphins (Diercks et al. 1971), we wondered if the behavior was accompa-

Figure 88. Beak-genital propulsion at Bateson's Bay, Sea Life Park Oceanarium, Oahu, Hawaii.

nied by acoustic stimulation of the recipient dolphin. Although we were occasionally able to hear low level click trains being emitted while this pattern was in progress, we could never be sure which dolphin in the school had emitted the sounds nor could we see any evident reaction by the propelled dolphin.

A much more active pattern is *pectoral whetting* (Bateson 1974), which was occasionally seen in active schools. In this pattern, one dolphin typically swims inverted under the other. The flippers of both are rapidly scissored back and forth over the surfaces of the other. On occasion, dolphins seen in this behavior swim rapidly and even spiral through the water, maintaining the activity.

Formation Swimming
Formation swimming is the predominant low energy pattern of daytime resting schools. It is typified by sedate movement of the various subgroups of the dolphin school, each animal swimming out of physical contact with its nearest neighbors. typically, about one to three body diameters intervene between the pectoral fin tips of the dolphins. The pattern has a range of expression dependent upon activity level. Looking from above, as we often did from our viewing vehicle, one could sometimes see an entire school move slowly just above the sandy ocean floor, the animals regularly arrayed but still maintaining the usual en

Figure 89. Resting spinner dolphins over the sand patch at Kealake'akua Bay.

echelon relationship. This formation was seen often enough that we gave it the name *carpet formation* (fig. 89). Subgroups, while muted, were always evident.

In more active states of rest, subgroups became more evident but the sedate movement continued. Not until arousal began did we note caressing patterns in such schools. The view from above then was of small clusters of dolphins often accompanied by the flash of belly-white as one animal tilted toward another. The longest formation swimming bout observed in captivity lasted 66 min.

Echolocation Bouts

The following description of this solitary behavior was observed in the captive school on February 24, 1980 (fig. 90):

> A lone dolphin circled the tank while directing loud high repetition rate click trains at various tank features, especially those beneath it. The dolphin typically circled very close to the bottom and tank walls. As it passed over cracks between the cement slabs of the tank bottom, it often deviated from its course to swim out along such features, echolocating into them. When the dolphin passed over the 1-m^2 recessed central drain box, it routinely paused and directed train after train of clicks into it.

Such click train bouts were moderately long; the average for five bouts was 27 min each (range, 15–46 min), during which time the animal swam alone and emitted train after train of sounds in a more or less continuous fashion.

At sea, because spinner (and other dolphin) schools travel in a constant cascade of clicks, it seems evident that, however the behavior is maintained, a number of schoolmembers are engaged in echolocation at any given time. It seems likely that trading of this duty occurs among schoolmates, as we noted in captivity, if only because it is a task having considerable physiological cost. The production of train after train of intense clicks might require rest because the tissues directly involved in echolocation may need repair from the intense act of click generation. Beyond those suppositions, we can bring no direct evidence to bear that bouts are involved.

DO DOLPHINS ECHOLOCATE ONE ANOTHER?

Does a clicking dolphin direct such click trains at other nearby dolphins? Was one reason for the staggered echelons of dolphins in a normal school to allow each animal an "open window" for its echolocation? Our observations of the captive school indicate that they did not direct intense echolocation trains against one another at short range, except under conditions of aggression. We made no measurements of sound pressure levels.

In fact, dolphins in echolocation bouts seemed to avoid ensonifying their tankmates. For example, on February 23–24, 1980, echolocation bouts were watched for about 4 hr to see if the echolocating animal would direct such trains against tank mates. In one example, an echolocating dolphin (Kehaulani) circled the tank 46 times and on each circuit came up behind the other two animals, who were caressing. In every case, either she contrived to swim well below the other dolphins if they were near the surface and continued click emission, or if they were near the bottom where she was swimming, she muted her clicking to a very faint train or else shut down clicking altogether until she had passed them. At the same time, she obviously turned away from a direct course coming up from behind them. Once on a different path than theirs, she quickly recommenced loud train emission. The pattern was quite obvious, and we came to call it *echolocation manners* (fig. 90).

It is incorrect, however, to assume from this that dolphins never direct clicks at one another. Low level clicks are apparently directed between animals during caressing bouts, in what we human observers anthropomorphically call a *tickle buzz*. These unrecorded signals are soft sounds very different from the whining crescendo of echolocation click trains. Also, as noted elsewhere, a small burst of clicks, often just a few, were heard from schools during turning maneuvers, and we have heard louder clicks apparently directed among dolphins in the captive school in various social contexts. As described in detail later (see section on Uses

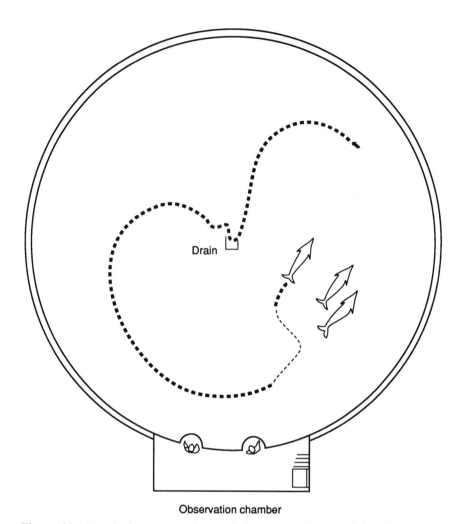

Observation chamber

Figure 90. A typical traverse of an echolocating spinner dolphin during an echolocation bout, at Bateson's Bay.

of Intense Sound), clicks or bursts of high intensity sound are often directed against adversaries in other species such as the bottlenose dolphin (Caldwell and Caldwell 1972a, Overstrom 1982). We have also had them directed at us, but have not heard them in use between spinners.

PARTNER SHARING AND EXCHANGE

It is generally difficult to observe patterns of behaviorial interaction among identified spinner dolphins in nature for longer than a few minutes because schools swirl in and leave both underwater viewing vehicles and surface craft. Sometimes identified animals will stay with the observer long enough for consecutive observations of modest length. Two examples are given here that outline partner sharing and exchange. These are given both to describe the observational problem cetacean observers face and to show something about how dolphins associate within a wild school.

On September 16, 1980, a school of four dolphins was followed from the surface for 101 min as it traveled between two points of land about 200 m (660 ft) from shore, south of Kealake'akua Bay (fig. 91). It is worth describing their movements and associations in detail since we were able to document the way in which these wild dolphins reacted to our presence. The school was the smallest we monitored during our entire study, which allowed us to sort out relationships. It was composed of two large adults, both with the high triangular dorsal fins we associate with males (fig. 92) (dolphins #26 and #165); an adult with a smaller, more falcate dorsal fin (#87); and a nearly grown subadult (#152). Later, one of the adults of this school threatened our diver who was able to film the event underwater.

The adult–subadult pair in this little school could have been a mother and young, but we were unable to determine the sex of either. The two stayed close to each other throughout the entire encounter and especially so during the first part of our observation. With one exception, the subadult was interposed between the two presumed adult males. The association between the juvenile (#152) and one of the presumed males (#26) became ever closer during our observation. At the same time, the subadult continued to maintain a close association with the small adult. (#87).

Proximity seemed to be correlated with a close synchrony of breathing in all the dolphins even if they were not directly adjacent to one another. In other words, there was a clear cohesion between these dolphins, as shown by our statistics. For example, #165 swam in rather close association with the other three dolphins during the first 40 min, but he swam alone and was usually separated by more than 8 m (26 ft) from the others during the latter part of the encounter. It frequently moved within 4 m (13 ft) of the others but did not join them, nor was it joined by others.

We could occasionally see behavioral signals among dolphins. We saw both #87 and #26 tilt their bellies toward the subadult, #152, twice during the latter half of the observation series. Once we saw #152 reciprocate, tilting toward #87, and then the two swam on their sides with stomachs touching for about 10 sec.

The closer animals (ones that swam less than two body lengths apart) showed their close association because they tended to surface together. Their surfacing was called synchronous when dolphins surfaced within

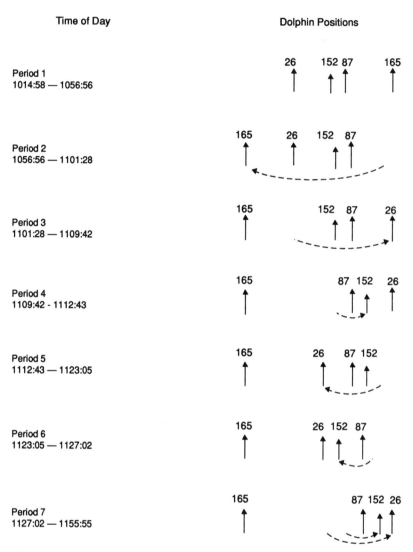

Figure 91. Relative positions of four dolphins traveling together, Kona coast, September 16, 1980.

2 sec or less of each other) (fig. 92). During the first 56 min when the dolphins were mostly within 4 m (13 ft) of one another, 140 of the 229 surfacings were made synchronously, while in the later 45-min period, the association loosened and only 32 of 272 surfacings could be called synchronous (the difference was highly significant; $\chi^2 = 134.4$, $p < 0.001$). The general impression was that at the start of observations,

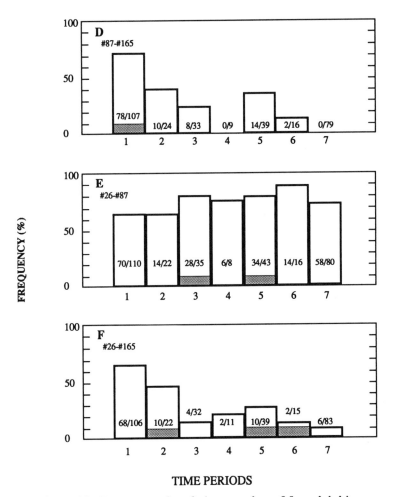

TIME PERIODS

Figure 92. Frequency of surfacing together of four dolphins.

the dolphins were afraid of the inflatable vessel and became more accommodated to it as time went on. The subadult stayed on the outside of the school away from the boat. As time went on, the school began to loosen slightly and synchronous surfacing became less evident. The boat moved at a constant slow pace throughout the encounter, matching the pace of the dolphins.

These surfacing patterns were instructive (fig. 92). During the first five observation periods, the juvenile and the small adult surfaced almost completely in synchrony, but even later on when they did not, there remained a regularity of how they surfaced in relation to each other. From 10:39 to 11:03, the small adult surfaced just before the juvenile on all five asynchronous surfacings, and these were interspersed among numerous synchronous surfacings. Then, from 11:05 to 11:12, the situation reversed, and when asynchrony occurred, the subadult surfaced first until the end of observation. Six runs of regular surfacing patterns lasted 24, 7, 12, 5, 18, and 8 min (mean = 12.3 ± 7.34 min), defined by which animal surfaced first. These runs showed a marked nonrandomness of serial order ($p < 0.001$). Dive times for each run were remarkably similar for all dolphins, with an overall mean for all 580 runs of 31.5 sec.

Similar fluctuating associations have been followed by underwater observation, but for shorter times (3–42 min). Frame-by-frame analysis of films allowed delineation of these changing association patterns. In figure 87 filmed on April 16, 1980, a single individual can be followed through association with four different partners. Each time the focal animal, named Triangle, split away from a partner at a surfacing. This same pattern of changing behavior around a breath was noted in other sequences.

In our encounters with dolphin schools from the *Maka Ala*, we were impressed with the high incidence of partner changes as compared to the long patterns seen in the small captive school. Even so, we could sometimes see that they were complete events, with a beginning solicitation and oscillations into a new pattern at the end, just as we had seen in the captive animals. Some *Maka Ala* sequences gave hints that one partner was sometimes ready to switch before the other was. We presume this left a bout incomplete for the latter animal. Such asymmetry sometimes seemed to lead to chase behavior. Thus, the tapering off of a bout seems to promote social symmetry. One partner repeatedly tested the readiness of its partner to move on to other patterns, and as a result, the two animals usually terminated the bout together and each went into other activities without disruption.

The responses of one dolphin to another can be very subtle. In our captive group, the adult male, Lioele, had injured himself by spinning out of the tank several years before we began our observations. He had a noticeable bend in his spine, which produced a peculiar wobbly swimming motion. It was interesting to watch Lioele's partners entrain to and imitate this swimming pattern when in a caressing bout with him. This imitation allowed us to see that there were solicitations to engage in a bout, as well as what might be called "acknowledgments" during a later shift of partners.

Lioele tended to spend more time than the others moving slowly at the surface during rest periods. A new bout partner usually circled him slowly or floated near him for short periods and then began to move with him. At times such a partner passed very close to him, imitating his wobbly swimming motions until he, too, started to move, and then the two dove and circled slowly together. Such imitation certainly occurs between healthy partners as well, although it is usually more difficult to see. Tayler and Saayman (1973) report on a number of examples of play and imitation similar to those we report here seen in captive Indian Ocean bottlenose dolphins. Many times we saw pairs swim near our viewing capsule, diving and twisting in such perfect synchrony that only careful film analysis could hope to separate out a leader and a follower.

During the group swimming of the rest period, the two females sometimes arrayed themselves on either side of Lioele, who because of his injury swam tilted to one side. In this circumstance, the dolphin on his left usually swam a little higher than he did and the dolphin on the right a little lower. One possible interpretation of this is that the two consort dolphins were using the signals and pattern of this tilted animal, and to view them properly, they had to take such staggered positions.

MALE COALITIONS

The subtle dimorphism of the Hawaiian spinner dolphin at first obscured the fact that often the first group of dolphins we would see upon *Maka Ala*'s approach to a school were male coalitions swimming between us and the remainder of the school (figs. 1 and 93). These subgroups consisting of a few to about a dozen dolphins could be discriminated from other age and sex groups by their thickened, muscular tail stocks, evident postanal humps, tall, black erect fins, and their "businesslike manner." Pryor and Shallenberger (1991) emphasized this same stolidity in groups of spotted dolphins seen in tuna seines, which they frequently saw weaving among other more quiescent rafting dolphins. Connor (1987, 1990) and Connor et al. (1992) have reported such coalitions of tight pairs or triplets of males for the Indian Ocean bottlenose dolphin. He describes these coalitions as engaging in aggressively maintained consortships with females or with other male coalitions.

In our observations of Hawaiian spinners, the males of a coalition showed little interaction with one another except for the interposition just noted and the purposeful orderly formation swimming noted by Pryor and Shallenberger (1991), both of which carried a sense of "threat" or perceived intrusiveness to the observer.

MATING

Sexual activity in Hawaiian spinner dolphins peaks in summer and fall, and this was concordant with both the hormonal cycle we determined for the captive animals and with the timing of sexual behavior as mon-

Figure 93. A proposed male coalition (foreground), Kealake'akua Bay.

itored in the captive group (Wells 1984) (see chap. 9). During this same reproductive period, it was not uncommon to see intromission between caressing animals watched at sea. Our records list 16 examples of complete mating observed between June 26 and October 14, 1980, and none outside that time, even though much underwater observation was carried out in other months. In such reproductive behavior, the observer noted either the erected penis or belly-to-belly contact with pelvic thrusting.

The matings appeared to be promiscuous, that is, no mate selection was obvious to us. Perhaps the terms *serial polyandry* or *polygynandry* better describe the situation, in the sense that more than one male was observed to achieve intromission in very short sequences with a single female and with no obvious mate selection (see chap. 9). For example, on August 31, 1980, two males were observed mating with a single female. During this short sequence (lasting only a few seconds), a male swam up under a female, achieved intromission with a short flurry of pelvic thrusts, then righted himself after intromission and remained close to the left side of the female (fig. 94). The second male turned belly-up, swimming upside down *above* the female, and then rubbed against her right side as he slid down into the mating position under her. As soon as the second male's penis became visible, just before intromission, the first

© 1991 Jenny Wardrip

Figure 94. Mating posture of two spinner dolphins; male is underneath.

male darted off to a nearby adult (unsexed), performed a belly-up to it, and then chased it as it tried to escape his attentions. A receptive female seemed rarely alone for long.

During the estrus season, we often observed mating behavior within pairs or trios of animals, but we also observed the dense interweaving masses of caressing dolphins (*wuzzles*). In wuzzles, intromission was observed. It seemed curious to us that there should be two such different circumstances in which mating takes place within the confines of a school. The one involves just a few animals in which mate choice would seem possible, while in the group process involving a significant part of the school, choice seemed impossible. At this point the question is unresolved.

BIRTH, NURTURE, AND JUVENILE GROUPS

We saw neither births nor identifiable newborn animals at sea during this study, although we did see small young frequently throughout the year (fig. 95). It may be that we simply had difficulty discriminating newborn animals from those a week or so old.

A previous Kealake'akua study (Norris and Dohl 1980a) listed newborn for every month but August on the basis of small size and the presence of an adult in attendance, but we now feel that these records did not discriminate between newborn and small juveniles and that the attending adult could have been either a parent or an alloparent. Newborn bottlenose dolphins are obvious for some time after birth because of a vertical pattern of light and dark gray striping referred to as "fetal fold marks" and on the day of birth by bent fluke and dorsal fin tips. These marks, however, are usually difficult to see in Hawaiian spinners.

Figure 95. A mother spinner dolphin and her young. (Photo courtesy of Sea Life Park, Makapuu Oceanic Center, Waimanlo, Hawaii; photo by Nicki Clancey.)

Although birth was not observed at sea, we are grateful to Ingrid Shallenberger, the head trainer at Sea Life Park Oceanarium, for a complete record of a spinner birth in captivity. It contains a number of points of interest to this work, so we quote from her notes at considerable length.

July 21, 1977
The birth took us by surprise since Kehaulani did not show the usual signs of imminent birth: the squareness of the belly or bulging mammary glands. Since the mother was a relatively new captive, the calf may have been conceived in the wild. These are notes:

15:40: Spinner dolphin Kehaulani discovered giving birth in the large rectangular tank where several other dolphins were held, including other spinners and a female roughtooth dolphin. At the time of discovery, the calf's tail was extruded to about the level of its anus. The calf was turned belly to the right and Kehaulani swam alone.
16:06: Calf out to half-stomach, two adult females in attendance, Lilinoe and Kahe.
16:08: Calf retracts to level of anus.

16:15: Calf born alive.

16:16: Calf began normal swimming but rapidly ran afoul of a gate. The very attentive mother assists it. The other dolphins swirl closely around the pair.

16:20: By this time, three alloparents had established themselves: one, Apiki, was a large juvenile male, another was Kahe, an adult female. Another adult female, Lilinoe, took up attendance from time to time. Of these three, Apiki most frequently attended the calf, although with less success than the other two. When calf runs afoul of a tank gate, the mother helps it with her dorsal fin.

16:28: Both alloparents (Apiki and Kahe) take the calf, with the mother trailing, and press the ventral surfaces of their tails toward calf, as if to induce it to nurse.

16:30: Mother assumes nursing position toward Kahe, the female alloparent.

16:35: Apiki positions calf on his temple, the assisted locomotion position, while the calf "porpoises" partway out of the water during a breath. The calf was born with a faint but visible fetal fold pattern. Its dorsal fin, which had been folded flat against its body during birth, is now erected about half way to the vertical.

16:36–17:30: Calf mostly travels with alloparents and attempts to nurse from them.

17:47: Calf showing signs of distress; mother attempts beak–genital propulsion with it; attempts to chase alloparents away.

18:05: Apiki swims off with calf, but returns calf to the mother at 18:14, and then mother, calf, and Kahe swim as a trio, with calf in the middle.

18:20: Apiki steals calf away; it hits tank wall solidly. Kahe competes with Apiki while mother pays no evident attention.

18:56: Another adult female, Lilinoe, begins to swim with calf.

19:06: Apiki seeks control of baby from Kahe. Calf hits wall again, has trouble at first swimming upright, then attempts to nurse from Lilinoe. Placenta passed by mother at 19:10.

19:07–02:35: Calf mostly swims with Apiki and frequently tries to nurse from him. Kahe beak–genital propels mother. Calf hits the tank wall again, amid much agitated swimming by the other dolphins. At end of period, hits wall again, squeaks, and then sometimes swims alone for a few moments, but is mostly attended by the solicitous if seemingly inept juvenile male.

02:50–04:57: Mother resumed control for a time, but quickly surrenders it to alloparents, especially Apiki, and then at end of period resumes control of calf, surrenders it again quickly to Lilinoe, who is "open-mouth threatened" by Apiki. The calf stays with alloparents until:

08:26: when mother returns and may have nursed the calf.

09:08: Calf attempting to nurse from mother. Mother rolls onto her side assisting the calf in nursing (fig. 96). The alloparents, especially Apiki, remain in attendance.

09:29: Calf rides along on mother's left pectoral fin. It then splits time with mother and alloparents.

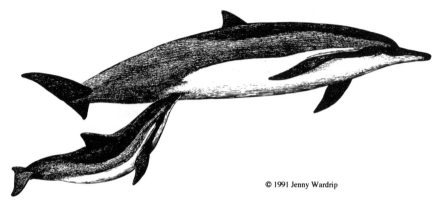

© 1991 Jenny Wardrip

Figure 96. Nursing posture of a pair of spinner dolphins.

10:01: Calf attempts nursing from mother; she controls young and threatens others who approach. By 10:38 she is obviously nursing the calf.
10:48: Mother threatens alloparent Kahe by hitting her on the head with her flukes. Apiki seems to try to take baby from the mother.
11:00: Cleary identified nursing by baby and mother.
11:13–11:28: Eighteen bouts of nursing, averaging 6.1 sec duration each, range 3–15 sec.
11:35: Apiki has partial erection, spins four times; mother tail-slaps other dolphins away.
12:06: Mother, calf, and Apiki swim together.
12:09: Eleven bouts of nursing, averaging 8.0 sec each, range 2–32 sec.
12:25: Mother defends calf against intrusion of Maile, the roughtooth dolphin. Kahe attends mother through event.
12:56: Nursing. Eight events averaging 7 sec each, arranged in series, range 3–15 sec.
13:35: Nursing. Four events averaging 8 sec each, range 5–10 sec.
14:34: Nursing. Eight events averaging 5.4 sec each, range 3–10 sec.
15:00: Nursing. Five events averaging 9 sec each, range 5–20 sec. Mother turns belly toward baby during nursing. Protects calf from running into wall.
16:45: Apiki takes up station with mother and calf. Calf hits wall.
16:46: Calf now with Apiki; hits wall repeatedly. Returns to mother at 19:17
19:18–24:00: Calf with Apiki on whom it tries to nurse. Kahe assists as well, and is much better than Apiki at keeping calf away from obstacles, nonetheless, the calf hits the wall at least five times. Tank is a froth of activity at midnight; Calf tries to nurse from Apiki and Kahe.
24:01–03:45: Calf alternates between Apiki and Kahe, with a long bout with just Apiki in attendance. Brief sortie with mother before Kahe takes calf again.
04:40: Kahe interposes herself between calf and wall. Long session with Apiki.

05:48: Mother nursing calf.

06:00–07:40: Calf mostly with Apiki, who once pushes it away from the wall.

08:11: With mother; no nursing.

09:00: Calf showing some signs of distress; labored swimming, bite marks noted around genital area.

09:31–09:58: Apiki threatens mother; hits her with flukes, calf tries unsuccessfully to nurse, Apiki aggressive to mother. Mother nudges calf to surface.

10:04: Apiki jaw-claps at mother, produces erection, makes loud "noises." Mother and calf move off alone; calf in trouble, being guided by mother.

10:36: Lilinoe jaw-claps at mother and calf.

11:16: Calf sinking; mother threatened aggressively by both Lilinoe and Apiki; long echolocation trains heard.

11:25: Apiki hits mother. Rammed her, almost knocking her out of the water; afterward jaw-clapping.

11:40: Mother places left fin under calf and lifts the largely inert calf. It finally sinks, the mother raises it, holds it briefly on her head, abandons it, while Apiki and Kahe stand by it. Once the calf is removed from the tank, Apiki circles near exit point while other animals circle farther away.

There are several elements of importance in this poignant birth observation, but we must not forget that it chronicles events that occurred in an enclosed tank and not in open water where there would be no obstacles. As is true for other open-ocean cetaceans, spinners are less adept at dealing with obstacles than are shore-dwelling species. Captive shore-dwelling species such as bottlenose dolphins have much less problem with confinement and obstacles during birth, and with modern husbandry practice, successful birth has become a routine event for them.

It is important to note that both male and female spinners other than the mother had seemingly accepted roles. The mother was more accepting of both male and female consorts than she was of other schoolmembers of the same or other species. Both accompanied the newborn during its first hours in a protective fashion, although in this case, the ability of older female consorts to "herd" the young away from dangerous obstructions seemed to be superior to that of the juvenile male, even though later he did protect the young in this fashion. But such protection was insufficient to ensure survival in this particular captive environment.

The behavior of male and female alloparents were somewhat different. The male was the most active of the alloparents and the one that showed the highest levels of agitation, especially approaching the time when the calf died. Both male and female alloparents harassed the mother at this time, even as she sought to support the calf. Once, during a time of high agitation by the dolphin school as a whole, the male showed a partial erection.

The mother surrendered control of the calf very soon after birth and only resumed it sporadically during the duration of these observations. Nonetheless, when the calf was in extremis, she resumed control even though harrassed by the alloparents.

What might these elements of the observation mean? First, a few comments about the mother's role. Spinner dolphins undergo a long gestation period (10–11 months) and produce very large precocial young. It is reasonable to suppose that this places special stresses on the mother and that she should rely upon alloparents to a considerable extent during these early hours, only rejoining the calf during difficulty or nursing events. The role of the female spinner alloparents seemed to be simply to help the mother through assistance with the calf, although toward the end when the calf was near death, the alloparent Lilinoe did seem to be in conflict with the mother.

The male alloparent's role seems to contain more apparent contradictions and conflict than those of mother and associated females. The male alloparent Apiki was more often in conflict with the mother than the female alloparents, even to the extent of giving threat and attack. He more overtly sought to "steal" the calf from the mother and was more in direct conflict with her during the final hours when the calf's declining condition seemed obvious. Such conflict reached a peak just before death when the young male attacked the mother.

If we view the male's role in terms of the male coalition behavior that it presumably will perform when it enters adulthood, a new light is cast on these events. Also, these events must be considered in relation to observations of dolphins trapped in tuna nets, when a segregation of schools seems to be sought by an active and aggressive cadre of adults, including the male coalitions that seem to be involved in attempts to order the school (Norris and Dohl 1980b, Pryor and Kang 1978). It may be that one aspects of the male's perceived role is to assist in a protective ordering of the birth school. Presumably a mother and calf, especially a calf in difficulty, requires such protection.

The partial erection that the male alloparent Apiki displayed toward the calf in difficulty is not an isolated observation. Oceanarium dolphin handlers know that males will both rake and nip at the genital areas of newborn dolphins and attempt to mate with them (see also Norris 1967). For this reason, the males are usually removed from the tanks when a birth by the main captive species *Tursiops truncatus* is in progress.

Dolphins have a fibroelastic penis that does not require engorgement, and hence erections are less tightly coupled to sexual arousal in these animals than in many other mammals (Slijper 1966). Hence, such pseudomating is perceived as a part of the male's threat signaling capability rather than as true sexual assault.

Such threats seem understandable in the context of a male attempting to order a school into a protective arrangement and being defeated by the constraints of a tank. It may be that under dangerous conditions, such as when the amniotic fluids and blood are introduced into the water, the rule may be this: *sensitive age and sex classes in the most protected sector of the school; any dolphin out of position will be met with aggression.* The result in the open ocean could be a protective exterior phalanx of aggressive animals through which a predator would have to pass to get at the newborn or the mother (see chap. 13).

The threats to the calf by Maile, the roughtooth dolphin, may have been real, since her species feeds on very large fishes such as MahiMahi (*Coryphaena hippurus*), which can weigh two to three times more than the calf (Brower and Curtsinger 1979).

The female alloparental role seemed nurturent, in that aggression was noted much less than for the male. Such alloparental role taking may reach its apex in delphinids such as the short-finned pilot whale (*Globicephala macrohynchus*) in which old post menopausal females still possess the capacity to produce milk (Kasuya and Marsh 1984, Marsh and Kasuya 1984, 1991). These females may be filling an alloparental role in schools that must sometimes dive deeply for their food, perhaps deeper than newborns can swim.

In Kealake'akua Bay, very small calves frequently traveled with other calves and larger juveniles, swimming in discrete groups within a school. The schools entering Kealake'akua Bay were obvious segregates of the total structure of the large offshore school. Schools with numerous young or adult–young pairs seemed to travel together, entering the rest coves together, and then after a time, being replaced by schools emphasizing other parts of the larger school. This is concordant with the observations of Miyazaki (1977) who found such segregated portions of a total school traveling more or less discretely within the large migratory front of *Stenella coeruleoalba* schools moving along the shore of Japan. In that case, migration seemed to spread out the elements of the larger school, while in our studies, elements of the larger school may have mixed together each night (see chap. 6).

CALF AND JUVENILE BEHAVIOR

Dolphin calves are frequently among the most active members of a school. From the viewing vehicle, it was not uncommon to see young spinners, especially small juveniles, dart away from adult companions for sorties covering perhaps 20–40 m (65–130 ft) and then rush back again to their companions. Such calves typically beat their tails about twice as fast as accompanying adults.

The movements of these calves are exaggerated, with much wiggling, tilting, and rapid darting course changes. We were able to discern what Bateson (1974) called "playpens," which are nursery areas where open volumes of water within the school envelope are roughly ringed by adult subgroups. The calves within these nurseries were noted chasing each other, rolling over, and touching bodies, fins, and genitals in play parodies of adult courtship and caressing.

Young calves occasionally rode the bows of our vessels, almost always in the company of an adult. On occasion these calves appeared to be inquisitive regarding us and approached our vessel head-on or turned and tilted to look in the window of our viewing vehicle.

Calves were sometimes remarkably aerial, even during times when the remainder of the school was resting. During one rest period, we watched such sorties during a span of about an hour when two calves repeatedly executed aerial patterns that were poorly enough defined that we could not fit them easily into our aerial behavior classification (chap. 5). In contrast, other calves performed well-coordinated spins. Aerial behavior is a practiced pattern that seems to require a learning period during the young and juvenile years to be perfected.

We noted that attending alloparents sometimes performed aerial patterns with a young animal, although more often they merely lingered nearby. Whether this represents "instruction" is unknown, but since the coordination between mother and young was sometimes precise, we tend to believe that it might. Such adult–young pairs were sometimes noted swimming in synchrony, breaking the surface in a nearly simultaneous leap. On one occasion, three adults and a young dolphin performed 13 simultaneous arching leaps in a row.

We were able to observe what we interpret as mother–young interactions in some detail when a well-marked adult female–young pair (dubbed Temple Baby and Mottled Mom) entered Kealake'akua Bay. The baby was given this name because the young animal spent so much time swimming above the mother's temple and forward of her largest diameter in the *assisted locomotion* position. These two were quite tolerant of the viewing vessel and were frequently observed a modest distance away from the main school but close to our craft. The following is a distillation of our notes from these encounters.

The pair was seen to leap together. During one morning activity period (prior to rest), a series of three tandem leaps was observed. Later the same afternoon, the two were noted spinning simultaneously, and then in one spectacular aerial display, they were seen in a high arcing tandem leap going in opposite directions, passing each other within a few centimeters at the apex of the leap. The mother was seen to belly tilt toward the young during the leap. On one occasion during rest when the remainder of the school was moving sedately in the carpet formation, the pair rose out of the formation toward the surface. The adult female leveled off about half way to the surface while the calf darted up alone, took a breath, and rejoined the waiting adult. Moments later the entire school surfaced and the mother–young pair rejoined it.

Young animals were noted on 16 occasions performing belly-up maneuvers relative to other larger dolphins. These mostly occurred within tight pairs. Most occurred beneath the mammary area of adult females and involved contact. One case involved a calf performing this behavior to an obvious juvenile.

A vignette of the process of behavioral maturation of a juvenile was seen on August 15, 1980. Temple Baby was noted swimming with Mottled Mom as before, but the young animal sometimes darted about, spinning alone. At 9:47 A.M., Mottled Mom was separated from the young animal in a swirling group of six adults in which much body tilting and contact was taking place. At 9:49, the young dolphin returned belly-up under its mother, its rostrum touching the adult's genital–mammary area. The mother moved off again into a tight subgroup of adults in which dorsum-up beak–genital propulsion was observed. A little later (at 10:05), the pair was seen again, side by side, trailing behind a swirling gaggle of dolphins engaged in group caressing. Bit by bit, independence between mother and the calf was appearing. Such juveniles were frequently seen without an obvious adult consort in groups of adults in which belly tilts, belly ups, and intromission were all occurring.

Juvenile or subadult behavior was in many respects like that of younger animals (Appendix B). Subadults were typically highly energetic, were often seen in pairs with an adult, and were sometimes noted accepting assisted locomotion from them. A distinguishing feature of these subadults was their greater independence. They were more apt to change swimming companions and to swim alone farther from a group than calves. Rostrum–genital contact (which may have been nursing) was a commonly observed pattern between calves and adults, but was seen much less often between adults and subadults.

Conversely, calves took predominantly one of two positions relative to an adult: either the temple position of assisted locomotion or the beneath-the-flukes position that is probably associated both with nursing (figs. 95 and 96) and the available defense of the adult's flukes. Calves also tended to associate more firmly with a single adult individual than did juveniles. For example, in an observation of a calf swimming with two adults, the calf was associated with one individual in 40 out of 45 serial instances, and in the remaining cases, the calf swam between the two

Figure 97. Dolphin (top) carrying plastic sheet on its flukes, Lanai Island.

adults. In undisturbed schools, adult–calf pairs were found anywhere within the school envelope and often at the periphery.

PLAY

We have already described some behavior patterns of calves that are best considered play—for example, the practicing of adult patterns such as leaping and spinning. Within the context of cultural transmission discussed in chapter 16, many more features of adult life must be transferred and practiced by young dolphins than we document here, including echolocation, reproductive patterns, social rules between individuals, and alloparental care. Earlier in this chapter, we described the actions of the juvenile male Apiki during the birth sequence, in which such role maturation was perhaps occurring.

But beyond these maturational features lies the honing of the patterns of muscular and other coordination essential to the lives of young and adults alike. A frequently observed example is the manipulation of play objects. This most commonly consisted of the carrying and trading of floating objects between dolphins (fig. 97), and it revealed to us some sense of the exquisite control dolphins have of their bodies in the three-dimensional weightless world in which they live.

Usually the object involved was a piece of flotsam, especially sections of black plastic sheeting used in the pineapple fields of the Hawaiian Islands to discourage weeds. Bits and pieces of this plastic blow out to sea and are picked up by the dolphins. Fragments of macroalgae are also common play objects.

One of our party (J. Solow), while swimming with a school of spinners off the south shore of Lanai Island, began to recognize an adult, whom she dubbed Linus and tentatively identified as a male. On one occasion, Linus was carrying a piece of plastic on his pectoral fin. He passed by her and let it go into the water, only to return to pick it up with his flukes and again with his rostrum. He circled playing this game of "catch" with himself, finally releasing the plastic fragment very close in front of the swimmer. The first time the dolphin attempted this, Solow dove down to retrieve it. Just as she was about to grasp it the dolphin swirled by, taking it away on a pectoral flipper.

Later that day, Linus was encountered again, still carrying a piece of plastic on a pectoral flipper. The dolphin let go of it near Solow, who succeeded in grasping it. She reports swimming with it while the dolphin zig-zagged back and forth in front of her, coming as close as 2 m (6 ft), emitting a sound like loud echolocation. Solow also reported seeing the dolphins trading pieces of plastic among themselves as they swam along.

Such play with floating debris was a common sight among the captives at Sea Life Park Oceanarium. Leaves that had blown into the tank, strands of grass, and strands of algae that we introduced for them all became play objects for all of the dolphins. Their "dexterity" in handling these light-weight floating objects was remarkable. On one occasion (February 22, 1980) Kahe had picked up a 4-cm (1.5-in.) leaf and precariously balanced it on the anterior base of her dorsal fin, where one could see it vibrate in the passing water stream, as if to slip off at any moment. Nonetheless, she balanced it there, even though it tended to slip from one side to the other as she swam. On one occasion, this slippage led to her spiraling off into a double roll in an apparent effort to keep the leaf from slipping away. Perhaps the most remarkable feat of balance we saw was when Kehaulani carried a bit of seaweed through an entire boisterous feeding period in which the dolphins competed for fish thrown to them by the handlers and never once lost control of her play object.

These play objects were often passed by a single individual from rostral tip to pectoral fin to dorsal fin and to flukes with great finesse and were traded among the dolphins with little evidence of competition between them. The considerable control needed to balance such objects was sometimes made obvious when the flipper not being used in balancing was held in the normal cruising position while the one bearing the play object was maneuvered constantly, sometimes being held out at a considerable angle from the body.

Pair swimming by spinners, which may be play, can be elegant in its precision. One such pair, watched in an active school in Kealake'akua Bay, traveled along the side of the capsule in a tight, precise spiral and then curved downward, still in the spiral, to disappear into the blue water below. It was not possible to discern a leader in this swift maneuver, nor was any other reason for the swimming pattern obvious. It was, however, a beautiful example of refined coordination by two animals.

AGGRESSION AND DEFENSE

Overtly aggressive interactions between spinner dolphins were found to be generally inconspicuous in undisturbed wild schools, although chases between two dolphins were relatively common. Tooth raking resulting from the mouthing of one animal by another seemed to be common, given the abundance of shallow scars seen on these same animals. In captives it seems coincident with reproductive hormonal peaks in both animals. But on the whole, aggressive interaction among spinners seems much less marked than in many other dolphin species (Norris 1967b).

The contrast between the generally peaceful interaction within Hawaiian spinner schools and that of dolphin schools trapped inside tuna nets is marked (Pryor and Shallenberger 1991, Norris et al. 1978). In the nets, groups of passive rafting animals were seen surrounded by highly active aggressive adults with much jaw snapping, insistent echolocation, bubble release, and scissoring of pectorals, all signs of agitation or aggression. The majority of these aggressive animals were spotted dolphins, but some were spinners. Human swimmers in the net often faced the S-shaped dolphin threat pattern (described later in this section). Traveling among the trapped animals were male coalitions of five to ten animals swimming purposefully through the school.

Thus, though day-to-day activities seem to include only modest elements of aggressive behavior, they clearly can be released under the threat. Let us examine these patterns in more detail and then discuss the more speculative possibility that they can be organized into a system of shark mimicry.

Open Mouth Contact

Spinner dolphins frequently mouth or nip one another. Their rostra and dentition are both very delicate, and hence these actions carry scant threat to a recipient. A typical example was recorded from the *Maka Ala* on August 31, 1980, in Kealake'akua Bay. At 10:10 A.M., an active subgroup of about 12 dolphins, mostly pairs and trios of adults but including one calf, was encountered.

One threesome was observed in very active interaction. One of the animals opened its mouth and pressed the side of its tooth row against the back of another member of the group, behind the dorsal fin, and then quickly touched the same area with its pectoral fin tip. The same pattern was repeated again but this time in the region of the external auditory meatus. This led to a chase by the recipient, and another pressing event by the first dolphin, this time with jaws closed. Such tooth pressing and raking frequently leaves rows of shallow whitish marks or grooves in the skin where the teeth have penetrated the epidermis into the hypodermis. These heal and disappear rather quickly, judging from observations on captives.

Later (at 10:18) in the same school and amid an extremely intense outburst of noises including loud whistles and buzzes, a caressing subgroup of about 12 tightly interweaving dolphins was noted. In this group, belly tilts were frequent and open mouth contact was noted. There was literally no space between the participants.

Open mouth contact was sometimes associated with chases. On September 1, 1980, in Kealake'akua Bay, a subgroup of five dolphins was seen traveling in front of an active school. One dolphin pressed its open mouth against the uppermost dolphin of a belly-to-belly pair. The pressing dolphin then swung under the upper animal in the inverted position. At this point, another dolphin nipped the flukes of still another schoolmate. This precipitated a rapid near-vertical dive by the nipped animal followed by the aggressive animal and one other. At one point in this chase, both pursuing animals swam belly-up and then disappeared from view. Directly afterward a cacophony of sounds was heard and the *Maka Ala* passed through a bubble field 3 m (10 ft) across, extending downward as far as the eye could see. We presumed this consisted of expressed air from the phonation event.

Scars from open mouth contact are frequently seen on many species of dolphins. Since the teeth of dolphins tend to splay outward in what Norris and Møhl (1983) have called the *cage jaw* (a means of trapping prey before excurrent water forces it away), such a raking motion can furrow the skin, although serious wounds do not appear to result even in large toothed species such as pilot and false killer whales. The wounds sometimes left by the very large teeth of sperm and beaked whales can be deep (see Norris and Prescott 1961, Heyning 1980, 1984, McCann 1974).

Threat Posture

One wonders how the seemingly defenseless spinner dolphin protects itself in an ocean where predators are numerous and well equipped. It is

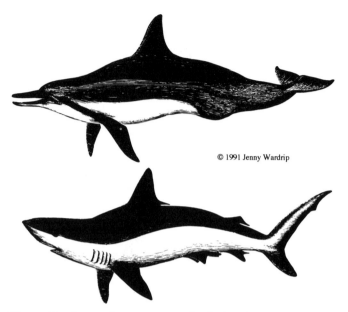

© 1991 Jenny Wardrip

Figure 98. Aggressive posture and mimicry by an adult male spinner dolphin compared to the aggressive posture of a gray reef shark.

not that they simply avoid the company of predators such as sharks altogether by use of echolocation, since our swimmers encountered sharks near spinner schools. On the tuna grounds, sharks are frequently seined along with tuna and dolphins (Leatherwood et al. 1973).

Although the teeth of spinner dolphins do not seem to be useful weapons, these dolphins do direct aggressive patterns at schoolmates and at intruders. Such aggressive behavior seems to vary greatly in intensity from mere negative hints to overtly threatening acts. In most cases fully adult dolphins were involved.

We have observed the spinner posture in detail from encounters with the captive male, from the tuna nets, and have once filmed it at sea off Kealake'akua. The spinner uses an S-shaped posture in which the dolphin faces the object of threat, usually with mouth agape or snapping, head down with a sharply arched neck and anterior back, and with the flukes flexed upward from the general arch of the back (fig. 98). When such a threatening dolphin faces an observer, the view is of the dark snapping snout and head facing the observer and contrasting strongly with the immaculate white of the belly over which it is bent. The pectoral fins are typically spread and may be scissored up and down in rapid slicing motions.

This posture may be given by an animal that stations itself in front of an offending subject, or the dolphin may swim rapidly in a circuitous path around an intruder (as was the case to be described in detail later in which our diver was threatened). In that case, the body curvature was less extreme than in more stationary threats, and the moving dolphin rolled from side to side as it circled our swimmer, with pectorals slicing.

Aggressive Encounters

Our first two observations of aggressive encounters were noted from the *Maka Ala* in Kealake'akua Bay. The first took place at 11:00 A.M. on March 3, 1980. Four spinners ran in front of *Maka Ala*'s bow. Two were identified as adult males because of their thickened tail stocks and evident postanal humps. The sex of these two was confirmed as male by direct observation of the genital area. The sex of the other adult animals could not be determined. Three of the dolphins curved away from the bow in a tight chevron, the middle animal ahead and slightly below the other two. All were within one body diameter of one another. A fourth dolphin approached this group from the side, coming in nearly perpendicular to them, with an exaggerated thrashing of the tail. The dolphin nearest the intruder thrashed moments later, still swimming in formation. The intruder may have struck its dorsal fin against the outside animal who then fled, the intruder sliding into its place in the chevron. The aggressor then swam in synchrony with the remaining two animals. The thrashing tails (which we describe in more detail later) and perpendicular approach were both unusual events in our observations of spinners.

On July 5, 1980, a topside observer noted a trio of animals, one of which exhibited the S-shaped body posture immediately adjacent to another animal that was swimming belly-up within pectoral touching distance of it. All three then swam rapidly down and away from the dolphin toward which the display had been directed.

A complete but brief sequence was filmed on June 23, 1980, by Wells. He dropped over the side of our inflatable and swam into a small school traveling along the island coast south of Kealake'akua Bay. An adult spinner raced toward him, zig-zagging back and forth in front of him. The dolphin assumed the aggressive S-shaped posture during this approach with arched back and pectoral fins flexed forward. As this animal swept around in front of the swimmer, it flexed its body rapidly in an almost rotary tail stroke, partially from side to side and partially up and down. So rapidly did its tail move that these movements were blurred on the frames of the film while no other aspects of the sequence were. This locomotion was different from the usual clean sinusoidal swimming of dolphins. As the dolphin swept around our diver, it rolled from side to side on its longitudinal axis. Our diver photographed the little school and then clambered back onto the inflatable; the dolphin returned to normal swimming.

Convergence Between Shark and Dolphin Behavior/Structure

In reviewing the previous film, we were struck by some aspects of the dolphin's locomotion that seemed reminiscent of the aggressive swimming patterns of the gray reef shark (*Carcharhinus amblyrhinchos*) as reported by Johnson and Nelson (1973). The very rapid sideways component of the dolphin's tail, while normal for shark locomotion, is at right angles to the tail movement normally used by dolphins, whose tails normally beat up and down (fig. 98).

Three possible interpretations for the lateral tail stroke presented themselves. First, it seemed possible that dolphins imitate shark aggressive patterns, including tail movement in a plane that they do not normally use. Second, it might be that dolphins use such a lateral stroke as an aggressive weapon, which would bring the thin edge of the flukes against an enemy. Such aggressive or defensive use of lateral swipes of the tail flukes is well known for both odontocetes (*Tursiops truncatus*) (Caldwell and Caldwell 1972a) and mysticetes (*Eschrichtius robustus*) (Norris and Prescott 1961). Third, the pattern could be an example of behavioral convergence.

To highlight the similarity between the shark and dolphin patterns, we quote from a description of Johnson and Nelson (1973, p. 79) for the gray reef shark:

> The display consists of two locomotor and four postural elements. The locomotor elements include (1) Laterally exaggerated swimming motion and (2) rolling, and later tilting of the body and/or spiral looping (spiral up and down movement through the water, more pronounced in the anterior region). Rolling, although often seen independently of spiral looping, appears indistinguishable from the initial phase of the latter, and both are frequently followed by resumption of laterally exaggerated swimming (the most common display mode). The postural elements include (1) lifting of the snout, in its most intense form resulting in a distinct bend between the chondrocranium and the spinal column, and slight opening of the jaws, (2) relatively prolonged and at times marked drooping of the pectoral fins, (3) arching of the back, and (4) lateral bending, often more pronounced at the posterior end of the body cavity.

At first glance, it is easy to mistake a displaying spinner dolphin for a shark. Both are swift and terete. In both the dorsal and pectoral fins are similar in shape and color. Only the tails and the snouts are markedly different between the two animals (fig. 98).

The overall body patterns of spinner dolphins and those of some sharks, such as the black-tipped shark (*Carcharhinus melanopterus*), are very similar. Both have dark fins, a dark cape, intermediate lateral body color, and a white venter. Both are about the same size as adults and have widely overlapping geographical and ecological distributions. Both come daily to or live in the passes into atoll lagoons (Stevens 1987).

There is yet an additional curious parallel between dolphins and sharks. Two tropical oceanic dolphins, the spinner dolphin and the roughtooth dolphin, (Norris 1967*b*) have an anomalous connective tissue swelling called the *postanal hump* on the tail stocks of adult males. This hump is very prominent on adult males of the eastern Pacific spinner (Perrin 1975a) and has been noted on pelagic specimens of spotted and common dolphins from the offshore eastern tropical Pacific (W. F. Perrin, pers. comm.) (fig. 5). The hump is much more subtle yet recognizable on adult male Hawaiian spinners and is nonetheless the major feature that allowed us to recognize such adult males at sea in Hawaii. What engaged our attention was that these humps are placed in exactly the same location as the claspers of adult male sharks. The concordance is even greater when one notes that the hump of the roughtooth dolphin is longitudinally folded, making the match with the paired claspers even more precise.

We now view this hump as a structure that may passively carry a communications message, that is, a signing organ. In this case, the message may be "dangerous animal." During aggressive posturing, the spinner dolphin's tail stock is bent downward and forward. This thrusts the swollen vent and genital area forward. In this same posture, the male shark displays its male intromittant organs, the claspers, as part of its aggressive display. Male gray reef sharks are defenders of territory, and such display precedes attack, so the signal has strong import.

A final feature of this proposed shark imitation is found in the dorsal fins of the shark and its imitator. Adult male spinner dolphins develop tall, erect black dorsal fins with age, and they hold them erect in the male coalitions within which they often travel. This is especially evident in eastern Pacific spinners in which the dorsal fins are large and tall and often cant slightly forward as if "put on backward." When such a dolphin arches its back in the aggressive posture, the effect should be to bring these fins into tall upright triangles.

It seems probable that the dolphin imitates the shark. Nearly every feature of the territorial display of the shark is contained in the dolphin pattern, including the lateral movements of the tail, which in the shark are basically locomotory but are in the wrong plane for the dolphin. The exaggerated swimming of the shark, the figure-eight path, the posturing, and the opening and closing of the mouth are all present. How might this imitation by dolphins be used? Clearly, the convergent patterns are used by both spinners and sharks against intruders and may have marked effect on conspecifics as well. It also seems likely, given the widespread evidence of shark attack on dolphins (chap. 15), that sharks are important dolphin predators. The pattern, then, might be used outside the dolphin school, being directed at the predators themselves.

USES OF INTENSE SOUND

One final and complete observation of the S-posture of an aggressive adult male spinner was observed at the Oceanic Institute on February 16, 1980, when two of us (Norris and Brownlee) attempted to elicit aggression from the adult male Lioele by tapping smartly with a metal pencil on the viewing port each time he swam in facing the window. We hoped, in effect, to "echolocate in his face." Finally, the dolphin turned from its two female consorts and paused about 1 m directly in front of the window. He arched his back, flexed his head, pointing his snout at us, opened his jaws slightly, and emitted a very loud double pulse of sound, so intense that Brownlee, listening through ear phones, was left with her ears ringing. The intensity of the dolphin's phonation effort caused a double bubble of air to leak from his blowhole.

As the sequence began and Lioele turned toward the window, the two adult females fled the scene, swimming directly away from him, and stopping only when they reached the far slope of the tank. They then stopped, resting on the sloping tank bottom with their snouts pointed directly away from Lioele, a very unusual posture. After he had emitted the burst of sound, the two females began circling the tank again.

This aggressive sequence contains two new elements: the production of a very loud sound, and the apparent knowledge by the females that an aggressive event was about to happen, whereupon they seemed to protect themselves from the loud sound. It also contains most of the elements of the shark display, except for the figure-eight swimming and tail thrashing, of which the crippled Lioele may not have been capable.

In 1983, Norris and Møhl presented the hypothesis that odontocetes might be able to debilitate prey by sound or use very loud sound signals in aversive social situations. The idea was that dolphins might emit sounds so intense that prey could be debilitated and capture made easier, or that such sounds might be used within the dolphin school in social ordering. This hypothesis was based on a number of suggestive observations, including a complete sequence in which two captive female spinner dolphins ensonified a school of akule (*Trachuropsis crumenophthalmus*) until some of the fish lost the ability to school and one was sucked down the drain of the tank (fig. 99). Since that paper was published, a number

Figure 99. Prey debilitation by the spinner dolphin Kehaulani, Bateson's Bay, Sea Life Park Oceanarium, Hawaii.

of new observations have been made that both support the production of very loud sounds and the contention of acoustic prey debilitation can happen. They do show that spinner dolphins can make exceptionally loud sounds and that they affect other members of the school. Workers have been able to kill schooling fish (Northern Pacific anchovy) with artificially generated sounds very similar to those produced by dolphins in feeding situations, and several examples have been collected of odontocetes producing similar sounds, sometimes volleys of them, in feeding situations (Marten et al. 1988) (fig 100).

We briefly summarize this evidence here because, if dolphins possess this capability, it casts new light on the defense they may use against predators at sea (see chap. 15) and because it may be of profound importance in obtaining food and in social relationships within their schools. The latter possibility stems from the proposition that any adaptation that could effectively deafen schoolmates should have importance for a species that depends completely upon the acute hearing needed for echolocation. For a wild dolphin to be deaf is probably to be dead.

The first evidence that suggested such a capability was that dolphins and other toothed whales have been found to be capable of generating remarkably intense sounds (Au et al. 1978). Bottlenose dolphins have been trained to echolocate on small, distant targets, and when they do

so, they are capable of emitting echolocation clicks of almost unbeliev-
able intensity; the highest recorded was 228.6 dB re 1 μPascal at 1 yard
(Au et al. 1978). Such a sound is verging on the finite limit of sound, the
region in which adding more energy to sound production produces
mostly heat and not higher levels of sound. These sounds, however, are
very brief (70-μsec to 1-msec clicks arranged in typical echolocation
trains) and predominantly of very high frequencies (generally above
100 kHz).

Other evidence comes from the sperm whale, which is known to produce loud bangs of very high intensity (Caldwell et al. 1966) and which
feed on the widest size range of prey known, ranging from 4-cm (1.5-in.)
lantern fish to very large squids (up to 3 m, 10 ft, in length) (Norris and
Mohl 1983). Sperm whales somehow catch prey even though they are
equipped with peglike teeth located only in their elongate, polelike lower
jaws. Even these teeth do not seem essential to feeding since teeth are
unerupted in young whales and deformed sperm whales whose jaws
were curled to the side and could not close have been caught with stomachs full of good-sized, still-living squids.

In the evolution of odontocetes, the rise of modern forms in the late
middle Miocene was accompanied by a widespread reduction in jaw
length and frequent tooth loss. Early Miocene species with very elongate
jaws in many cases gave way in the middle Miocene to modern forms
with blunt snouts and reduced teeth. Some modern species are essentially edentulous, and the adult males of one form (*Mesoplodon layardi*)
are unable to open their jaws more than about 8 cm (3 in.) because two
peculiar straplike teeth curve over the upper jaw, holding it nearly
closed. Without a new means of subduing prey, this would appear to be
a terminal adaptation.

Evidence that sharp mechanical disturbances can kill fish comes from
the pistol shrimps of the family Alphaeidae, which produce very loud
clicks by rapid closure of one hypertropied claw. These shrimps have
been observed to immobilize small fish by clicking this claw near the
heads of their victims (Marten et al. 1988).

A case of probable acoustic prey debilitation by the captive spinner
dolphins used in this study was cited by Norris and Møhl (1983). It is
worth quoting the description of this event (p. 95) in detail here:

> A direct test was carried out at the Oceanic Institute, Oahu, Hawaii, with
> three captive Hawaiian spinner dolphins (*Stenella longirostris*). Its results
> were also inconclusive from the standpoint of proving acoustic prey debil-
> itation, but they do provide new data regarding ensonification of prey by
> dolphins. The test, performed on September 10–11, 1980, utilized an un-
> derwater viewing room equipped with listening gear, but no means of de-
> termining [acoustic] source levels. The spinner dolphin is a moderate-sized

animal that forms large schools and feeds on small mesopelagic fishes, squids, and shrimps, mostly from the deep scattering layer. . . . Its prey items are typically small, including postjuvenile forms; the largest is about 10 cm in length. The akule used in this test were about twice that length and therefore not normal food. Another 12 akule from the same collection were placed in a nearby oceanarium tank, serving as a control group.

Shortly after introduction of the fish, first one dolphin and then two began to make click train runs on the fish school. They did so in quite a stereotyped manner; they always approached the school from its side, racing at it while emitting a whining crescendo of clicks. The school typically parted in front of the onrushing dolphins, both parts turning tightly toward the passing dolphin's tail [our fig. 100]. The mammal was unable to turn so tightly and had to make a broad circle before coming in on another run on the now reformed fish school. Such runs continued for approximately an hour without noticeable effect on the fish. After two hours, involving perhaps two dozen runs by the dolphins, a noticeable depolarization of the fish school began to appear. All the fish no longer pointed in the same direction. They began to wander in the school. A wandering fish usually drew the attention of a dolphin, who pursued it relentlessly, emitting train after train of rapid clicks. The fish was typically placed within a few cm, often 1 or 2, of the rostral tip, or well within the expected acoustic near-field. During such pursuit the dolphin scanned with its rostrum, in small rapid excursions, presumably playing the sound across the fleeing fish. By this time the fish were unable to avoid the dolphins, either in their main school or as separated individuals.

One such wandering fish was noted to change from a silvery color to a pale lemon yellow. The school traversed the central drain box, which was drawing a modest vortex. The fish struggled and dipped down as they passed over the drain; behavior that would not be expected of a normal school. One fish was spun down and sucked from the tank. No fish were eaten.

The control group of fish in the oceanarium tank remained alert and active though they were repeatedly attacked by jacks (*Caranx*) kept in the tank. Although the dolphins finally succeeded in debilitating the fish, it was not known exactly how this was achieved. The most obvious cause was the repeated ensonification, but lactic acid debt from the repeated chases could not be completely ruled out, even though control animals were also chased and did not suffer in the same way. We could say that the debilitation of the fish appeared to involve dysfunction of the labyrinth of the inner ear rather than simple fatigue. The affected fish lost their orientation ability, often traveling in long arcs while at the same time showing pronounced difficulties of vertical orientation. So, the probable cause of the debilitation was an effect on the inner ear. Nonetheless, it remained impossible to say categorically that ensonification had caused the effects noted. It was also curious that debilitation had taken so long. Certainly an activity that took 2 hr to complete would do little good in food getting.

It remains possible that a sex difference exists in the ability to produce such loud sounds and that they may be given in various contexts. Only females had participated in the long, slow sequence of prey debilitation reported here, while the rather immobile male was not involved. But only he demonstrated a sound having the characteristics we now associate with rapid prey debilitation when we irritated him.

Since the Norris and Møhl (1983) hypothesis was first published, a number of new observations have brought proof of the theory closer (Marten et al. 1988). First, a new class of sounds has been described, recorded on several occasions and only during odontocete feeding events. The sounds are 200–700 times as long in duration as a single echolocation click and are composed of many transients in an unorganized burst that could possibly consume much or all of the air supply normally used to produce an entire train of clicks (fig. 100). These very loud rifle shot–like sounds have been recorded in volleys when dolphins or killer whales encounter a school of fish. The sounds seem to be much like the single threat sound that was directed at Brownlee. Similar blasts of sound have been called "jaw claps" (Overstrom 1982).

The sounds are low frequency-rich blasts centering within the general hearing range of fishes (0.8–1.5 kHz), and sounds of similar characteristics can kill fish. Carl Schilt (1991) of Long Marine Laboratory at the University of California, Santa Cruz, has directed similar artificially produced sounds at schooling fish (northern anchovy), and those fishes located within a few centimeters of the sound source have been killed or seriously disoriented.

It is still not clear whether only dolphin males produce the sounds we implicate in prey debilitation. Jaw claps, which are loud, long duration, multipulsed sounds associated with rapid jaw closure, are produced by both males and females and may be similar to sounds that we implicate in debilitation of prey. These relationships need further study. Note that these sounds are very different from the brief, high frequency echolocation clicks previously suggested to be involved in prey debilitation.

The scenario for the way these sounds might kill or debilitate fish or squid centers around the fact that fish apparently do not have a protected inner ear. Most have no middle ear ossicles or middle ear reflex, which in mammals and birds blocks out intense sounds after about the first 1 msec. Possible exceptions are the ostariophysan fishes, which possess Webberian ossicles, a chain of bones connecting the ear and the swim bladder (van Bergeijk 1966). We propose that the unprotected fish receives the sound, the ear then goes into overload, recovers after about 1 msec, is hit again by the same continuing sound, recovers again, and

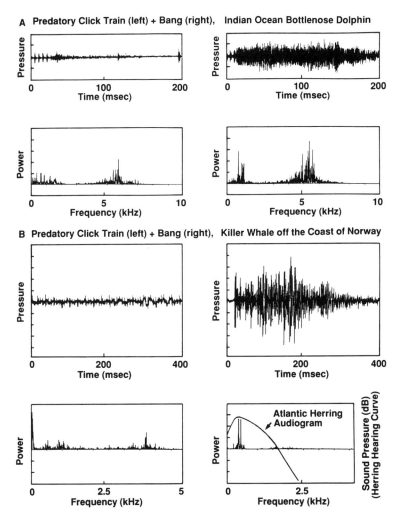

Figure 100. Long duration impulse sounds emitted during odontocete feeding episodes. Oscillogram and power spectrum of an echolocation click train followed by a "bang" from (A) a wild Indian Ocean bottlenose feeding on Perth herring, and (B) a wild killer whale off the coast of Norway feeding on Atlantic herring. The low frequency component seen in clicks is made loud in bangs. This manipulation of the power spectrum of clicks to make bangs suggests that clicks and bangs are produced by the same mechanism.

For each species the oscillograms of the click train and bang have the same vertical scale, so the intensity of the bang relative to the click can be read. The bang power spectrum vertical scale is 20X the click vertical scale in A and 40X it in B. Absolute intensities are unknown. The killer whale bang power spectrum closely matches the audiogram of its prey (see Marten et al. 1988).

on and on, perhaps for hundreds of times before the sound finishes. The psychophysical assault of a single 700-msec sound may therefore be very great (Marten et al. 1988). With each such overload sequence, the psychophysical effect of the sound is thought to summate (temporal summation), and shortly the fish simply collapses from the effect. Hair cell damage is likely to be involved.

We move on now to a consideration of the species aggregations in which spinner dolphins travel at sea, a review of their parasites, and a review of what is known about the predators of spinner dolphins.

FIFTEEN

Predators, Parasites, and Multispecies Aggregations

Kenneth S. Norris

Attacks or successful kills by predators on healthy wild dolphins have almost never been observed. This is not particularly surprising in view of the long lives of dolphins and the limited observations that have been made of them in nature. Although they may be stalked frequently, their predators seldom succeed in catching them. Nonetheless, the high incidence of shark scars in some dolphin populations indicates the reality of predation as a force in dolphin societies.

Stewart Springer, a well-known shark biologist, has provided an especially interesting record of sharks attacking a dolphin school, even though no dolphin was taken while he watched (American Institute of Biological Sciences 1967). This record is of special interest to us because of the close evolutionary relationship between the dolphins Springer observed, which were apparently common dolphins (*Delphinus delphis*) and spinner dolphins. Both form large schools that frequent the open sea in parts of their ranges. They are about the same size as adults. The account gives some idea of the problem of predation for an open sea dolphin. Springer records the following observations:

> We were loafing along at three or four knots washing down the deck after a frustrating early morning purse seine set that caught little except jelly fish when about 50 porpoises came and maneuvered close against the ship's side. This was unusual because we were moving too slowly to give them any sport. They were an unusual kind for the mid-Gulf of Mexico, *Delphinus*, and their behavior was startling. It was obvious that these porpoises were dead tired and some were injured. By looking closely, we began to see the dim shapes of sharks staying perhaps 30 yards away from our boat and the porpoises. There were six or eight quite young porpoises, not more than three feet long, and these were herded in close to the boat

while some of the large ones patrolled the fringe of the porpoise school that now had our vessel to protect one flank.

The porpoises trusted us with their young and our 250 ton tub of a fishing vessel responded as well as it could. We tried stopping but this seemed to bring the sharks in and to disturb the porpoises so we went ahead as slowly as possible, slower than big diesel engines like to go. Someone thawed some frozen fish and tried to feed the porpoises but they appeared to be too exhausted to be interested. Now and then a shark would come closer and one or more of the patrol porpoises would burst into a series of dashes. We could not see whether any contact was made between the porpoises and the sharks. The sharks always retreated but did not go away. Some of the patrol porpoises had been injured. Several were scarred and one had a badly shredded tail fin. It was easy to see that the porpoises were cooperating and that they were not only protecting their young but also were protecting each other. One shark would not have given the porpoises much trouble but a large group of sharks working with the same objectives had seriously endangered the entire school of porpoises and possibly had already decimated it. There was no real evidence that the sharks were cooperating. Given a little intelligence or some pattern of cooperative attack, there seems little doubt that the sharks could have finished the action in short order, and at least could have eaten the young porpoises as well as any last ditch defenders.

We felt badly because we could not do anything about the sharks and we didn't feel like abandoning the porpoises. Our dilemma was resolved in about an hour by a small but violent rain squall. When we came out of it the porpoises and sharks were gone. Perhaps, the short rest and the surface disturbance caused by the squall allowed the porpoises to escape with their young.

PREDATION IN SHALLOW VERSUS OPEN SEAS

It may seem anomalous that workers who have looked for shark scars on both nearshore and oceanic dolphin species find many more on those living in shallow water. Bottlenose dolphin populations living in the shallow waters along the Florida and Texas coasts have been found to be heavily scarred; between about 20 and 50% of the dolphins living in shallow waters along the coast bear unmistakable scars of shark attacks (Wood et al. 1970, Irvine et al. 1973). Yet few authors report shark scars from dolphins living in oceanic habitats (Wood et al. 1970). To be sure, reports are few where these might be noted. Is the reason for this difference because predation upon oceanic dolphins is somehow less or because once a dolphin is injured at sea its chances of escape are scant? Or does it mean that the protection of the school at sea is so complete that no predator can breach such a defense?

The contention that predation is somehow less in the open sea rests uneasily. Why should the dolphins in Springer's example seek the side of a ship for surcease? What change in the balance did the sudden squall provide? Why should Hawaiian spinners come once a day to rest over shallow sand, when open water is near at hand, unless it offers some respite?

Predation in the Open Sea

A review of records of predation upon dolphins in the open sea shows that a dolphin away from its school may be in imminent danger of death even if reasonably capable of flight. For instance, Leatherwood et al. (1973, p. 3) describe what frequently happens to live spotted and spinner dolphins trapped in tuna seines when they are thrown back, alone, into the sea:

> In the first instance, a 4.5 foot male, which had been hauled aboard with the fish, was recovered and thrown over the side. Though he was active on deck and ventilated well once in the water, he blew only once and began to swim away sluggishly. When he was less than 35 yards off the bow, he was bitten in two by a shark which showed only its dorsal fin, so its size and species were undetermined. The remainder of the porpoise sank through the bloody water and out of sight. . . .
>
> In the second instance, fishermen pulled a live adult female from the brailer and threw her over the side. As she was swimming away, a large shark (about the same size as the 16 foot chaser skiffs used in the fishing operation), with a robust body and sharply pointed head, came up from deep below the boat and bit the porpoise's midsection. The porpoise sank slowly and was hit several more times by smaller sharks before what was left of her sank out of sight.

These descriptions are of normal, not exceptional, events on board these seiners. Even if healthy dolphins of these oceanic species can evade the attentions of sharks, it is clear that the protection is transitory and disappears once a dolphin has left the protection of its school. This also suggests that predation pressure upon oceanic dolphins may well be important, even if we seldom record the scars of encounters that failed to kill the dolphin. Most of the dolphins seem to be able to evade the sharks so long as they are healthy, but when any debility befalls them that causes them to straggle outside the school's envelope, they are probably taken shortly thereafter. To be sure, the unusual circumstance of a net being set, crowding dolphins and tuna together, clearly attracts sharks. But it also indicates that they are nearby to be attracted. So the paucity of scars notwithstanding, dolphin societies at sea seem likely to have been shaped to an important degree by the presence of these predators.

Evidence from Hawaiian Spinners

Some individual Hawaiian spinner dolphins we examined showed evidence of shark attack. The male, Four-Nip, seemed obviously to have had a close call with an attacking shark, and other dolphins having nicked fins, tails, or flukes could also have undergone shark attack. When we first saw Four-Nip, the posterior margin of his fin was tattered with four pennants of tissue that clearly were left when a shark bit the edge of the fin (see fig. 56). From the study preliminary to this one, Norris and Dohl (1980*a*, p. 486) report the following:

> Hawaiian spinner dolphins seem to be attacked with some frequency by sharks. Several of the scarred animals we catalogued had obviously been wounded by large sharks. Lunate rows of tooth marks, especially on the tail region, some apparently from sharks with a 12–15 in. (31–38 cm) gape were noted. In one case it seemed that the entire tailstock had once been in a shark's mouth. Nicked and tattered dorsal fins may also have been produced by shark bite.

Perhaps because the Hawaiian spinners with which we worked come into shallow water to rest, they bore abundant evidence of shark attack, just as seems to be the case in the scarring of Florida shallow water bottlenose dolphin populations. Does this provide evidence of higher predation rates nearshore than at sea? The modified view, which we favor, is that in shallow water there is increased chance of a wounded animal being able to escape after being attacked by a shark, while in the open sea any animal that is seriously debilitated by a shark bite is effectively dead. Thus, fewer dolphins should bear scars in the open sea as compared to populations in shallow shore environments.

This view squares with Springer's observation that the dolphins sought the protection of his boat. In doing so, the dolphins had reduced the number of directions from which attack might come. In this way the dolphins had temporarily simplified the process of keeping track of and responding to their predators by reducing these possible directions.

The association of dolphins with objects that reduce the problem of surveillance may be more prevalent than is at first apparent. This may be the central reason Hawaiian spinner dolphins come nearshore to rest over shallow sandy-bottomed coves and roadsteads. Dolphins resting over a sand patch can look both up and down and be assured that no predators are near in those directions. Perhaps this is why they so assiduously turn away from the dark bottom that surrounds these patches— these are places where visual discrimination of a shark from the bottom is presumably difficult. It may also be an important reason for the establishment of multispecies aggregations, the topic of our next section.

MULTISPECIES AGGREGATIONS

Multispecies aggregations are ecological assemblages of various species, both predators and prey, that travel or aggregate together in the open

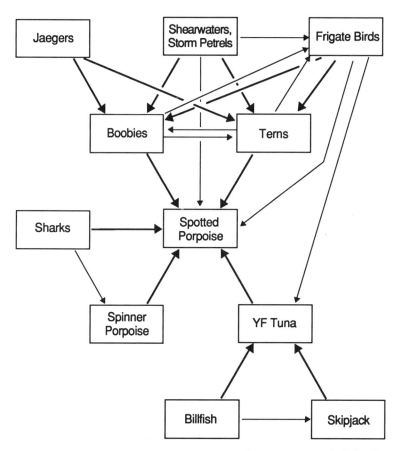

Figure 101. Multispecies aggregations on the eastern tropical Pacific tuna grounds. (Courtesy of W. F. Perrin.)

sea (see chap. 12). The possibility of these interesting aggregations forming with dolphin schools was first mentioned to me by William Perrin and expanded upon by Au (1991) and Scott (1991) (fig. 101).

Such aggregations may form either around stationary floating objects or with certain swimming species, including dolphins. Sportfishermen in Hawaii and elsewhere in the warm water world know that if they can find floating debris, even insignificant items such as styrofoam coffee cups and small blocks of wood, they are apt to find large predatory fish such as the mahi mahi, or dolphinfish (Coryphaena hippurus), circling nearby. If one dip nets the block of wood or other debris from the water, it is often found to carry a melange of species with it. Small prejuvenile fish of various species and swimming crustaceans may have hovered under the floating object. On its surface may be goose neck barnacles, crabs, and simpler encrusting life such as bryozoan colonies.

Do the logs and the dolphins serve the same function? That is, are each of them nuclear to the association they come to be part of? It appears so. If an aggregation formed around floating objects is encircled by a seine and the large fish taken away, others of their kind and their smaller associates will shortly reaggregate. The same is true of dolphin schools. The aggregations that travel with the dolphins, including large tunas, may be stripped away, apparently over and over again, by fishermen during a given year.

Why do such aggregations form in the open sea? I perceive that there is no single cause, but instead the process of aggregation, whether around dolphins or coffee cups, is ecological. For an individual member of such an aggregation, the cause may be protection from predation, availability of food, removal of ectoparasites, reproduction, or more than one of these reasons. That is, the niches of the participants differ.

Cover is the scarcest of commodities in the open sea (Williams 1964). For a small fish in the open sea, a knot of rope may provide protection simply because few oceanic predators are designed to turn sharply enough to catch a small fish hovering behind debris (Gooding and Magnuson 1967). Individuals of some species may hover behind one another as the aggregation grows, which provides a measure of protection. The game in the open ocean is not to avoid predators altogether; that is not possible. For the prey organism, it is instead to make the presence and position of predators predictable enough that avoidance can be carried out.

What assembles itself is a transitory and fluid webwork of organisms, not a rigid interacting entity that comes to maintain its integrity once formed. It is instead a marine community that one can observe assemble itself. The idea and the process are extremely instructive to consider when one tries to understand ecological dynamics, ashore or at sea. To give an idea of both causes of assembly and how the various organisms find their niches, here are some observations that have been made of these multispecies aggregations. Such aggregations form either around stationary floating objects or with certain species of constantly swimming dolphins. A different suite of species forms the aggregation in each of these two cases, apparently relating to the species ability to stay up with the swimming dolphins. This limits the retinue following a dolphin school to large species that do not suffer excessive energy expenditure in the process and to birds.

Included in multispecies aggregations is a considerable variety of species ranging from small fishes and invertebrates to dolphins, billfish, tuna, sharks, and several species of birds that move with the aquatic species in the air above (Murphy and Ikehara 1955). It is as if a dust particle had collected a drop of water and begun a rain storm. Each species, as it finds its place, expresses its individual patterns of life within a larger matrix. The process seems like an oceanic equivalent of the successional process ashore.

Tuna fishermen know this behavior well. Whenever floating logs are present, they may set their seines around them in what is called *log fishing* (Cosgrove 1991). They can make excellent catches of many tons of tuna from seemingly insignificant logs and other debris (Greenblatt 1976, Hunter 1968b, Hunter and Mitchell 1966, 1968).

Are these logs somehow a substitute for the dolphins in the bond between fish and mammal? The dolphins, after all, do not collect around the logs, while many fish species, including tuna, do. Tuna associate with logs in many parts of the ocean where they do not associate with dolphins.

I propose that such aggregations build from the initial nucleus of the object, log, or dolphin, then animals begin to associate with the nuclear object, probably because at first it provides partial protection for them. The combined mass itself then represents a larger object. This process, which Williams (1964) called *cover seeking*, tends to accumulate more members.

Plunging birds and their kleptoparasites, the frigate birds, are part of the multispecies aggregations, but they may still roost on land. The frigate bird has wettable feathers that prevent it from even landing directly on the sea surface (Kielhorn et al. 1963). Thus, these avian members leave the aggregations at dusk and seek out new ones again the next day. Dolphins, who must seek out scattered patches of mesopelagic prey, I propose, are nuclear in moving aggregations but are absent when aggregations stand still, such as around a floating log.

Gooding and Magnuson (1967) performed a direct experiment involving the assembly of multispecies aggregations at sea. They floated in a raft from which an underwater viewing chamber was suspended, working in three different open sea areas, and kept track of the numbers and species of fish that accumulated around their chamber. The first Gooding and Magnuson tests were a series of 11 drifts, the longest for 52 hours off the Kona Coast of the island of Hawaii. Two longer drifts lasted 8 and 9 days in the equatorial current system between 150° and 155°W longitude. They chronicled a class of transient fishes, including flying fishes that did not stay with the raft, and the aggregation of two classes of resident fish species. The smaller of these resident fishes stayed in view most of the time, and the larger was composed of predators who tended to take lengthy sorties around the raft but who returned again and again.

One small species, the freckled drift fish (*Psenes cyanophrys*), was the first to appear and had the highest rate of accumulation. They noted that by the end of the second equatorial drift, 729 had been caught in a purse seine and several hundred had escaped. The fish arrived at the raft singly or in small groups. A total of 27 fish species were involved, and these ranged from a variety of juvenile forms, including four typical reef species, triggerfish, filefish, puffers, and four large predators including mahi mahi (*Coryphaena hippurus*), the wahoo (*Acanthocybium solandri*), and the whitetip shark (*Carcharhinus longimanus*). Ecological interactions such as attempted predation, parasite picking, and chafing (rubbing the body against the raft) were observed. Gooding and Magnuson (1967, p. 495) noted the following:

> At least nine species of fish, both large and small, reacted to the raft in a way that made them less vulnerable to predation. Typically, when a predator approached the raft, the prey formed a compact group very close to the understructure. When the predator left or ceased harrassments the prey again dispersed about the raft. The value of the raft to the prey was demonstrated by the fact that only one species, the amberjack, frequently caught fishes that had taken shelter under the raft.

When Gooding and Magnuson (1967) describe parasite picking, chafing, and varying levels and successes of predation, they describe, it seems, the first stages of such an emerging socioecological system. This appears to be the same protective use of an obstacle that Springer's dolphins showed when they collected against the side of his vessel. In both cases, the floating object reduced the options of the predator and simplified the environmental surveillance problem of the prey.

Multispecies aggregations also seem to emerge as weblike assemblages that include various ecological categories of participants (fig. 101). The niche partitioning that emerges corresponds to the ecological attributes of participants and their interactions.

Multispecies aggregations, including those containing dolphins, have been seen underwater. Divers from the Cousteau Society vessel *Calypso* reported that in dives under porpoise schools conducted over several years' time tuna and sharks were frequently associated with moving dolphin schools (Leatherwood et al. 1973).

Elements of multispecies aggregations were sometimes seen in Hawaii. Frigate birds were used by the dolphin collector at Sea Life Park Oceanarium as a means not only of locating dolphin schools but also of estimating the number of dolphins from a distance. From the number of such birds circling, he could estimate approximately how large the dolphin school was likely to be underneath (Norris 1974, Norris and Dohl 1980*b*). As noted earlier, on occasion our divers in Hawaii saw sharks in close proximity to open water spinner dolphin schools. It is a common observation on the Kona coast that schools of pilot whales (*Globicephala macrorhynchus*) are accompanied by ocean white-tipped sharks (*Carcharhinus longimanus*), the association being constant enough to be reasonably predictable. The tuna in Hawaii, though present, swim deep in relation to a very deep thermocline, and any association between dolphins and these fish, if it exists, was out of our sight.

Unlike the associations of Hawaiian spinners, some oceanic multispecies aggregations, including some that involve spinner dolphins, live in "downstream ecosystems" in which the basic supply of nutrients derive from upwelling, ridging, or divergence zones associated with currents or even with deep water topography that causes nutrient-rich water to reach the surface and then to drift with surface currents, sometimes far from the original source (Au et al. 1979, McGowan 1972).

A reasonable terrestrial parallel to these aggregations has been described for migratory African ungulate herds, which follow scattered rainstorms and the grass it grows and are themselves accompanied by a retinue of carrion-eating birds and mammals, as well as by a complex of predators and prey (Jarman 1974). The grazing ecosystem of ungulates and grassland resources on the plains of Tanzania and Kenya is itself a reaction system of animals of various sizes and digestive efficiencies who crop the grassland system in a regular order of exploitation and who prepare the grassland itself for each other (Bell 1971). The emergent system includes social responses to the energetics of the various participants and to the nutritional value of the quality of food available for various seasons.

Cody (1971) has described another such system in the American deserts, in which multispecies flocks of finches selectively crop seeds in relation to varying annual rainfall patterns. These birds widen or tighten the random paths of seed search that optimize food availability for the flock members under differing conditions of drought. Each member takes its food from a slightly different part of the available ecosystem.

It also seems obvious that these moving terrestrial aggregations affect ecological conditions in ways that extend beyond their immediate presence, for example, through cropping of grass resources or the fertilization of the soil. Communities, such as those in the soil in the terrestrial example, can be influenced by the passage of a multispecies aggregation. This in turn affects food resource production.

Spinner dolphins associated with islands, as in Hawaii, are partially locked in place compared to populations of their species living in the tropical current systems or in the northward-moving Kuroshio Current off the coast of Japan. Their island, in effect, becomes their floating log, and aggregating to it are not only the dolphins but also their resources for life and a complex of ecological associates. Perhaps because some of the dolphins retreat daily to rest coves along shore, the elements of multispecies aggregations are not so evident in them as they seem to be for populations living wholly at sea. Yet we know precious little about their associates during nighttime feeding or about the fluid population that seems to remain offshore.

It could be that the very simplicity of assembly of a socioecological system in open three-dimensional water can tell us trenchant facts about how similar systems are assembled elsewhere. All one has to do is study the progress of assembly of species, just as Gooding and Magnuson (1967) began to do. But a longer term observer faces difficult observational problems in the open sea, and these have yet to be surmounted sufficiently well that multispecies aggregations and the way they are assembled can be fully understood. Nonetheless, one can imagine carrying out direct experiments on floating objects that might show the dynamics of ecosystems with unusual ease and clarity.

DOLPHIN PREDATORS

Two major classes of large predators seem of importance to Hawaiian spinner dolphins: certain large sharks and large predatory marine cetaceans, including the false killer whale (*Pseudorca crassidens*), the pygmy killer whale (*Feresa attenuata*), possibly the short-finned pilot whale (*Globicephala macrorhynchus*), and more rarely, the killer whale (*Orcinus orca*).

Shark predation on dolphins, especially the bottlenose and spotted dolphins, has been described in some detail by Wood et al. (1970), who note that porpoise remains are often found in the stomachs of the tiger shark (*Galeocerdo cuvieri*), the dusky shark (*Carcharhinus obscurus*), and less often, the bull shark (*C. leucas*). In tuna nets, other oceanic sharks such as the oceanic white tip shark (*C. longimanus*) and the hammerhead shark (*Sphyrna zygaena*) are encountered. For Hawaiian waters, Gosline and Brock (1960) list the tiger, the white tip, and hammerhead shark, and all of these are doubtless familiar species to the spinner schools that travel the same waters.

The great white shark (*Carcharodon carcharias*) has been implicated in attacks on cetaceans, especially in cool waters (Arnold 1972). The species has been taken on the Kona coast of the Island of Hawaii on nearshore shark lines baited with dolphin flesh, although there is no direct evidence of their predation on dolphins there (Albert Tester, pers. comm.).

The presence of cetaceans predatory on dolphins brings an entirely new dimension to the spinner dolphin's predator–prey equation. While sharks are at a sensory disadvantage relative to spinner dolphins, this is not the case for mammalian predators. Sharks have to rely upon sight, probably low frequency hearing, and perhaps their sense of smell to locate dolphin schools. Large predatory cetaceans, however, have at least sensory parity with the dolphins, for they, too, can seek out dolphin schools by their own echolocation and they can listen to the range of sounds the dolphins emit, while the sharks cannot. All three of the cetacean species implicated by Perryman and Foster (1980) in cetacean predation are commonly seen on the Kona coast feeding grounds of the Hawaiian spinner dolphin (Shallenberger 1991).

Perryman and Foster (1980) describe the relationships of the false killer, the pygmy killer, and the pilot whale to dolphin schools on the yellowfin tuna grounds, including obvious fear reactions and attacks, especially from the false killer whale. One is led to believe that these species may all be dangerous adversaries to dolphin schools. This is supported by observations in captivity where two species have been held together. Note, however, that these captivity data indicate only possible relationships because two such animals that might otherwise avoid one another are being forced together, and other relationships, such as those between humans and captive cetaceans, are similarly distorted.

Pryor et al. (1965) have described experiences with a captive pygmy killer whale at Sea Life Park Oceanarium in Oahu, Hawaii. After describing the initial aggressive reactions of the pygmy killer whale upon being transported and introduced to captivity, they noted the following (p. 453):

> Ten days after capture the animal was moved to a tank containing two pilot whales (*Globicephala sp.*). It became much more active, swimming almost ceaselessly. It began swimming with the smaller immature pilot whale. Occasionally it would approach this whale at right angles; the pilot whale always evaded actual contact with a little spurt of speed. In the evenings, the pygmy killer whale was seen occasionally chasing the small pilot whale. One morning the small pilot whale was found dead. Autopsy revealed that it had been killed by a single powerful blow to the temporal region of the cranium, possibly a lethal butt from the pygmy killer whale, since the only other occupant of the tank, another pilot whale, was very weak and lethargic at this time. There were no other significant bruises or tooth marks.

Later the pygmy killer was moved in with a dolphin (our records list it only as *Stenella* sp., which could have been either a spotted or a spinner dolphin). This dolphin (referred to as a porpoise in the following quote) seemed unafraid of the pygmy killer whale until the whale became clearly aggressive, whereupon

> . . .the porpoise very evidently became frightened. This aggressive behavior consisted of short dashes from the center of the tank at the porpoise,

which action set the porpoise swimming in fast circles around the tank while the little whale pivoted in the center. In the adjoining tank several other individuals of *Stenella* also became excited and dashed about, possibly due to distress signals reaching them through the wooden gate that separated the two tanks.

Beyond these hints of aggressive temperment toward other dolphins, nothing is known of the ecological position occupied by this little whale.

Pilot whales are capable of eating quite large prey and may be a threat to spinner dolphins, especially to their newborn. A record of a common dolphin giving birth to a stillborn calf in a tank containing both false killer whales and pilot whales is of interest (Brown et al. 1966). The authors report the following (p. 12):

> The striped dolphins and the false killer whale followed the laboring female. The dolphins showed particular interest and nosed the female's abdominal region on several occasions.
>
> The dorsal fin of the calf appeared to obstruct its further passage. In normal births the dorsal fin folds at its base either to the right or left, but in this case it remained erect and caught internally at the apex of the vaginal introitus.
>
> At 12:15 PM one of the striped dolphins grasped the fetal tail flukes in its mouth and withdrew the infant from the parental birth canal. A discard of amniotic fluid and a little blood followed delivery. . . . Our common dolphin, attended by the striped dolphins, carried her dead infant's body to the surface. These efforts were, however, terminated by the male pilot whale, who seized the body by its head. The pilot whale devoured the small cadaver, entire, after carrying it to and from the surface for 38 minutes.

THE COOKIE CUTTER SHARK

Another predator feeds upon discs of flesh and blubber cut from the sides of dolphins and whales throughout the warm water world. This is the curious little bioluminescent squaloid shark (*Isistius brasiliensis*), colloquially called the "cookie cutter shark" because of the shape of hole it makes in its prey. It is thought to be a squid mimic, attracting feeding dolphins to it (Jones 1971). It has teeth only in the lower jaw, which are used in its attacks to scoop out discs of blubber and flesh. Perhaps the shark's peculiar dental and branchiostegal apparati are used in conjunction with the flow of water past the two animals to twist the shark and thereby cut the crater wound.

Scars from this shark are frequently seen on the cetaceans of the Hawaiian area, especially the offshore dolphins and toothed whales including the false and pygmy killer whales and the roughtooth dolphin. Spinner dolphins are also common targets, and many dolphins were seen during our study bearing healed pinkish white scars from attacks by the animal or sometimes open, circular 5-cm (2-in.) diameter wounds that had yet to heal.

PARASITES

We will not attempt a review of the parasites known to afflict spinner dolphins, but instead refer the reader to Dailey and Brownell (1972) for a list of records and citations. These records include eight genera of trematode, cestode, nematode, and acanthocephalan worms as parasites of the spinner dolphin. Dailey and Perrin (1973) describe a new species of spirurid nematode and a trematode from the spinner dolphin.

The life patterns of spinner dolphins seem to promote the transmission of parasites, viruses, and bacteria through the air, water, and milk or via the placenta from the mother (recognizing that endoparasites typically require vectors for transmission, some quite complicated). Not only is it certain that schooling dolphins breathe each other's exhalate as they surface near one another but also they swim regularly through each other's fecal material. It was our regular observation that as spinners began to settle into rest, trails of semiliquid greenish brown fecal material could be seen issuing from the dolphins as they swam along, many animals passing through the clouds of material as it dispersed. Geraci et al. (1978) have investigated the question of transmission of dolphin parasites and pathogens and have reported on school-wide infestations, especially those of the mammary glands and adjacent tissue.

Perrin and Powers (1980) have investigated the role of the nematode *Crassicauda* sp. in natural mortality of spotted dolphins in the eastern tropical Pacific and have found that it could be responsible for as much as 11–14% of the projected mortality. This parasite causes bone lesions associated with the air sac systems of the skull that can become so severe that the brain is invaded. They also note (p. 962):

> Although related spirurid nematodes use intermediate hosts, potential transplacental transmission or transmission via milk of the nursing calf should be investigated. A closely related form (or even possibly the same form) may cause reproductive failure in Atlantic white-sided dolphins (*Lagenorhynchus acutus*) . . . through invasion of the mammary glands.

Ridgway and Dailey (1972) have investigated the role of trematode parasites that entered the brain in strandings of the common dolphin near Point Mugu, California. They found adult trematodes in the bile and pancreatic ducts, severe liver damage, and massive brain necrosis due to the presence of numerous trematode ova in the brain tissue.

Although these records are from the closely related spotted and common dolphins, Cowan and Walker (1979) have shown that the spinner dolphin seems to face similar problems from parasitism. These authors also investigated the importance of lungworms (*Halocercus delphini*) in spinner and spotted dolphins on the tuna grounds and found that "it is probable that no adult animal is without scars on the lungs." They found that a substantial portion of lesions associated with natural disease processes could be attributed to this organism, by recognition of viable worms or degenerated fragments of worms in the foci of inflammation or scarring.

We move on now to a comparative consideration of spinner dolphin society, especially looking at the commonalities and differences seen in behavioral and socioecological studies of higher terrestrial mammals. From this comparison, we attempt to assess the evolutionary place of delphinids such as the spinner dolphin in the sweep of the Mammalia as a whole. Another thrust is to assess the causes and results of dolphins having reached what we consider to be the cultural level of organization. In essence, we return to a question posed early in the book: what might dolphins, as large-brained mammals, do with the capability such features may impart to them?

SIXTEEN

Comparative View of Cetacean Social Ecology, Culture, and Evolution

Kenneth S. Norris

INTRODUCTION

This chapter covers what I consider to be the primary questions underlying this study. First, what do cetaceans do in their wild societies with their obviously high order cognitive development? What does it mean to the dolphin lineage that some species, at least, can perform mental abstractions and cross-modal transfers (that is, to be asked questions by visual means and indicate the answer by use of sound, an uncommon faculty in animals)? What does it mean that dolphins can remember long strings of unrelated numbers, perform high speed vocal mimicry, and be successfully trained to use symbols of speech, even if they do not seem to use them themselves (Herman 1980a, Pilleri and Gihr 1971)?

Second, in terms of evolution, how did they get that way? How can it be that cooperative and even altruistic patterns are so widespread in dolphin society, apparently among nonkin and kin alike, when theory argues that it is difficult to explain how animals that are not direct relatives can develop such patterns? Where does the idea of reciprocity fit into all this (Trivers 1971)? Do we need to think of the nonrelated giver of cooperation being locked in a system in which it must receive a relatively direct payback for any cooperation it gives? For reasons outlined in this chapter, I do not think either of these problems present evolutionary difficulties, given that dolphins, and apparently many other mammalian species and some birds, have achieved the cultural level of social organization.

In the course of this evolutionary discussion I will compare the population structure and social arrangements of the spinner dolphin with ecologically equivalent primate taxa ashore.

Third, from the natural historian's view of spinner dolphins developed throughout this work, I will attempt to present the major descriptors of their lives. What has been required or has developed that allows these mammals to nurture only a few young and live long lives at sea in an environment where nearly every other taxon persists by the opposite strategy—that of producing large numbers of rapidly growing young?

Defining Some Terms

First, three terms require definition, as used in this chapter: socioecology, culture, and phenotype. By *socioecology,* I mean a collection of species occupying a given area, which through the evolutionary process, have apportioned the available food resources among themselves *and* which respond to environmental change in this area in a variety of ways that ramify through many aspects of social arrangements and behavior of the component species. In other words a socioecology is an ecosystem with behavior and environmental change taken into account. I see socioecological systems as having been arrived at by a long period of coevolution between their component species, resulting in dynamic arrangements that allow the participants to extract the food resources they need at any place and time. Each of the member species has its own labile geographic range. A socioecology is, therefore, not a "thing" with circumscribed boundaries and participants but instead a reaction system composed of semi-independent players.

One example of a socioecology is the grazing ungulate-centered system of plant and animal species that utilize the African savannah, which includes grazing animals, carnivores, carrion feeders, parasites, and their environment. Another is the ecologically partitioned traveling association of oceanic dolphins and various other species of animals that accompany them—the birds, sharks, billfish, tunas, and food species on which they collectively feed. I will use the terrestrial ungulate system for comparison throughout this chapter.

The hallmark of these systemic states is that the composition, movements, social arrangements, and deployment of the component species are responsive to local ecological change within given limits and that the component species and details of their interactions can be expected to vary geographically and temporally.

Some reviewers have confused my use of this term with the similar sounding but greatly different term *sociobiology.* The sociobiological doctrine (Wilson 1975), which emphasizes the linkage between the genome and both social structure and behavior of animals, is now seen by many workers (including myself) to be a considerable oversimplification. Even when sociobiology was proposed as a new synthesis, it was understood by its proponents that a shadowy gulf existed between the genome and behavior, which remained largely unassessed. Since then, studies of molecular biology and development have made it clear that the gene–behavior gulf can be wide indeed, masking all the complexities of gene expression during development. So wide can be the gap that a considerable amount of evolutionary process can and does go on without immediate functional linkage between environmental selection and mutation (Borchert and Zihlman 1990). Nonetheless, I see sociobiology as having provided an important viewpoint that has pushed our understanding of the evolutionary control of behavior sharply forward.

By *culture*, I mean the collective concourse of ideas or concepts within a species; their origin, passage, and storage; and their consequences in higher animal society. Because ideas are constructed in the mind of the conceiver, they cannot be linked directly to the genome. Instead, they pass from individual to individual by instruction and observation, and they may change with each transmittal. They can be stored as memory, as traditions, or in iconography. I perceive the elements of culture in many higher mammals and some birds.

Later, I discuss at length the concept of the phenotype. I believe a misapplication of this concept has frequently hampered understanding of evolutionary process, particularly for cultural species such as dolphins. The problem, succinctly put, is that the entire ecological reach of a species defines the feedbacks from nature that shape it. This is what I call the *phenotype*. When we humans release compounds into the atmosphere that allow increased ultraviolet light to enter, and hence to cause increased cancer rates among us, we have altered the selectional environment upon our own species (and a host of others). I include that ecological reach in my definition of the phenotype, and do not restrict the term to the immediate exterior of the organism itself, as has so often been done.

Dolphin Variation

Not all dolphins are nearly alike. Is it correct to discuss these concepts with reference to spinner dolphins when most of the cognitive measures have been determined for bottlenose dolphins, a species that adapts rather easily to captivity, is easily trained, and shows singular flexibility of behavior (see Gawain 1981)? The spinner *is* a considerably different animal. It is much more completely a group animal. Held alone it is lost, frightened, and totally dependent upon its handlers. But we have had experience enough to know that if spinners are placed with sufficient members of their own kind, they learn easily, they engage in imitation, and their memories are excellent and not unlike the bottlenose dolphin (see Bateson 1974). If perhaps not a perfect match for the bottlenose dolphin, they certainly classify as large-brained mammals living in highly cooperative societies that show evidence of a cultural level of organization. Why and how might have this animal evolved in this higher mammalian salient in the sea that is almost completely isolated from other mammalian groups?

Rhythmic or Oscillatory Behavior

One more point should be raised before beginning this discussion. This chapter ends with a description of how I see a spinner dolphin's day-to-day and year-to-year life as being organized by a series of contrapuntal behavioral and physiological rhythms that may be the natural result of a society of mammals organizing itself in three-dimensional open space.

Oscillation, or repetitive change within defined limits, is definitional of life. Every living process, every living organism, and every living aggregation oscillates at various periods about equilibrium points that can shift over time. An animal's life is organized into a complex of features such as sleep, migration, sociality, and temperature regulation, none of which are totally rigid all-or-nothing processes or events. A migration is eased into over time, with the animals showing increasing signs (or intention movements) of the impending migration, followed by the event itself. A dolphin social exchange can oscillate between an old and a new pattern as behavioral states change. This oscillation is required for life to persist because the environments in which living things occur also fluctuate. For example, the body temperatures of animals oscillate about central points during the processes of temperature selection, acclimation, and acclimatization (see Bullock 1955, Fry 1947, Norris 1963). The lives of spinner dolphins are clearly organized by such cyclic change. They oscillate into and out of nearly every behavior pattern that has been examined in any detail.

To approach these ideas, I will first set up a framework for how I believe cetacean and other mammalian socioecologies and cultures have been shaped. The socioecology of most concern here consists of the spinner dolphin itself, its other cetacean associates, its predators, parasites, food, and scavengers that occur together in the tropical current systems and around tropical islands of the major oceans. This is no casual gathering of animals. I imply that the component species have coevolved with relation to each other, allowing the species to fit together easily into a system of species that has a great deal of coherence even though its composition may change from place to place and time to time.

SELF-COMPENSATORY SYSTEMS

There is a growing understanding that the living world is shaped according to the rules of systems theory; a world of interaction among the many living species comprising natural systems, their aggregations, and the environment. I perceive that for a given animal, a socioecology is but one of several different self-compensatory or oscillatory systems that either directly alter the state of the organism or involve it (see Hofstadter 1979). Physiologies, for example, being within individual organisms, are at one level of organization, and socioecologies, being composed of various species of organisms, are at another level. What happens at one epigenetic level of organization is only indirectly related by measures of fitness to events on other levels.

A direct consequence of oscillation is that life processes at all levels are arranged in self-compensatory systems. The entire balance of every socioecology is in constant flux relative to the environment. For example, if rainfall is heavy over the area occupied by a group of species, one pattern of species distribution may appear, whereas if aridity sweeps in, the component animals may shift toward another pattern, involving differential movements by the component species and for the species to persist these patterns must not only change, but oscillate.

I contend that socioecologies, physiologies, and the design of an animal's multicellular architecture are *emergent* cybernetic or compensatory systems that have arisen because as an animal, a physiological process, or a structure adapts to its environment, it carries with it this capacity to regulate its short-term circumstances within set limits. The collective capacities for change of the component organisms of a socioecology, the systems of a physiology, or the patterns of a structure thus come to produce collective response systems that, in the short term, *can temporarily change independently of the underlying genome.*

ARE SOCIOECOLOGIES EVOLUTIONARY ENTITIES?

The degree to which socioecologies are found to be evolutionary entities may help to resolve such arguments as, "Are communities of organisms epiphenomena?" (Hoffman 1979, 1983, Webb et al. 1987, Webb 1988). Hoffman (1979, p. 357), for one, denies the reality of communities on the basis of paleontological evidence with these words:

> The basic assumption of community paleoecology is that communities or biocoenoses represent a distinct, real level of biotic organization achieved through ecological integration of a coevolution among the species. This assumption seems invalid for two reasons. (1) The actual degree of community integration is in general insufficient to induce any driving forces for a structural development as predicted by systems theory. (2) The concept of biological reality and distinctness of the community level of biotic organization implies assignment of a significant role to group selection.

I disagree. My answer, which I argue in the pages that follow, is that "Yes, these fluid groupings of species are in fact, epiphenomena, but they are not without evolutionary coherence and that one need not invoke group selection to explain their structure." I will attempt to show that their component organisms have fit their tropic patterns together and that these systems involve a great deal of behavioral integration, a feature notably difficult to interpret from paleontological evidence. This does not mean that the component organisms cannot be found elsewhere but only that the fit between organisms in a community or socioecology is the result of coevolutionary process and has involved considerable fitting together. One animal in any such grouping becomes an aspect of the environment of the others, and hence coevolution of both occurs without requiring recourse to the ideas of group selection.

THE ORIGIN OF CULTURES

As mentioned earlier, I believe cultures to be the result of the evolutionary arrival in various lineages of higher vertebrates of the capacity to form mental concepts or ideas. This follows the arrival of the capacity for mental abstraction in evolution and dictates its considerable consequences. By *abstraction,* I mean the act of "putting two and two together" by an animal; the ability to make predictions. A spinner dolphin listens to the signals of the outlying dolphins of its school, moving out of sight in the dark, and constructs a mental picture of the shape of the school, its speed, and its direction. It may make a projection about where a good rest cove is and how long it will take to get there. The mental constructions of cultural animals are incorporated into the life patterns of their species through teaching of and observational learning by the young.

I see the elements of culture as widespread in homeothermic animals, including both birds and mammals. I contend that spinner dolphin society involves both socioecology and culture and that both have had major roles in shaping the general genic evolution of the group.

Understanding is often clouded by attempting to assign a direct genetic encodification to the states of epigenetic systems, such as socioecologies and cultures, because each is a semiautomous system and systems are, as I have outlined, responsive entities that change constantly within set bounds. Outside influences, such as that from the genome, act largely to change the set points of oscillation or the players in such systems, but not the momentary state of the system itself. It is for this reason that difficult to explain levels of cooperation are found in many higher animal societies, a point I return to later.

It took me a long time before I was comfortable with the idea that an epigenetic system such as a socioecology could function as an analog computer does; that what we see of the players in a socioecology at any given moment is one state among a graded series of possible states; and that it is almost irrelevant to attempt to calculate gene frequencies or to link such momentary states directly with the genome. The analog design of socioecologies everywhere makes it especially difficult to point to causes since they shift with changes in the environment. To understand the momentary state of a species, one needs to form a broader overview of the adaptive circumstances of an animal's life rather that to seek unitary cause and effect correlations.

COMPARISON OF MARINE AND TERRESTRIAL MAMMALS

When we began this study we could not answer the question: Is a mammalian society living wholly in the ocean organized differently from one at a similar level of social complexity ashore? My discussion is condensed by comparing the society of spinner dolphins with just three ecologically equivalent primate species. Major sources for my various comparisons include an excellent summary volume on primate social behavior by Smuts et al. (1987) and such works as Cheney and Wrangham (1987), Eisenberg (1981), M. Jarman (1979), P. Jarman (1974), Mech (1970), Geist (1971), Kavanau (1987), and Walther (1984), as well as the cetacean literature cited throughout this book.

It is notable how broadly applicable the same major behavioral and structural descriptors are to both the marine and terrestrial species and how only the degrees of emphasis of certain features are obviously different or unique. This is true in spite of the fact that cetaceans and terrestrial mammals have evolved wholly isolated from each other since early in the Cenozoic radiation of mammals (archaeocete cetaceans were established at sea by early Eocene time, about 50 million years ago) (Gingerich 1977, Gingerich and Russell 1981, Gingerich et al. 1983, Barnes and Mitchell 1978, Barnes 1984).

The mating systems of terrestrial and marine mammals show a similar range of architecture even though the cetacean lineage has evolved mostly in water completely away from any significant ecological contact with land. There are indeed a great many features of the "structural grade of mammalness" that have played themselves out, whether they developed on land or in the remote open sea.

Obviously not everything is the same between the two groups. I perceive four major sources of difference between terrestrial and aquatic lineages. First, there are large *scaling*-related differences between terrestrial and marine mammals (Peters 1983, Schmidt-Nielsen 1984), especially related to body size and body architecture. But where the size or surface area of a structure, such as the brain, is less directly controlled by the environment, such scaling is less clear.

Scaling differences between terrestrial and marine mammals spring especially from the more rapid heat loss of a warm-blooded mammal in water as compared to air and from the high viscosity and density of the water in which dolphins live compared to air-dwelling creatures. Water buoys up animals (whose density is near 1), while they are not significantly supported in air, which has a density that is 800× less (Huntley et al. 1987). Also, crucially involved are the hydrodynamic "packaging" both of body form and arrangement of parts in marine mammals and the differences in size-related drag upon a moving animal between air and water (see chap. 10). These factors act together to shift the size range of cetaceans far upward from that of terrestrial mammals filling comparable ecological niches.

Socioecological parallels exist between spinner dolphins and small primates, especially tamarins, marmosets, and titi monkeys (*Saguinus, Callithrix,* and *Callicebus*) of the tropical forest canopy, although their average adult body weights differ by a factor of about 90–420× (160–750 g for tamarins and marmosets [Goldizen 1987] versus 67.7 kg for an adult male spinner dolphin [Hampton and Whittow 1976]). The tamarin, an arboreal monkey with many socioecological parallels to spinner dolphins, has a brain weight of a few grams, while that of the ecologically equivalent spinner weighs about 400–540 g (Perrin et al. 1977*b*).

A second major difference between land and aquatic mammals is that dolphins have an especially heavy reliance upon the acoustic modality for navigation and communication (Nachtigall and Moore 1988). The acoustic sense seems crucial to dolphins for the operation of open water schools and for maintaining a favorable balance between predator and prey, especially in these long-lived mammals that cannot afford to sacrifice members very often. Equally important has been the development of the schooling mode of deployment, which imparts protection to the dolphins in times of direct threat once the acoustic shield has been breached (see chap. 12).

This is not to say that the acoustic sense is not important for many terrestrial mammals in much the same way, only that the odontocetes appear to have refined echolocation and listening capacity to a high level. Without such capability, it seems improbable that dolphin societies could have invaded an open sea environment populated by predators larger than themselves and there set up a long-lived mammalian society (Norris and Evans 1988).

Third, an essential problem is for prey species (the dolphins) to carry out extended lives, frequently more than 30 years in length, in the three-dimensional accompaniment of predators larger than they and sometimes having the advantage of greater speed (but perhaps not greater elusiveness or endurance; see chap. 9). I suspect that this unusual pop-

ulational arrangement reflects the past terrestrial history of cetaceans, a
holdover from the formation of these patterns on land. By this I mean
that a mammalian society with a very long life cycle, much involved with
instruction of young, that lives all its life at sea, is a great anomaly in
open water, and seems to me likely to have carried with it many features
established on land. The open ocean dolphins appear to have achieved
this by partitioning their lives between times of immediate predator
avoidance, when they act like fish, and times of sociality, when their essential mammalness emerges most clearly. The social patterns of dolphins appear to be expressed most fully when their echolocation and
sensory integration systems tell them no predators are a direct threat.

You might note that many open water fish form schools and ask,
"How do they survive as species?" The difference seems to lie in the divergence of reproductive strategies. Schooling fish typically produce
large numbers of eggs, and many live short lives (5 years is old for a yellowfin tuna, for example) (Cosgrove 1991). The equation shifts from the
importance of the individual and its nurture, in the dolphins, to the statistics of numbers and survival rates in such fish.

Most land mammals or even flocking birds that seek protected roosts
at night do not have to deal constantly with the third dimension of *unimpeded* predator approach to the degree that dolphins do. Nonetheless,
terrestrial animals face the reverse problem more fully than do cetaceans: their predators can use terrain for surprise attacks much more
than can predators upon open water dolphin schools. The equation becomes similar for the two groups when we consider shallow water cetaceans and terrestrial animals or ones like the Hawaiian spinner dolphin
that traverse shallow water daily. Without the sensory advantage of
echolocation, which has allowed oceanic dolphins to detect predators
early and hence make their capture energetically expensive, long-lived
open water dolphin societies might not exist.

The closest terrestrial example to the open water cetacean condition
appears in the mammals of the forest canopy who are subject to predation in three dimensions. However, these animals do derive some protection from limbs and leaves and at the same time face stealth to a degree
unlikely in open water. This is not to say that there are no aspects of protective topography at sea, only that they seem much less pronounced
than on land. As described earlier, waves and turbulence produce an underwater topography of bubbles, thermoclines reflect sound, and rain
storms may greatly increase a clutter of background noise (see chap. 7).

When an antelope herd localizes the position of a predator outside
the periphery of the herd, the problem of predator avoidance is two-
dimensional rather than three-dimensional. I was impressed with this
difference when watching a lion stalk a zebra herd at Tarangiri Park,

Tanzania. By a higher than usual repetition rate of its barking, the zebra herd localized the lion for all its members, and the vast majority of herd members went about their grazing and socializing, even though the lion was remarkably close by. Predators that operate in coordinated groups, such as lions often do, may negate this advantage to a degree by distributing attack at more than one locus. Such group predation is a strategy also seen at sea (Major 1978).

Finally, the fourth difference is that the terrestrial environment can be very fine grained compared to environments at sea. In the open sea, large scale, even global processes are often regulators of pelagic populations and their distributions. This is not to say that the environmental responses of spinner dolphins do not vary locally; they do in some places. Around the earth, the spinner dolphin maintains some populations wholly in the open sea current systems, while others associate with islands such as described in this book. Still others move back and forth on annual migrations within north–south currents, such as the spinners of the Kuroshio system off Japan (Kasuya et al. 1974, Miyazaki and Nishiwaki 1978), and still others live in shallow water, such as in the Gulf of Thailand. (Perrin et al. 1987). The spread in maximum spinner body size seems directly related to the warmth of the environment in which they swim, that is, to their energetic balance. The tiny Gulf of Thailand spinner, perhaps the smallest of all dolphins, lives in very warm shallow water, while the much larger northwest Hawaiian animals live in much cooler, often much deeper water.

ADAPTIVE RADIATION OF WARM WATER CETACEANS

When one reviews the social arrangements found among the marine members of the 78 species of the modern Cetacea, it becomes obvious that a broad range of patterns exists. These range from the highly cooperative almost cryptically dimorphic societies of Hawaiian spinner dolphins to the large and highly polygynous sperm whale, in which adult males can be about twice the length of adult females (Norris and Møhl 1983), among the largest sexual difference of any mammalian species.

Resource partitioning is also obvious among the various cetacean species that make up the warm water pelagic cetacean fauna of which spinners are a part. Spinner dolphins are feeders on small abundant prey items that are continually renewed throughout the year, either by islands interdicting subsurface water movements or by current dynamics in the open sea. Some cetacean species from the same tropical assemblage of species feed on deep sea squids, others on large fish, and still others on cetaceans. Sharks may occupy a trophic role similar to that of hyenas in ungulate-centered systems—at times predatory but perhaps more often feeding on weakening or dead animals.

The range of trophic and socioecological patterns associated with these cetacean faunas are probably about as great as those found among the major terrestrial radiations such as those of ungulates and primates (see Gaskin 1982). In other words, the ecological constraints upon population parameters may extend over comparable ranges on land and at sea, and the range of social patterns expressed may also be similar. Where does the spinner dolphin fit into this range of socioecological expression?

SPINNER DOLPHIN SOCIETY

One can typify Hawaiian spinner dolphin society by saying that it is a mammalian group process–oriented society in which sexual dimorphism is minimal and that feeds on small, vertical migrant prey. It takes experience to be able to differentiate Hawaiian spinner dolphin sexes quickly in the field, and the skill is based mainly on differences in behavior between the sexes. Of such behavioral dimorphism as can be observed, much is voluntary, hence allowing great uniformity between sexes when needed (see chap. 14). A direct correlate of such physical and social symmetry is that males and females can each take a considerable range of roles within the school, some that are relegated wholly to one sex in more polygynous societies. Both sexes of Hawaiian spinners are involved in birth, nurture, instruction, and the cooperative defense of the school. Another correlate of social interdependence is strong expression of the dolphin's version of grooming—caressing.

Cooperation has reached a high level in spinners (and probably also in other smaller open sea cetaceans), matching any terrestrial group, including the highly cooperative marmosets, tamarins, and titi monkeys. I believe this to be a direct correlate of the high potential for predation faced by these dolphins, both from sharks and predatory cetaceans such as false and pygmy killer whales (see chap. 13). It also reflects the fact that oceanic cetaceans seem to have little surcease from the proximity of predators, living permanently as they do in three-dimensional open space, often in multispecies aggregations with their predators. Their lives appear to revolve around maintaining a geometry with predators that avoids nearly all successful attack. By virtue of their echolocation capability, the dolphins are energetically too expensive to pursue, that is, predators are kept far enough away that attack sequences are probably not often undertaken. Echolocation and the integration of sensory input provided by the school, which together provide an "early warning system" for the dolphins, are crucial in maintaining this equilibrium (Norris and Schilt 1988).

As is true in other species of small open ocean cetaceans that form large schools, the result seems to be strong submergence of individuality in spinners in favor of a group process when under attack. Under such circumstances, decisionmaking appears to pass from the individual to the superorganism, or protective mechanisms of the school, mechanisms which allow predator evasion (see chap. 13).

This subordinance of mammalian patterns during defense from predators seems to reduce spinners when under direct threat to a social level comparable to that of schooling fish. Only when predators are kept an energetically expensive distance away is it likely that most of the basic mammalian behavior of spinner society can emerge.

The part of the protective behavioral armament of the spinner dolphin for which I know of no parallel among terrestrial mammals is the apparent division of the basic protective duties of the school into *bouts,* in which nearly all members, apparently regardless of familial relationship, seem to share certain duties of the school on a rotational basis (see chap. 14). Because in spinners this pattern seems to occur in schools that are "societies of friends" and not primarily kin groups and because it depends upon the ability to generalize, bout behavior it is most likely a cultural phenomenon, a point to which I will return.

In this assessment of dolphin society as a predation–pressure driven system, we must bear in mind that the events of successful predation are just as hard to observe with these dolphins as they have been with primates and other large mammals (Cheney and Wrangham 1987). In the 35-year life span of a spinner dolphin, actual attack must be rather rare, and yet I believe that it continues to exert much pressure in the shaping of dolphin society. The common scars noted on Hawaiian spinners that seem attributable primarily to medium to large sharks (Norris and Dohl 1980a) attest to a significant level of "near misses," at least in the relationship between an island-frequenting dolphin and sharks.

The availability of food also seems to have been a shaping force in spinner society. A reliable, constantly renewed food supply seems available for Hawaiian spinners, though perhaps of considerable patchiness. While we have not chronicled seasonal shifts in feeding strategy, we have shown that spinners spend significantly less time in bays in winters than in summer (chap. 3), which may indicate a lower winter replenishment rate for food species and the necessity of longer feeding times. Nonetheless, repetitive use of the same coves by relatively fluid assemblages of dolphins occurs throughout the year. Overlain on this pattern seems to be a continual flux of new occupants to and from the immediate coast.

Accumulations of food are probably found by the group process of forming large coalesced feeding schools. A wide front of echolocating dolphins certainly has a better chance of finding food patches on the island slopes than does a smaller group.

The circumisland shifts of spinner populations we observed probably occur because rest sites nearshore are not useful for occupancy when water becomes turbulent and dirty, thus causing the dolphins to seek a lee shore. Visually oriented resting dolphins may not be able to countenance dirty water, probably because it gives an advantage to predatory sharks who can thus approach close enough to afford attack.

THREE-DIMENSIONAL SOCIETIES

Let me now propose some additional parallels with arboreal social mammalian populations ashore. As I have mentioned, the most striking parallels seem to be with the new world tamarins (*Saguinus*) (Goldizen and Terborgh 1987), marmosets (*Callithrix*), titi monkeys (*Callicebus*) and perhaps the lesser known old world (African) talapoins (*Cercopithecus talapoin*) (Gautier-Hion 1973).

The concordances between the spinner dolphin and the small primates—the marmosets, tamarins, and titis—are more interesting than similarities between spinners and other primates since they are correlated with similar important descriptors of their respective socioecologies. For example, both the spinner dolphins and these primates live in three-dimensional environments with predators capable of coming from most directions, and both appear to live with predation pressure as a constant factor in their lives. The food supply of both appears to occur in rather large patches relative to the needs of a given individual, and these supplies are scattered widely across the feeding areas and are replaced regularly. All of these features, it seems to me, feed into the muting of sexual dimorphism and into the promotion of group processes in these animal societies, including various aspects of cooperation. Now let's consider these similarities in more detail.

These primates are all the smallest, or among the smallest, of all the primates living in their given habitat, which is the forest canopy (Goldizen 1987), just as the Hawaiian spinner dolphin is in its environment. Scaling factors, which were described earlier, may come close to equating these animals of such different absolute size. Similar parallels are found in the slightly larger capuchins and squirrel monkeys (Robinson and Janson 1987).

The forest canopy is clearly the terrestrial habitat having the closest parallels to the open water environment. Events take place in three dimensions, and the primates or dolphins living in each can respond in terms of the three-dimensional geometry and placement of their members. Predators can come from multiple directions. Aerial predation seems important for these primates, and attacks also come from within the canopy, requiring vigilance in various directions.

Spinners certainly require similar vigilance, although predators do not usually come from above (except for humans). The sea surface may be something of a trap for dolphins and other open ocean prey species because they cannot maneuver farther to escape a predator once they have leaped through it, perhaps to be met on the way down. A marmoset on the ground is probably similarly disadvantaged except that an aerial predator must avoid crashing into the earth. At any rate, within some constraints, the habitat of both types is three dimensional.

At least for new world marmosets and tamarins for which estimates of attack frequency have been made, predation pressure seems high, and these primates are constantly wary (Goldizen 1987). Eagles, wild cats, and snakes are all involved, and attacks averaging once every two weeks were recorded by Goldizen. Associated defensive formations with adult sentinels constantly on guard were also noted.

Resting of tamarins, like that of spinner dolphins, takes place in small, tight, silent groups, as if vocalizations can lead predators to them and as if visual mediation of group cohesion has supplanted contact calling during this period. Both spinners and tamarins occur at certain locales in mixed species groups but retain recognizable separation between the associated species. In both cases, this interspecific association may be related to feeding patterns.

It was suggested by Norris and Schilt (1988) that the need for silent rest by a dolphin may be partly due to the unusual stresses placed on the dolphin's vocalization system by the necessary production of tens of thousands of high intensity echolocation clicks during a single active period. Hence, a daily time for tissue restitution of the sound system could be necessary. However, such dolphin silence may also be related to avoiding being located during a time when the school stays in one place and can be approached with a minimum of energetic expense by a predator.

Both forest canopy primates and spinner dolphins appear to station adults close to juveniles, perhaps both as guards and instructors (see Goldizen 1987). We made frequent observations of the adult male spinners Four-Nip and Finger Dorsal in juvenile groups, as if this were a permanent or common role for such adult males, and we often encountered patroling male coalitions.

Elements of fission–fusion societies are seen in both these small primates and spinner dolphins (Ramirez et al. 1978). Both appear to fragment during the day into small groups scattered throughout the territory, only to reassemble at other times. In these terms, our study of

Hawaiian spinner dolphins may have dealt with a single "home range" (the entire island and the unknown offshore range), although we were unable to discern the boundaries and interactions with animals in adjacent home ranges.

The minimum group size relationships between tamarins and dolphins are also suggestive. The minimum group size is reported as about 2 to 13 individual for new world tamarins by Goldizen (1987), figures that are comparable to the basic reaction units of dolphin schools (our subgroups) (Norris and Schilt 1988). The lower numbers (2 to 4) match those found among shore-dwelling dolphins living in relatively two-dimensional habitats (shallow water and embayments) (Wells et al. 1980), but perhaps in auditory contact with other such groups, while groups of a minimum of 4 to 12 fit our observations of the open water spinner dolphin.

Evans and Norris (1988) have speculated that the minimum number of individuals in a functional school reflects a basic function of the short term memory in which the environment is subdivided for a given sensory modality into "seven plus or minus two" sectors of surveillance (Miller 1956). In other words, a dolphin can track only about this number of sectors or items at any one time, by short term memory operating in the "cognitive present"—a period of 1 or 2 sec into the future. It is within these time and perceptual limits that dolphins maneuver through their environment and attempt to evade attacking predators. This relationship could well feed back into the number of animals that can effectively take part in the smallest protective unit of a school or a troop.

Terbrough and Goldizen (1985) describe the mating system of tamarins as "cooperative polyandry." Nonreproductive matings have been reported for tamarins by Goldizen (1987). We have also seen that quasi-reproductive patterns are a marked feature of spinner dolphin society outside the mating season, so much so that we have come to perceive a nonreproductive function for such behavior in these estrous animals—that of "reaffirmation of relationships." Juvenile callitrichid monkeys have been noted to help with infant care, and we have given an example of a juvenile male spinner assisting during a captive birth (see chap. 14).

The carrying patterns of the young of forest canopy primates and spinner dolphins may be thought of as similar, although after a time young dolphins seem to wander more widely from the mother than primates apparently do. This is easier to do in three-dimensional open water than in a tree. Perhaps it is also related to the distance at which predators can be perceived. Among the primates, carrying is shared by males and females, but heavily by males, including juveniles (Terbrough and Goldizen 1985). Spinner calves use assisted locomotion, which involves drafting alongside an adult, while the inboard pectoral fin touches the adult and allows the young to be carried along without locomotory effort of its own (Kelley 1959, Norris and Prescott (1961). As in primate carrying behavior the costs mostly fall upon the adult animals. We have no data indicating the age or sex relations of such carrying, except to say that very young dolphins are often seen in this behavior with their mothers, and some adult males frequently or habitually act as alloparents (as described in chap. 14).

New world tamarin adults have been noted sharing large insects with infants. Such behavior is known for cetaceans, such as false killer whales and the rough-toothed dolphin, that feed on large prey, but it has never been observed in spinner dolphins, which feed on relatively small prey far out of sight below the surface. Nonetheless, spinners do share playthings such as floating algae and pieces of plastic blown into the sea (see chap. 12). It will be interesting to see if spinners provision young dolphins; an opportunity for observing this has not yet arisen.

Goldizen (1987, p. 42) makes an interesting comment about helping behavior among mammals generally:

> The most striking aspects of helping behavior (not exhibited by all helping species) are delayed reproduction by helpers (or permanent sterility), donation of food to young, [and] significant loss of foraging or hunting time due to "baby sitting" of young. These aspects of helping behavior are interesting because they are hard to explain in terms of individual selection; all at least potentially reduce the helper's reproductive success.

ARTHROPOD SOCIETIES

I found it deeply suggestive to find parallels between the organization of communal arthropod systems (Gordon 1989) and those of social mammals. The structures and dynamics of the societies of arthropods and higher vertebrates show important, even remarkable commonalities at the level of colonies (both ants and dolphins appear to live in group decision-making systems). However, it is my contention that at the level of individuals, trenchant differences that reflect the emergence of concept formation at the vertebrate level and its apparent absence at the arthropod level remain evident.

An example of commonality is ecological role taking in these two kinds of societies. For many years, it was believed that social insects live in rigid castes, each individual devoted to an essential task such as foraging, and that the behavior of a colony is the result of the numerical distribution of workers set to such essential tasks (Oster and Wilson 1978). Within a colony, the proportion of colony members belonging to a given caste was considered to be a fixed quantity that was set by the ecological milieu of the species. It is now emerging that this is widely untrue. For taxa in which workers are monomorphic (216 out of 263 ant genera), task switching is the rule, and it even occurs in some polymorphic genera (Wilson 1984, Gordon 1989).

In spinner dolphins and forest canopy monkeys, the distribution and mutability of trophically related roles may be determined by cultural rules or physiological state and, in broadest outline, appear to be much like those described by Gordon (1989) for most ants. For colonies of ants, herds of savannah antelopes, or schools of dolphins, many aspects of trophic role taking may vary in relation to environmental stress upon individual organisms.

TOWARD A MODEL FOR SOCIOECOLOGICAL AND CULTURAL EVOLUTION

When I attempted to compare the characteristics of mammalian taxa across wide phylogenetic distances, I found that a clear theoretical model was lacking for what might have shaped them in the two different environments. I also found a long-standing muddle about the various evolutionary relationships of the genotype, phenotype, and niche. These problems existed during my early training in the 1940s and still seem to plague understanding in the 1990s. Most workers continue to agree, as they had during the seminal debates about the "new synthesis" in the 1940s and 1950s, that the forces of natural selection act upon phenotypes and not genotypes. However, as far as I can find, there is still no broad concurrence on what the phenotype is. As a graduate student, I was much engaged in the debates of the great evolutionary protagonists of that time, such as Ernst Mayr, Theodosius Dobzhansky, Sewall Wright, G. G. Simpson, and R. A. Fisher. I was later introduced to the ideas that underlie the cybernetic or systems theory revolution. To me, these rules say that when the fitness of an organism is determined, such determination usually occurs not at the level of discrete genes, but among the set of epigenetic systems that make up or include the organism. I now believe that in this view lies a means of envisioning how evolution shapes populations, socioecological systems, and cultures, and how they together shape what happens to the genomes that ultimately come to regulate their expression.

My subsequent experience owes much to my contacts with Gregory Bateson and George Bartholomew, who perceived that social and evolutionary systems seemed to respond to the same set of systemic, cybernetic, or self-organizing rules—those of feedback, feed forward, and so on (Wiener 1948).

When one considers the diverse but ordered collection of species with which a given species of ant or monkey or dolphin is associated in its socioecology, another level of trophic organization is found, one in which the different guilds occur that apportion the available resources. These take resources from the total array of food resources in different ways, such as predators, herbivores, and those who consume the dead. In a socioecology experiencing environmental stress, migrations may change the kinds of food eaten, may shift its composition and social roles, and may change among and within the component species. So the term *socioecology* is based on the concept of the community, but includes a great deal of adaptive social ordering and response to environmental change.

Socioecologies that include mammals are apt to include true concept-based culture as a factor in response to environmental change, and such response can be very rapid indeed. Culture is at its foundation a commerce of ideas and is likely to require much central processing capability. Arthropod societies apparently do without concept formation but, even so, build complex societies by more direct genetic means, such as haplodiploidy (Trivers and Hare 1976).

Before returning a final time at the end of this chapter to make specific comparisons between the socioecologies of spinner dolphins and terrestrial species, I want to discuss my view of the evolution of such systems. I especially want to reflect on how the phenotype might interact with the genome and what its role is within a socioecology. My frame of reference draws especially from the works of Whitehead and Russell (1910–1913), Waddington (1960, 1968, 1975), Bateson (1963, 1979), Bartholomew (1970), Hofstadter (1979), Axelrod and Hamilton (1981), Dawkins (1976, 1982), Bateson and Bateson (1988), Slobodchikoff and Schulz (1988), and Stearns (1989).

Cultures are emerging as a new systemic level from within the socioecological framework. I view culture as a direct consequence of sociality in group-dwelling higher vertebrates, where members of such a society require the capacity for mental abstraction. This includes the ability to conceive of the deployment of a school made up of many animals from assessments of their various sounds and then, by constructing a mental image (or "gestalt") of the positions of the various members, to make mental predictions about things in the future and past (which are also abstractions), such as the predicted time of arrival of the school at a rest cove.

Probably the best indication that a species has cultural elements is the presence of a juvenile period interposed between the time of early nurture by a mother and adulthood, during which instruction and imitation (often called play) are dominant behavior patterns.

SYSTEMIC NATURE OF THE PHENOTYPE AND THE ORIGIN OF COOPERATION

I do not wish to digitize the phenotype and its effects, as the concept of memes seems to do (see Dawkins 1982, Bonner 1980), but rather to conceive of the compensatory systems of an individual, its population, socio-ecology, or culture as responding to environmental change by an alter-ation of both geographic distribution and behavioral responses. For ex-ample, when the seasonal weather pattern undergoes a shift in the area occupied by a socioecology of savannah animals, a complex set of com-pensatory changes in the ecological patterns and behavior of the constit-uent species can be expected to occur. Each species responds in relation to the others in the system and to the changes in the environment, es-pecially food resources. The same kind of change appears to occur in open ocean populations of spinner dolphins faced with the vagaries of global current systems during the cyclic events of El Niño (Reilly and Fiedler 1991). Cetacean and fish populations move, and food-getting strategies appear to change. The tuna seine fishery itself responds, alter-ing strategies for tuna capture (National Research Council 1992).

Only if such change were to be sustained beyond the capability of the plains mammals' or dolphins' compensatory systems to respond (as, for example, during a prolonged drought or large scale oceanic climate change) would fitness of the constituent species be affected. They are al-ready adapted to short term cyclic change. Another way of expressing this idea is that, through evolutionary process, the constituent species have built a compensatory system within which they live that buffers them from a normal range of transitory shifts in fitness. Only when the limits of such compensation are approached or exceeded does the basic system itself change.

The same sort of shifts can occur within any of the hierarchy of sys-tems that comprise the organism and its environmental relationships, and in all cases, the organism has a wellspring of responsiveness that generally allows it to survive. At the physiological level, such compensa-tion might be called acclimation; at the socioecological level, the effect is apt to involve changes in patterns related to food getting and predator avoidance.

In the past, failure to perceive this hierarchy of systems seems to have produced the many attempts by sociobiologists to show the direct control of the genome over behavioral events. Especially perplexing have been the frequent examples of altruistic behavior among animals. How can one animal cooperate with another when there is a cost in fitness to the giver of such cooperation (in other words, altruism)? The conception arose that the cost of altruism had to be paid back, but not necessarily immediately. Trivers (1971) indicated that reciprocity could be delayed for a time, allowing a later "fitness accounting" to be carried out by se-lection. This has been difficult to prove. I believe at least part of the an-swer lies in the analog dynamics of systemic levels above the genome and the means by which they interact with the genome.

When one attempts to square cooperative or altruistic events in mammalian and avian societies observed at the socioecological level with direct genic control, one is often unable to find the expected concordance between the predictions of either kin or reciprocal altruism theory and the behavior observed. Instead, one usually finds a perplexing melange of patterns, ones for which predictions from these theories seem to be borne out and others for which the theories seem to fail. For example, after reviewing social patterns in primates from such an evolutionary perspective, Silk (1987, p. 329) concludes warily

> . . . that individual selection has shaped the pattern of competition and aggression among primate groups and that kin selection and reciprocal altruism have favored the evolution of altruistic interactions. In general, successful competitors seem to obtain reproductive advantages over their less successful peers, and altruism is selectively directed toward kin or reciprocating partners. Nonetheless, few firm conclusions can be reached about the adaptive consequences of such behaviors. Much of the evidence that we need to adequately evaluate adaptive consequences of altruism, competition, and aggression has not yet been collected. Moreover, much of the evidence that has been cited in support of predictions derived from evolutionary theories could also be interpreted differently.

In addition, after more than a decade of testing, a recent symposium held to examine evidence relating to reciprocal altruism theory produced few unequivocal cases (*Ethology and Sociobiology*, 1988, vol. 9). Such theory attempts to explain the prevalence of altruistic behavior in standard selectional cost–benefit terms and as related to direct genomic fixation. The theory states that an altruistic event performed by one animal toward another typically has a clear cost in terms of the fitness of the performer, and if it is to be fixed in the genome, there must be a mechanism by which the recipient "pays back" this cost in fitness. Axelrod and Hamilton (1981) show how one animal can afford to cooperate with one another at the outset of an exchange using a tit-for-tat model, which relies on a direct genetic accounting of the costs in terms of fitness.

Such theory fails to explain the evolution of societies of mammals, such as the arboreal primates or the dolphins, in which cooperation is a way of life and which edge over into altruism through various sorts of assistance or epimeletic behavior (Caldwell and Caldwell 1966). Like the social arthropods, tamarin troops may contain members whose role precludes breeding altogether. Whether this condition exists in spinner dolphins remains a tantalizing possibility. I suspect that the answer to how these societies could have evolved will be found not in some sort of group selection but in the selectional environment of epigenetic systems such as the socioecologies and cultures in which these societies occur. It will probably be related to their long life spans, in which inclusive fitness involves the systemic outcome of the activities of whole societies over long periods of time. For both dolphins and monkeys, it must involve the cultural level of organization that operates by the remotest of epigenetic means: teaching and observational learning by individual animals.

The theory of reciprocal altruism is tightly locked to the fate of the individual animal, while at the level of the population or socioecology, the individual is a cipher in a statistical game. There, individuals occupy as much space and coopt as large a share of resources as the genome of the species will allow. Frequently, a large percentage of individuals live in suboptimal conditions and a considerable number may not survive. This ensures that available ecological space will be fully occupied to the physiological limits of a species. What, then, happens evolutionarily when the fitness equation of the individual runs counter to the needs of the species as the statistical game of the population is played out? If cooperation or altruism among its members allows a population to occupy more ecological space, can this run counter to the fitness equation of the individual where the ability to maximize the number of successful offspring remains determinative?

Most important, because a socioecology is an analog response system capable of a myriad of graded changes, the specific costs of cooperation or altruism become all but impossible to separate out on an individual basis. Other related factors in the complex survival equation of a cooperating species may blunt any such direct accounting. This may explain why in social animals from insects to tamarins, and perhaps dolphins and humans, there are cadres of individuals who make the supreme surrender of their fitness and do not reproduce at all.

In open ocean dolphins, cooperation to the point of altruism appears to have become necessary for population survival. It seems likely that once the track toward cooperation was entered in dolphin society, it began to assume the nature of an evolutionarily stable strategy (Parker 1984) and became an integral part of the social design largely without regard for relationships. Thus, the need for reciprocation may never have been an issue (Norris and Schilt 1988).

I regard the pressures toward cooperation in open ocean dolphin socioecologies to be so strong and so unavoidable that it is probable that these species would never have succeeded in invading the open ocean without it. Because the sensory integration system of a dolphin school is vital to the survival of the dolphins that make it up and because it requires school members to subordinate themselves to it to make the SIS work, I believe that cooperation in dolphin society is constantly crucial to survival. A result seems to be that they have become cooperating animals to such an extent that when alone they are truly less than complete.

It seems clear to me that many of the features we perceive today in ecological relationships and behavior are shaped heavily from the "top down," that is, they are the phenotypic result of the social and ecological environments within which the animals and plants live (Futuyama and Slatkin 1983). Such shaping occurs in part as a systemic coevolutionary response of a species to the state of a population socioecology or culture in which it lives. When I speak of systemic events, I mean this in the true cybernetic sense of self-regulation, bias, feedback loops, and oscillation (Wiener 1948).

In the 1940s and 1950s, a few voices in the debates of the emerging new synthesis, especially that of C. H. Waddington, attempted unsuccessfully to swing the discussion away from heavy emphasis on the genotype to include development and the phenotype, insisting that the face an organism presented to nature was what was selected and that linking all evolution directly to the genes obscured how the process actually worked. Within this phenotypic face was included a whole realm of phenomena that were not directly genetic but that were all part of the environment of the phenotype.

My view is that from the evolutionary process there has emerged an linked hierarchy of systems, nested in relation to the generation time of their species, each influencing the others. If a fitness change at any one of these systemic levels is sustained sufficiently long and with sufficient strength to influence inclusive fitness so that the fitness of the genome is biased, then genic evolution may occur.

The hierarchical systems of nature seem clearly not to be the cleanly separated logical systems of Whitehead and Russell's (1910–1913). Instead, such effects as the pleiotropic nature of many genes or the diffuse effects of adaptation at various levels of organization may blur such systemic boundaries, making the arrangements even harder to perceive than they might otherwise be.

Stearns (1989) attempts to grapple with systemic effects within physiological and populational systems of organisms. He refers to them as "trade-offs" and states that they need not necessarily have an immediate and direct effect on the genome. He also perceives (Stearns 1989, p. 266) that focusing tightly upon such unitary trade-offs as events in a single physiological process may obscure the larger issues of the whole organism or of events at the population level:

If interactions there [of populations] are as dynamic and flexible as [I] would suggest, and the relationship between two traits could easily change from positive to negative over the normally encountered environmental range, then the more general phrase "dynamic linkage" might be preferable to "trade-off" as a descriptor of the underlying mechanisms. Population level trade-offs differ qualitatively from trade-offs among functions within individuals.

His figures 1–4 describe the genetic covariance of growth in various environments, and at various growth rates, the relationships of reproductive investment and survival, and the relationship between energy investment and reproduction.

The organism itself embodies a number of internal systems, by means of which architectural and physiological variation and development of the individual is ordered according to the resultant inclusive fitness. The populations of single species represent another systemic level, and embracing that are socioecologies, and in some cases embracing all, an emerging system of cultural adaptation.

In other words, built into the operation of every organism are a series of analog processes (continuously varying along various axes) that taken together determine the momentary state of the organism. Then within groups of such organisms is another series of analog systems operating in the same way. Adjudicating such analog processes is the digital system of the genome, which, given a generation's worth of time, may impose long term order. At its simplest, reproduction at the level of the chromosomes either succeeds or it does not. At the level of the organism, multiple implantations may quickly make reproductive success statistical.

Similarly, the systems within the individual, the physiology, and growth processes are "judged for fitness" if they fall out of kilter. The organism gets "sick" in some way and may not survive. Its systemic state is immediately judged; the individual *and* its genome lives or dies. Thus, a given individual may fail, but at higher levels of organization, the stochastic processes that operate in the socioecology or culture within which it lived may allow others of its species to survive because they are somewhat different.

I will return later to discuss a few more details of how I see these systems shaping their constituent organisms, but first I want to make some comments on the phenotype.

DEFINING THE PHENOTYPE

In my view, part of the reason the New Synthesists failed to include the phenotype adequately in their discussions was that there never emerged from the debates a clear and accepted definition of *phenotype*, and it became defined in exclusional terms almost as another directly genetic construct. The usual definition was as follows: the appearance of an organism resulting from an interaction of the genotype and the environment (*Random House Dictionary of the English Language* 1967). Such def-

initions shaped the debate of that time and did not allow the evolutionary theoreticians to include in their constructs the fact that a great deal
besides the direct expression of genes or the exterior surface of an animal shaped an organism.

While there was much effort at the time given to dealing with the role
of populations in evolutionary change (Dobzhansky 1941), there was
scant understanding that associations of populations of many species or
their cultures were also operant. It is emerging that socioecological arrangements that assemble for energetic reasons (Bartholomew 1970), or
cultural systems that arise from the injection of ideas into the evolutionary milieu, have had profound effects upon the selectional environments of their component organisms. Waddington (1960, 1968) called
repeatedly for "a general theory of the phenotype" to go along with the
then well-developed constructs of how the genome behaved in evolutionary terms, but he did so at a time when only the barest understandings of either socioecologies or cultures of wild animals existed.

The Extended Phenotype

This lack of definition has often left the impact of the phenotype on evolution unassessed in the behavioral literature. This has happened even
though the phenotype has been profoundly important in shaping the
course of evolution, while the genome produces variation, encodifies the
most systemically fit results, and regulates the timing of their expression.
To sustain such epigenesis, the genome must also encodify the allowable
limits of change or oscillation. Most evolutionary scholars realize that
when we attempt to find genes that directly control cultural features of
animal lives, or ones that describe direct genic regulation of the ecological responses of animal populations, we gloss over unspoken levels of
complexity. But a new understanding of the phenotypic evolution is
emerging. Dawkins' (1982) book *The Extended Phenotype* seems to carry a
considerable measure of understanding about this matter, although I
find that effort still tied down by the need for direct linkages to the genome. I am likewise impressed with Bonner's (1980) book *The Evolution
of Culture in Animals,* although it, too, shifts the debate from the systemic
nature of socioecologies and cultures to a search for a unit of culture,
the meme. If an idea or a concept can be a meme, then we agree, but I
do not believe memes are assorted like genes. They are reactants in the
analog system of culture.

Because the existing use of the term *phenotype* has frequently been re-
strictive, I have searched for a term that would express these relationships better. I have informally used the term *inclusive phenotype* in recognition of the same need that Hamilton (1964*a,b,*) seemed to have in
extending the reach of the term *fitness* to *inclusive fitness.*

But I did not want to coin a new term if I could avoid it. Such coinage should not be a casual act. So perhaps modification of an old term would better serve the discussion. I therefore use Dawkins' term *extended phenotype*, but imply in my usage that phenotype includes the entire ecological reach of the organism. More precisely, my definition of the extended phenotype *of an organism* is the selectional milieu within which its species is shaped by evolution, as defined by such environmental interactions of structure, development, physiology, behavior, and culture as affect its inclusive fitness. This means that if the organism is a modest creature whose environmental interactions are modest, so too will be its phenotype. But if it is like the human race, its phenotype can involve its cultural artifacts, and much of the globe and many of its processes are, perforce, involved. For example, I perceive of our release of chemicals into the atmosphere and its likely evolutionary results upon our species as parts of our extended phenotype.

I could also have used the ecological term *niche* for what I intend. My use of extended phenotype and niche are similar, except that the normal usage of niche has strong geographic intent. What I mean by extended phenotype is more linked with the functional attributes of organisms as determinants of their competitive ability in natural systems, and their geographic range is a consequence of these things.

THE SOCIOECOLOGY:
AN EXAMPLE OF AN EPIGENETIC SYSTEM

I consider the socioecology of an organism as an *emergent* cybernetic (that is, systemic) entity that has developed repeatedly throughout the living world and seemingly inevitably throughout evolution. It develops as a range of species has come to fill in the available niches of a habitat and whose component species have then come to interact ever more closely with one another by coevolutionary processes.

One of the most illustrative socioecologies, to which we have already alluded, is the grazing-centered system of African antelopes and their plant and animal associates. This group of organisms has filled the ecological niches of a broad range of scrubland and savanna habitats, with each species being fitted to a part of the total possible adaptive range. Some small species of antelopes such as the dik diks have come to exploit rich localized food sources, while others such as the African buffalo concentrate on lower grade but more ubiquitous sources. Some species exploit ephemeral food supplies and migrate to find them, while others eat food that is available year-round, such as that along water courses (Jarman 1974).

For example, if a series of years of late rains occurs, the effects ramify throughout the system. The shifted rainfall pattern affects the constituents of the vegetation differently, according to the individual plant's phenology and tolerances. Some species are hardly affected, while others may all but disappear. There are corresponding changes in animal distributions, survival, and fecundity. The entire system shifts into a different pattern of biomasses, numbers, behavior, and distributions. This new arrangement feeds back upon the vegetation, the microorganisms, the predators, the carrion eaters, and others, impinging heavily on some constituents and less on others.

This process of adjustment, while ultimately dependent upon the adaptations of individual socioecology members, occurs regardless of what species they are. The socioecology is of a different level of organization. It is open ended, and the fates of its constituent organisms are each the result of coevolved individual tolerances and abilities at ecological compensation.

Among dolphins I think we are seeing the assembly of socioecology in their daily gatherings into a multispecies aggregation (see chap. 15). Given food patches, open space, predators, and a series of animal types, these oceanic animals seem to build a socioecology that functions according to the attributes of the available organisms but that is more than the sum of its parts. It includes various trophic levels and feeding types, and ultimately its form function is dictated by energetic considerations, that is, it is a socioecology. Because dolphins are a keystone species, members must be able to keep up with them, which automatically imposes a size minimum for membership because of swimming efficiency considerations. Note that young dolphins are assisted by their parents in keeping up with such a school (see chap. 10).

Each of the various species that play roles in a socioecology retains a capacity for variable or oscillatory responsiveness to environmental difference or change, frequently related to an annual or other cycle. As Waddington (1968) repeatedly emphasized, many organisms have indeterminate development at some stage in ontogeny, and this may respond to environmental change. That is, the environment may be able to "reach in" to the developing organism and change its adult outcome. Once again, a phenotypic relationship ultimately determines the fitness of the adult and not a direct genetic fitness. It appears that long term evolutionary change is typically concentrated at times when a necessary response by a component organism approaches or exceeds its genetically established limits.

All organisms possess a range of capacity for adaptive physiological response. Each process, such as metabolic rate, has a range of environmental circumstances within which it can function, and these limits may be reversibly changed by the process of *acclimatization*. Beyond this, with time and exposure to environmental stress, this range of response itself may change through the more permanent *acclimation* of the process (Brett 1956, Bullock 1955, Fry 1947).

It is probable that all mobile animals are capable of behaviorally selecting the portions of the environment in which they come to live. In aquatic environments, for example, many species select the thermal environment in which they grow, hence, they behaviorally alter their own selectional environment (Norris 1963). Thus, at every organizational level, there is the requirement that the constituent organisms retain flexibility to respond to systemic events. In the jargon of cybernetics, there is a *bias setting* for every natural system that regulates the excursions of its oscillations.

OSCILLATION OF BEHAVIOR STATES

Oscillation is definitive of life itself since no living thing can persist in a single, unvarying state. It is not surprising, then, that in recent years the study of behavior has gone from being a typological exercise—in which one described a given behavior pattern as if it were cut in stone—to one that recognizes that most patterns are labile. After years of work, I began to perceive that everything a spinner dolphin does is encased in and ordered by a series of oscillations, a point to which I will return later.

Lott (1984) has shown that many of the behavioral descriptors of socioecological systems are dynamic. For example, a species may change from being territorial under one set of circumstances to social under another, and this change may happen very rapidly in some species. For example, sun birds are territorial at one time of day and social at another. Lott has catalogued more than 100 such documented shifts in vertebrate social systems, mostly in birds and mammals. It seems that whenever the larger environment shifts, seasonally or secularly, a whole complex of physiological and behavioral changes can be expected to occur (Bartholomew 1970). Every organism in a socioecology can be described by the curvilinear relationships of its energetic efficiencies in relation to various environmental states, and these in turn are related to behavioral responses (see Carpenter 1976).

If this scenario is correct, we should not wonder that a given species is territorial under one environmental state and living in cooperative groups in another, nor should we be surprised to find that monogamy is practiced at one time and polygyny at another. Neither should we be surprised to find reversible changes in metabolic efficiency curves and acclimation state. Nor should we be surprised to find that cooperation among nonkin is common in socioecologies.

Because the cyclic fluctuations in socioecological systems are typically more rapid than the reproduction (generation time) of the large mammal genetic sys-

tems, such fluctuations may or may not affect fitness. No average change in fitness need occur and no further genetic change need result as long as the oscillation remains truly cyclic, does not bias system response, and occurs within the generation time of the species.

Genetic Control of Oscillation

What remained unclear to me for a long time was how a genetic system typified by a random assortment of genes between each generation could support an epigenetic system over geological time. How could the gene complement lay down a pattern long lasting enough to define a physiological or socioecological system within which an animal's physiology or society might oscillate? It seemed to me that if all of the genome was subject to changes due to mutation and genetic rearrangements, it would be unable to specify the basic design of an organism for very long. Yet, it is obvious that it does do this. Brachyopods, for example, have had the same basic body plan since the early Paleozoic. A high degree of such constancy appears throughout all mutlicellular lineages.

How, I wondered, is the long-lived message of a genome conserved? This message must somehow define body plans and the ranges of their function. In this way, the boundaries of an animal's structure or socioecology might be specified from the genome. Such control should be uncommonly stable. While the species of antelopes may come and go through geological time, their basic antelope plan has changed only very slowly and the basic features of their cellular chemistry at a far slower rate still.

Somewhere in the genome the allowable range of variation (the amplitude of oscillatory events) must be specified. If the genic system itself were labile (had a high mutation or cross-over rate), such control of the boundaries of oscillation would be defeated and any system built up above the level of direct genic control would fall apart. Such genetic control of oscillation must also underlie epigenetic systems beyond the individual since such entities as breeding systems and trophic relationships of various species must be specified and must involve a definition of limits. How that might be accomplished has long been a shadowy matter.

There seems to be two possible ways in which such long term stability of parts of the genome can be achieved. First, islands of structural stability might be defined in the structure of the genome, whether within genes or chromosomes. Second, once a variable pattern is achieved, for example, one that defines the basic patterns of multicellularity, this successful pattern might be stabilized simply because it becomes basic to so much else.

I then learned of one example of what I was searching for—linked collections of sequences of nucleotide pairs called *homeoboxes*. These are found in a special class of about 50 control genes (homeotic selector genes) in both invertebrates and vertebrates (Alberts et al. 1989), and their job is to specify positional information in a developing organism. Such information is of a higher logical type (or abstraction) than the direct expression of a gene. It deals with the patterned arrangement of the expressions of a gene or genes, just as a socioecology is an abstraction based upon the physiologies and other features of its component species.

These homeoboxes, which consist of a sequence of about 180 nucleotide pairs, contain parts that are remarkably stable and conserved through the course of evolution. Alberts et al. (1989, p. 938) describes them as follows:

> The homeobox sequence is conserved with astonishing faithfulness at the protein level; one of the homeobox proteins of *Xenopus* [the clawed frog], for example, has a sequence in which 59 of 60 amino acids are identical with the homeobox of the *Antennapedia* protein of *Drosophila,* despite more than 500 million years of independent evolution. Such conservation suggests that there may be fundamental similarities between insects and vertebrates in the mechanisms that control the development of the basic body plan.

In my terms, homeoboxes must contain instructions that cause regulation of the pattern of body plans during development. There may well be other such genetic instructions that serve a similar function for other systems of higher abstraction within organisms, for example, in physiology and behavior and in the control of epigenetic systems. At each of these levels of organization, there are architectural features that need to be defined in broadest outline and to be conserved. In each case, an allowable range of oscillation must be specified. This setting of limits that bound the allowable range of a structure or process is an example of a broader and wholly fundamental biological phenomenon—that of defining the limits of oscillation through which a living structure, process, or behavior may range. Life vanishes if it is not allowed to oscillate and also if such oscillation is uncontrolled.

Fowler (1984) has alluded to the same split as I propose here, between the features of an organism that are directly controlled by the genome and those that are indirect genetic systemic (or oscillatory) responses. In discussing the dynamics of populations, he separates the determination of R (the fraction of the carrying capacity at which maximum production or population growth is realized) from the short term environmental interactions of species. Fowler (1984, p. 378) says:

It is especially important to note the possibility that R is more of an inherent characteristic of the species than it is a property of short term interactions between species and their environments. This would explain the difficulty experienced in attempts to relate R to such factors as predator–prey relationships and competition.

I aver that these features are parts of different systemic response systems. It is no surprise, therefore, that features of one level of abstraction are difficult to relate directly to those of another.

Coexistence of Organizational Levels
It is worth emphasizing that all levels of organization seem capable of coexisting within a single species simultaneously, producing an intermixed melange of adaptive patterns that is extremely difficult to tease apart or to understand in terms of direct cause and effect. That is, humans and other social mammals such as spinner dolphins appear to be regulated simultaneously by adaptation as mediated by populations and by events in socioecologies and cultures. The indirect organizational contribution to the adaptive complex at levels above the genome is what Bateson (1963) refers to as *soma*. The interplay among the different systemic levels of organization, the soma, and the genome is what, in sum, makes up the adaptive complex of the organism and ultimately determines the fitness of its extended phenotype.

The genetic systems have received extensive treatment (Hamilton 1964a,b, West-Eberhardt 1975, Williams 1966, 1975), while the epigenetic systems have remained less clearly understood. A major interpretive problem has been that many authors have mixed organizational levels when thinking about adaptation, even though each organizational level may operate by its own internal systemic rules and can only be related indirectly to the others. This is what Whitehead and Russell (1910–1913), who were concerned with logical systems not genetics, were saying. A logical system has its own coherent internal rules, and one such level of organization only relates to the next through biasing other systems. An adaptation expressed in a socioecology is tested directly in terms of its effect within the socioecology and, outside it, only by the bias it produces upon the physiological or other systems of the organism. That is, if the balance of a socioecology is shifted by climatic change, its various component organisms may be variously pressed against their allowable limits of response. The effect in other systems of these organisms, such as their physiologies, may be expressed in terms of stress and compensatory responses, for example, the mobilization of food reserves.

Bateson (1979, p. 158) inverts my perceived order of emergence of evolutionary complexity, in which I think of culture as the highest and most indirect expression of evolutionary process. Yet, he sees the relationships as I do when he separates soma (my epigenetic systems) from genetic change:

> . . . it is important to note
> a. That somatic change is hierarchic in structure.
> b. That genetic change is, in a sense, the highest component of that hierarchy (i.e., the most abstract and the least reversible).
> c. That genetic change can at least partly avoid the price of imposing rigidity on the system by being delayed until it is probable that the circumstance which was coped with by the soma at a reversible level is indeed permanent and by acting only indirectly on the phenotypic variable.

I want to reverse Bateson's order by suggesting that the "hard wiring" of genetic control came before an epigenetic system in evolution. As the structural and perceptual complexity of organisms expanded (as they began to build bodies and then societies organized around learning and then, finally, to develop the capacity for concept formation or ideas), these pressed against the time constraints of the genome, altering it. As a result, in cultural organisms, a long juvenile period was produced that allows cultural transmission to occur between generations. Bateson seems correct, however, in that as such a process proceeds, the instructions of the genome that control these epigenetic systems become more and more abstract. In terms of oscillation, the genome can only control the determinants of an epigenetic system, not its instantaneous state. The genome becomes once or twice or more removed from events in the epigenetic systems themselves.

ADAPTIVE RADIATION

We have been looking at the present face of socioecologies. In the geological sense, where did their parts come from and how did they come to fit together? Why is it that there are players in the warm open ocean that can assemble, almost lock and key, into a functional multispecies aggregation (such as that of spotted and spinner dolphins) and seemingly do it over and over again, perhaps every day? The process by which the parts of such a socioecology have evolved is called *adaptive radiation* by paleontologists and is usually described for a single lineage of animals or plants. In the sense of this discussion, we should think instead of the entire socioecology as the unit to be filled in, that is, the fitting together of parts of an environmental reaction system.

There is imprecision in trying to point to the primary adaptations that may have spurred such events even for a single lineage, although it is a favorite parlor game of evolutionists to attempt it. The focal adaptation in ungulates is considered to be the development of the fermentation stomach (Brody 1945). For primates, large brain size, binocular vision, the freeing of hands for manipulation, and aspects of locomotion have all been suggested. For the odontocete Cetacea, a focal adaptation may have been development of echolocation (Norris 1967a, Fleischer 1976).

But such a unitary adaptationist's approach is enmeshed in the problem of understanding such events in the larger systemic sense that I emphasize here. I aver that one class of features falls within an energy-regulated epigenetic system—the socioecology—and that the importance of these features varies according to the immediate balance of the system. True, some adaptations are much more important, more focal, than others, and the most important ones are probably related directly to the conferral of superior energetic advantage for their bearers. For example, the development of the ungulate fermentation stomach gave access to an enormous suite of easily available food of many morphotypes, just as the dolphin's development of food-finding echolocation might have done.

It is also clear from the geological record that various key adaptations or circumstances have provided opportunities for adaptive radiation to fill in. These include the development of multicellularity, the proliferation of adaptive body plans and protective anatomy (Gould 1989), the emergence of terrestrial vertebrates onto land, the subdivision of Pangaea (which spurred Permian extinctions and left much ecological space behind), the emergence of airborne pollination, the emergence of flowering plants, the emergence of homeothermic animals, collisions of meteorites on the earth, and many more.

I contend that in each case, socioecologies were assembled during the process of adaptive radiation that progressively took advantage of these new energetic opportunities. In each case, an adaptive radiation of body and behavioral types emerged from the crucible of coevolution, being assembled into systems with their own *emergent rules*. I expect that such a radiation first deals with species primarily by adjusting them to aspects of the physical environment, such as heat, light, and water. Then as the ecology fills, the new adaptations are increasingly social, relating to the activities of other players in the system.

Gradually as such coevolution became tighter and more intricate, socioecological systems began to emerge and began to obey their own internal rules, just as those multispecies aggregations seem to do out in the central Pacific. The multispecies aggregations to which the dolphins belong are, I suspect, ancient assemblages. The mathematics of such emergent systems is clear—it is the mathematics of cybernetics or systems (Wiener 1948, Hofstadter 1979).

Note, however, that not everything evolution produces need have a long term positive effect upon the species. This is necessarily true since the process is not predictive. The same adaptation that confers great immediate advantage may in the long run come to feed back negatively upon the organism. Once again the example of humans' effect on the ozone layer through our technology comes to mind. The great conceptual power of our language and ability to store and summate the ideas of many of us have led to an intricate culture and technology and to the enormous ecological expansion of our species. At the same time, however, we have begun to threaten our very life support systems themselves, and we may now be watching an evolutionary oscillation in progress.

THE CULTURAL SYSTEM

The *cultural system* is an interesting and frequently misunderstood concept. I will not attempt to review the numerous studies of what a cultural system is thought to be or how its evolution might proceed. Instead, I outline what I conceive it to be and how it might be integrated into the activities of organisms. For an excellent recent review of this flux of opinion, see Boyd and Richerson (1985).

Unlike events at the socioecological level of organization where a regular supply of food is focal, culture is, at its core, truly noncorporeal. As alluded to earlier, I contend that a culture is based on *ideas* or *concepts* produced in the minds of cultural animals and transmitted by observation or instruction within the context of an animal's society or, in the case of spinners, their fluid schools. I purposely do not attribute culture to the transmission of simple *information* (genes do that), but suggest that its hallmark is the transmission of second-order mental phenomenon, that is, ideas or concepts.

Such concept formation appears to be fairly widespread in mammals and birds. Cognitive experiments designed to determine if an animal is a concept former have uncovered the capability in a number of group-dwelling birds and mammals, including the dolphin, the sea lion, several primates (Itani 1958), the elephant, and the African gray parrot (see, for example, Herman 1980b, Savage-Rumbaugh et al. 1980, Seyfarth et al. 1980, Schusterman and Kreiger 1984, Pepperberg 1987).

I feel that Bateson (1963) was wise when he used the term *soma* to indicate epigenetic processes that were distinct from those defined by direct genetic determination. By this usage, he lumped what I call socioecology and culture together. I can see no clear line between most animals at the socioecology level and those that I call cultural. I bow to typology by my classification, the very thing I decry elsewhere, recogniz-

ing that the emergence of culture is a process of gradual increase in social and cognitive complexity. But I also perceive a "touchstone," a single
definable new development in the mental grasp of an animal at the cultural grade as opposed to those left behind at the socioecological grade.
It is, as I have said, the capacity for abstraction that allows concept
formation.

Each definitive feature of an animal culture, such as the prevalence of
instruction of young during a juvenile period or epimeletic (assistance)
behavior (see Caldwell and Caldwell 1966), requires that the participant
be able to form concepts. Bateson (1966) associates this with refinement
of communication systems and the brain. He says (1966, pp. 571–572):

> My first expectation in studying dolphin communication is that it will
> prove to have the general mammalian characteristic of being primarily
> about relationship. This premise is in itself perhaps sufficient to account
> for the sporadic development of large brains among mammals. We need
> not complain that, as elephants do not talk and whales invent no mouse-
> traps, these animals are not overtly intelligent. All that is needed is to sup-
> pose that large-brained creatures were, at some evolutionary stage, unwise
> enough to get into the game of relationship and that, once the species was
> caught in this game of interpreting its members' behavior toward one an-
> other as relevant to this complex and vital subject, there was survival value
> for those individuals who could play the game with greater ingenuity or
> greater wisdom.

It is Schilt's and my proposition (Norris and Schilt 1988) that the spinner school is a sensory integration system (SIS) allowing each member to
sense the spatial disposition of its comrades and thus to define the
school's protective envelope for its members. This implies the capabilities that underlie the emergence of animal culture. That is, each dolphin
must construct in its own mind a gestalt of the shape of the school at any
instant from just a few data points It must mentally fill in the spaces in
three dimensions. At the same time, it must also make predictions about
the immediate future. Receiving information that some members are indicating the imminence of a turn tells the dolphin what it must do. The
dolphin is not following a leader *seriatim* but is instead behaving in terms
of a more abstract conception—the state of the school. I suspect that the
primordea of such capacity must also exist in other organisms that form
schools, flocks, or herds. Instantaneous spatial information about a series of neighbors or larger spatial states may well be a *sine qua non* for fish
or flocking birds just as it appears to be for the dolphins in a school (see
Tyack 1986, 1991).

I agree with Bateson, that beyond these simple geometric determinants lies the whole sphere of social relationships that seems to deter-

mine so much of what spinners do inside their school envelope whenever immediate conditions are right for their mammalian sociality to
emerge.

Cultural change is initiated, I contend, in *the mind of the individual.*
This view may be implicit in the idea of the *meme* as put forward by
Dawkins (1976) and discussed by Bonner (1980), although as noted earlier, I do not perceive the need for a unit of culture beyond the notion
that a culture is based upon concepts or ideas. As I see it, culture is transmitted in the only ways that minds can exchange information—by one
animal observing and learning from another (observational learning),
by instruction in the context of a given animal's society, or by time bind-
ing (coding and decoding in human written or oral traditions). These capabilities allow us to store and summate our individual experiences and
hence greatly extend the intellectual reach of the human species beyond
any other.

My statement that cultural transmission (tradition) is carried out in
the context of an animal's society is not trivial. A young animal coming
of age in a cultural society is shaped in many ways by the form of that
society. Rules of living are not only received through instruction or observed but are also imposed by the structure of the group within which
the animal comes of age. In the case of the spinner dolphin, the young
animal takes its place in the sensory integration system, with whatever
complexity and nuance about the world that that system may provide.

My student Carol Howard described what seems to be a crucial aspect
of cultural transmission when she contended that the process of cultural
exchange between two minds is unlike a message going along a tele-
phone line. Instead of an open wire along which the message passes with
uniform quality between stations, cultural transmission depends upon
the transmitting organism sending a message formed on the basis of its
experience, received and interpreted by another organism on the basis
of its *inevitably different experience.* In this way, cultural transmission involves the individual experiences and intimate and immediate differences of all participants. The learned overlay thus impressed on all culturally generated messages is the substratum for every rapid changes to
occur in the group perceptions of cultural societies, as opposed to the
"deeply frozen" transmission of information by genes. It is also the
mechanism that allows very fine grained and rapid response of a cultural species to the patterns of its environment. It also requires the control messages of the genome to be remote and abstract. Furthermore, to
make use of what is in such a message, the receiving animal must be able
to interpret the message. Thus, it is economical and necessary to possess
the same broad descriptors of the communication process as the sender.
When a predator interprets the intentions of its prey from its signal, or

when we read the moods of our pets, this relationship seems to be occurring. For example, I could read, with considerable predictive success, the moods of dolphins by listening to their sounds, even though our lineages diverged more than 60 million years ago.

Goodenough (1957, p. 167) much earlier perceived some of these relationships. He stated that ". . . culture is not a material phenomenon; it does not consist of things, people, behavior or emotions. . . . It is the forms of things that people have in their mind, their models for perceiving, relating and otherwise interpreting them." Boyd and Richerson (1985, p. 33) take much the same view, although we differ on the issue of information versus ideas as a basis of culture. They state that "culture is information capable of affecting individuals' phenotypes which they acquire from other conspecifics by teaching or imitation."

Probably such storage of learned ideas, teaching, and observational learning underlie the flexibility of environmental responses of spinner dolphins. These features are why dolphins can circle any island to find a lee where there are good places to rest. These are why dolphins seek out coves nearest to nighttime food sources, and why they can shift from cove to cove wherever their night's peregrinations leave them. A cultural animal such as a dolphin can probably track local events and circumstances in far more detail and more rapidly than can an animal limited to the socioecological level of abstraction.

CULTURAL ASPECTS OF SEXUAL ROLES

There are hints that in spinner dolphins, male sexual roles might vary along a continuum, with some individuals at one end heavily concerned with juveniles. These males may deal with cultural transmission and instruction, which is known for such social predators as wolves and mongooses and for some male tamarins who apparently do not reproduce at all. Perhaps we were observing such an arrangement when the adult males Four-Nip and Finger Dorsal were repeatedly found in juvenile subgroups. Perhaps this is also what Miyazaki and Nishiwaki (1978) were sampling when a very large school (almost 1000 animals) of the striped dolphin (*Stenella coeruleoalba*) was caught in the Japanese drive fishery and was found to be almost all juveniles except for a small group of adults, nearly all of which were males. This possibility is certainly a priority for future investigation in spinner dolphins.

In addition, a partitioning of reproductive roles is suspected to occur in pilot whales (*Globicephala*). Postreproductive females continue to lactate and perhaps to assist in an alloparental role in their schools long after their last reproduction (Marsh and Kasuya 1984).

THE FITNESS OF CULTURES

The cultural components of social patterns in most higher mammals such as spinner dolphins seem to be highly positive features, allowing quick and fine-grained responses to environmental texture and fluctuation. Hence, they provide much increased flexibility compared to organisms whose response remains more tightly tied to the genome. Nonetheless, the loose linkage between the cultural system and the genome can make heavily cultural organisms especially vulnerable to reductions of fitness (as in the case of atmospheric pollution and our part in producing it). The problem stems from the increasingly abstract control of the genome over epigenetic systems as one moves from genic systems to socioecologies and to cultures.

The cultural mode of thinking allows the construction of both past and future, which, after all, are abstract concepts (neither actually exists). A spinner dolphin can probably use environmental cues and memory to predict where a given rest cove will be and to move toward it, and if it is found full, it can move onto the next one held in its cognitive map (Premack and Woodruff 1978). In the human case, such abstraction, such sense of past and future, leads to the construction of explanatory cultural systems by means of which life is ordered and regulated. Explanation becomes an important means of social control, and consequently, we find the emergence of ethical systems and cosmologies in most human societies. Note that these can be and usually are based on incomplete or even fallacious information. *We humans will explain our world whether we can or not.* We all live within such explanatory systems. And once explained, we build a set of ethical or social rules upon the conception to regulate our behavior. How far along such a path spinner dolphins, chimpanzees, or elephants are is presently guesswork. My guess is that because spinners are still tied firmly to the socioecological needs of their species and because their social flexibility is so limited by their three-dimensional open space world, they have not played out this equation very far.

Play and Juvenility

I have come to believe that the hallmark of a cultural society is a well-defined juvenile period in which "play" (that is, cultural learning) is interposed between the time of weaning and sexual maturity. This sometimes long period may greatly extend the life cycle of cultural species. In such societies, there is apparently much to learn, and it is valuable enough to spend a long and often dangerous time learning it.

The juvenile period, as Gentry (1974) has emphasized, is a time when young animals practice physical patterns at the same time as their bodies are maturing and when they form the social bonds that may last throughout their lifetimes. It is a time when they observe and practice the social patterns and relationships of coming adulthood. During this time, their actions may be accompanied by metasignals of the "play state" that tell adults and one another that "this is play." I assert that the existence of a juvenile period characterized by play can be used to typify the presence of a cultural system in a species. If so, cultural adaptation is an important feature of spinner dolphin societies, as we have described (see chap. 14). By the same token, some level of culture is surely widespread throughout social mammals and at least some social birds.

Walters (1987) has emphasized that play peaks during the juvenile period of primate societies but that it is not wholly restricted to this period. He notes that even earlier, during the period of infancy, aspects of play between mother or alloparents and young can be noted that later give way to play between juveniles in a society. Then, as adulthood is approached, play patterns wane.

Play patterns are obvious in spinner dolphins. Most juvenile spinner dolphin activity is marked by a well-defined locomotory metasignal. As far away as one can see a juvenile spinner underwater, its exaggerated swimming movements carry the clear message (to put it in human terms) "I am a juvenile; my behavior is not important in an adult context." Such signals of juvenility are metaphoric class markers that distinguish juveniles from mature dolphins, as Bateson (1979) has repeatedly emphasized.

The existence of culture lies in the minds of the participants and is transmitted from one animal to another by its signals. Therefore, we find that culturally transmitted behavior patterns are often marked by such metasignals. If the message is complex, such cultural signals contain two or more levels of information and this must be marked for the receiver *in the same message* if a distinction is to be made (Bain 1986, Evans and Norris, 1988). For instance, when a canid pack closes on prey during a chase, a yip may indicate pain in one member, and in the context of the chase, this means that actual attack is taking place and the prey is dangerous. The cadence and pitch of the howls of others, which has been rising and becoming more rapid, also carries the message of closure on prey. Taken all together, the call sequence from the start of the chase to the kill is a complex analogic representation of the entire event that keys all participants together whether they can see one another or not.

The appearance of a well-defined juvenile period takes up a significant portion of the life span of cultural animals. The delay of adulthood, as compared to noncultural species, is marked in elephants, primates, and cetaceans. It may even begin to crowd against and define the possibly more rigid limits of life span. The time required by our society to produce an experienced and qualified neurosurgeon, for example, presses against the earliest indications of approaching senility, a sobering thought.

The importance of this new period in the ontogeny of a species is indicated by the risks associated with juvenility, which are often considerable. For instance, the exploratory nature of juvenile activity often places these young animals at considerable risk. What they do is not just "play"; it is in fact part of the serious business of ensuring cultural continuity between generations. The long generation time that results must be buffered with excellent predator avoidance and other safety mechanisms or the species will fail.

Adolescence in Primates and Dolphins

There is a difference between the terminology used in studies of maturation in primates and in cetaceans that may have obscured commonalities. Primatologists speak of a period of *adolescence* in apes, which is the period following puberty before the first successful reproduction (Walters 1987). Adolescence can be a very long period—about one-half the period from birth to sexual maturity in marmosets (10 months) and about one-third of this period (53 months) in chimpanzees (Walters 1987). It is approximately as long in dolphins of the genus *Stenella* as in chimpanzees. About 30% of a male spotted dolphin's life span occurs before sexual maturity, and both this period and that afterward are clearly marked by a series of pattern changes (Hohn et al. 1985) that indicate various levels of sexual and social maturity.

Dolphins and primates, however, differ profoundly in one respect—the time of birth during ontogeny. To gloss over considerable complexity, dolphins are physically precocial, giving birth to very large young, capable of independent movement on the day of birth, while primates give birth to large but wholly dependent young that may remain physically dependent on a parent for a matter of months following birth. The period of pregnancy in spinner dolphins is long, 10.5 months (Perrin and Gilpatrick, in press).

The spinner dolphin juvenile period is clearly marked. Some examples of the cultural transmission seem reasonably clear. Both aerial behavior and echolocation are demonstrated and practiced during the juvenile period (Caldwell and Caldwell 1979). Reproductive and relational patterns are practiced extensively by spinner dolphins. In other species of dolphins, the protocols of food manipulation and sharing are also known to be learned patterns (Brown and Norris 1956, Brower and Curtsinger 1979), as are both vocal and physical mimicry (Richards et al. 1984, Tayler and Saayman 1973). It seems clear that we know only the barest outlines of what is being taught.

COST ACCOUNTING

During the debates on kin selection and the origins of altruism, it has been hard to understand why altruistic acts occur since an altruistic animal should be reducing its own fitness in the process (Trivers 1971, 1985, Axelrod and Hamilton 1981, Connor and Norris 1982). Most of the behavior of this sort that we see in dolphin schools can be classed as cooperation, in which both partners benefit, but there are examples of what appear to be true altruism as well. A dolphin that supports a sick schoolmate so that it can breathe, and does so for days while it loses weight and seems to teeter on the brink of death itself, clearly seems to be involved in altruistic behavior (Brown and Norris 1956, Caldwell and Caldwell 1966). Another example is dolphins of one species that help those of another species (Norris and Prescott 1961).

I think that the indirect (systemic) way in which fitness is arrived at in socioecologies and cultures can postpone immediate genetic accounting, and altruistic patterns may arise in them in the absence of kin selection. Bateson (1963, p. 538) in a deductive study of the role of somatic change in evolution concludes much the same thing. He says:

> . . . there must be an economics of somatic flexibility and . . . this economics must, in the long run, be coercive upon the evolutionary process. External adaptation by mutation or genotypic reshuffling, as ordinarily thought of, will inevitably use up available somatic flexibility. It follows—if evolution is to be continuous—that there must also be a class of genotypic changes which will confer a bonus of somatic flexibility.

This helps to explain my excitement about homeoboxes, which set the limits of flexibility.

THE EMERGING SYSTEMIC VIEW OF EVOLUTION

In sum, it seems to me that we are passing into a new conceptual stage about how evolution has worked to produce higher animal societies. Our older attempts to tie every social adaptation closely to the genome are being seen more and more not to be the sole explanation of the facts. A new paradigm, which I have described here, is emerging. It recognizes that different organizational levels governing parts of the total behavioral events of a species' life can coexist as semiindependent oscillatory systems nested within the effective time boundary of a species' generation time (see Boulding 1968, Feibleman 1954, and Capitanio 1991 for another approach to these questions). This generation time may even be shorter than the time that is evolutionarily effective since the effective genetic response time of a population or species may actually be longer than a generation.

Because the patterns expressed in these semiautonomous epigenetic systems typically cycle at shorter periods than the generation time, what happens in them may or may not impinge directly upon fitness. Therein lies one bit of loosening of evolutionary constraints that may allow the wide range of social patterning we actually see in nature. Other expressions of flexibility include the postponement of developmental fixity noted in many organisms, the selection of environments by the organisms themselves, and the processes of socioecological adjustment and cultural change. To me the question of whether genetic relatedness is required to support the emergence of altruistic acts no longer seems a central issue to understanding how higher vertebrate societies may be built and how they function.

CONCLUDING THOUGHTS ABOUT SPINNER DOLPHINS

Finally, I want to summarize my thoughts about what membership in socioecological and cultural systems may have conferred upon Hawaiian spinner dolphins. At first glance, it seems to be surprisingly little. Their lives are quite regular; they do much the same thing on most days—rest, awaken, travel, feed, and socialize when they are not threatened and behave like a fish school when they are. They seem locked in a life-long cat-and-mouse game with their predators. And yet a closer look reveals several cultural tendencies that may have provided crucial advantages allowing them to carry out long lives in three-dimensional open space. Most important, a considerable degree of behavioral flexibility is shown by spinners. Flexibility is the key word here.

Their bout behavior seems to allow their societies to operate as a society of friends, extending social groupings in a flexible way, far beyond the limits allowed by kin-related society. They can apportion themselves out after a night's fishing into the nearest available rest coves. Bout behavior never leaves them without a protective school structure (the SIS), whose operation is probably another expression of the capacity to think abstractly. Perhaps the large numbers of animals involved in a dolphin school make group foraging more successful. It is easy to imagine that a widely spread feeding school could better locate shoals of mesopelagic prey being bourne in upon the island slopes than the much smaller family groups could. Especially, the development of the SIS of dolphin schools incorporates and demands much involvement of cultural features. Unspoken and uncomprehended levels of complexity are probably hidden within this simple statement of the outer bounds of school organization.

The flexible behavior of spinner schools seems to allow them to function in schools of constantly changing membership, while a more rigid society might settle for a single place and a single restricted set of associates. They then reassemble into working and feeding schools at night, even though the membership seems clearly to exceed the boundaries of kin relationship. To a significant degree, they have become a society of remarkably cooperative friends.

There seem to be features of the worldwide spinner dolphin species that also reflect unusual flexibility. They seem able to set up their societies wholly at sea and, where opportunity presents itself, to use islands and atolls wherever they occur.

I see no evidence that they use a syntactic language, as has sometimes been suggested. Quite the contrary, the oscillatory chorusing their schools engage in as they go out to sea in the afternoon is typical of group-dwelling mammals whose communication is heavily analogic, that is, it is emotionally based signaling of the moment. If they had language, they should be able to produce succinct instructions to guide a school during the events of the coming night. Instead, they chorus and tune up like a symphony orchestra, and it seems to take hours for proper synchrony to be achieved. They oscillate back and forth between an old behavioral state and a new one until schoolwide synchrony is achieved.

The fact that they can understand symbols and syntactic arrangements when trained must have its origins elsewhere in their lives than in anything like a language. I suspect it lies deep in the cultural level of organization itself. The ability to conceptualize seems to have come first and then to have played itself out in various ways in higher animal societies, only one of which (humans) seems to have developed a full blown shorthand syntactic language.

I see Hawaiian spinner dolphins as group or social animals to such an extent that they seem to have surrendered much of their individuality to the school in which they spend their entire lives. I think this is reflected by their uncommonly cooperative behavior within the school, their docility in the hands of captors, and their disconnection when held alone. Another aspect of their group nature is that their entire society seems to be mediated by behavioral oscillations that are group processes.

It was not until near the close of this work that I perceived how pervasive a feature of spinner dolphin life the oscillation of behavioral states is. The major behavior patterns of their day all have oscillatory features. A nighttime school accomplishes its move toward shore by a series of back-and-forth movements, each a little closer to shore. They oscillate in both time and space into rest. Their awakening and leaving a rest cove is accomplished by a series of oscillations—fast–slow, out–in, back–forth, and so on—that we have come to call zig-zag swimming. At the same time as zig-zag swimming is occurring, their vocalizations and social activity are undergoing concordant oscillations of abundant sound and high activity that alternated with near silence and low activity.

Within such oscillations, and apparently occurring at different rates, are other exquisite oscillatory events, with tapering beginnings and ends, such as the bouts we have described. Such oscillations allow the various schoolmembers to achieve social symmetry as well as synchronous timing. A pair of dolphins in a caressing bout oscillates in and out toward each other, both performing precisely together. If symmetry is not achieved, one partner may show a flash of anger and begin a chase (see chap. 14).

It took the whole span of this study for me to understand that at dusk when the whole dolphin school suddenly burst into magnificent leaping locomotion this was just one more oscillation and not an abrupt, new event. During these cycles of oscillation, locomotion had increased in velocity and the dolphins had finally reached the cross-over speed (chap. 10) at which it is more efficient to leap than to swim below the surface. The result was that the school suddenly leaped in abrupt unanimous bounding flight toward the offshore sea, only to stop suddenly a minute or so later as the oscillation toward higher speed reversed into a slower state; the school then reversed direction back toward shore again. Finally, the old state was left behind and the dolphins had reached the next behavioral state. They had oscillated successfully from rest to feeding.

Is this ubiquity of behavioral oscillation as marked as it is because dolphins live in unimpeded three-dimensional open space and because they are animals that never leave their school? Perhaps so. Oscillations are frequently seen in the behavioral states of animals ashore, but seeming never in such pervasive form. Does the relative lack of environmental complexity at sea give the oscillatory process full rein so that nearly always the dolphins in a school can be found keeping time and space in some fashion?

In the last analysis, these oscillations regulate the ranges through which the living processes of dolphins can fluctuate. Homeoboxes and probably other stable collections of genes may determine the range of excursion of processes operating at other genetic levels and these may allow epigenetic systems to function in self compensatory way. Oscillation is, in fact paradigmatic of life, and no other animal that I know shows it as well as the spinner dolphin does.

There is much more to understand about these thoroughly group-dwelling animals, the spinner dolphins. This is not too surprising when it took on the order of 50 million years for them to evolve and we have been at work trying to understand them for only a little over two decades. Only now do we feel ready to make a reasonable description of their social patterns. We can now tell their sexes reasonably quickly, we can identify individuals, and we can follow them underwater at sea. Future workers will have to listen in three dimensions in order to unravel the complexity of their social signals, communication that remains largely a babble to us but which obviously contains keys to much that goes on in their lives.

Overview

Now that we have done our naturalist's best to examine the spinner dolphin with every tool and viewpoint we could contrive, what kind of animal emerges? Is the dolphin a truly different animal from its mammalian kin ashore, as some have imagined?

The answer is, no, it is not different from other mammals ashore, except in ways that fit it to its special environment of the sea. In fact, one of the most important insights that this work highlights is that, in spite of the dolphin's long separate history from other mammals reaching back almost to the origins of modern mammals, no major design feature of its life patterns seems to differ from mammalian societies ashore. It is a mammal, good and true. It is this very fact of commonality in spite of its nearly complete separation from other mammals that is cause for wonder. Separated from the mammalian stock at least as long ago as the early Eocene, when some mammalian lineages were just being defined, it has played out the features that make it a mammal just like the rest. But, at the same time, it has gone beyond its simple roots and has "explored the potentials of "mammalness" through the development of a high order of sociality, including the evolution of a culture like that found in other high order social mammals such as the primates, social carnivores, and elephants. It is the shaping of this path without the ecological influence of the others that makes one stop and ponder. It makes one realize that, for those mammals ashore, probably the same phenomenon has occurred. Much was determined for mammals very early and can be expected to produce certain kinds of results. The old argument about determinism in evolution has at least a little new life in it.

What emerges is the canalization of evolution by the very assumption of the mammalian condition, even from early in its modern history. The essentials of mammals such as endothermy and a constant body temperature were there. Large body size was probably dictated when aquatic life began because surface-to-volume ratio required it through the energetic accounting of life in cool water. Although spinners live in the tropics, they spend much time below the thermocline, and heat loss in water is much faster per unit of area than in air. Thus, the forcing function to large size is present even in such warm water habitats.

Another determinant of evolution that dolphins seem to have faced in greater degree than other mammals is life in an "unobstructed" three-dimensional world. True, forest canopy mammals such as monkeys and tamarins face a world that is similar in important ways. We have seen that some of their societies bear remarkable similarities to those of oceanic dolphins, once one applies scaling functions that drive size. We have also seen that this human presumption that dolphins live in unobstructed open space is, to some extent, just that—a presumption. Theirs is, in fact, a world of bubble clouds, wavetrains, violent surface storms, thermal and salinity stratification, plankton blooms, and the blackness of night.

However, it also seems true that these mammals survive in open space and are available to their predators to a degree unmatched by other mammals, even including the herding animals of open plains. What defends them is what defends a Thompson's gazelle. First, they have an early warning system that can pinpoint predators and help keep them far enough away that energetics dictates that these carnivores should not waste their precious energy in a chase. Second, if they are attacked, all seem to rely upon the group processes that define herds, flocks, and schools to defeat the attack sequences of their predators. At such times, all seem to surrender their individuality to the group and all become ciphers in a larger supraindividual reaction system that can, by fomenting illusions through the vision of the predator, utterly destroy the attack sequences of single predators.

Such surrender has its costs. To be a member of a herd, flock, or school is to raise cooperation among the individuals in a society to a high level. To take individual defensive action in the face of the maneuvers of the group is to pinpoint one's self and to be singled out instantly from the rest. In other words, assumption of individuality is to die. The evolutionarily stable strategy of the school is the result, an absolute in the life of those who live by it.

It is in this equation of obligate cooperation that the "personalities" of dolphins (or antelopes or flocking shorebirds) are defined. Alone they are bereft of defense, and when alone they exhibit their own form of "cooperation" with those who might destroy them. In the case of the dolphin, it surrenders to us, a thing we mistake for gentleness. But it seems to be better described as the reaction of a mammal without its protective superorganism, with nothing left to react to in its own defense. Having surrendered vital aspects of its survival equation when taken from the group, it is much less than a complete animal.

The surrender of the individual to the group has many positive implications. That is, many of the aspects of "dolphinness" seem to emerge only in the group, and we must look at whole schools to discern them. What we see is what others have seen when unraveling the societies of monkeys, apes, elephants, and other societal mammals.

Such societies seem to be knitted together by communication systems that carry a rich freight of information. Various tricks are used in such communication systems, all multiplexed together, to identify the emotional states of members and their relationships of location, family, age, and status by multiple modulations of the communication stream, thereby defining how the group as a whole is doing at any instant. From such a milieu emerge subtler nuances such as the politics of the great apes or the disinformation and managerial communication of ground squirrels faced with a rattlesnake predator who cannot hear their voices, such that one message is sent by the defender's behavior to the snake, while another is transmitted to other squirrels by voice. No one knows if there is a dolphin parallel to this kind of management. This could well be, since few predators hear the highest sounds dolphins make.

Cultural societies typically gain flexibility in response to the local variations of their environments by relying upon the teaching and learning of members. Important time is given in the life of spinner dolphins to the transmission of culture. This is best seen as the juvenile period, although it extends both ways from that clear time in the ontogeny of a dolphin, finally to include about a third of a dolphin's life.

For spinners all these things seem to occur in a "society of friends"—a social organization in which the bonds of family are not lost but new imperatives of cooperation with those beyond the familial boundaries become superimcumbent. Having gone beyond the imperatives of kin relations, the defensive school must be defined by cultural roles that can make it function. We think we have seen these in the behavioral bouts of spinning, caressing, and echolocation to which all spinners seem to contribute and which, together, allow the school to function.

The juvenile time is well defined in spinner dolphins. It occupies the time between nurture and adulthood. We have had glimpses that suggest it is a time in the life of a dolphin supported by adult teachers, in particular, certain adults in our observations, males. It also occupies space in their schools. If the dolphins migrate, as the closely related striped dolphin does, juvenile groups may travel as almost wholly separate entities within a migrating front of dolphins. It is within the workings of juvenile groups with their teachers that we expect the cultural traditions of these dolphins to be passed down.

What are these traditions? They do not seem to be dramatic features, but rather are found in the local knowledge of coves to rest in, places to find food, and the distance and direction from shore to feeding grounds. What else might be shrouded in our ignorance about the complexity of their culture is anyone's guess at this point.

The features of sexual activity and its physiological substrates of hormonal activity and gonadal development and its consequences of gestation and birth are all tied together in an exquisite reaction system that defines how spinner dolphins go about reproducing. What is remarkable is that dolphins violate no special arrangement seen ashore but play out the same kinds of arrangements, depending upon food sources and the predation equation, that land mammals do.

Over-lying these arrangements, and probably both allowing and driving them, has been the development of echolocation. The ability to sense the relatively distant world amidst darkness, or beyond the limits of daytime vision, seems an utterly necessary precursor to invasion of the three-dimensional open sea. This is especially likely for a long-lived mammal that in the statistical game of species survival cannot "afford" to lose its members to predation at a very rapid rate. It has to win nearly all of the time in order to persist, while other marine animals such as fish can afford to lose more often because their reproductive equation and growth patterns are so different.

The cognitive capacities that have been demonstrated for dolphins seem to include most of the mental equipment that humans and other higher mammals ashore possess. The mental agility and breadth to form gestalts seems to be present and is probably most commonly expressed in keeping track of the school's envelope and internal structure, a feature made evident only by the characteristic sounds of schoolmates. It is also expressed in the role taking that supports such schools, and in the flexibility of association that allows these mammals to form protective groups with other dolphins far beyond family boundaries.

What seems to be different from, say, chimpanzees is experimentation. Dolphins are the quintessential group animals and seem to have given much of their individuality to the group equation, quite in contrast to some primates who violate group conformity by tinkering.

These dolphins seem to live in traveling socioecologies whose parts are available for assembly and disassembly even day by day, the lock-and-key arrangements having been developed over long geological time. This is an exciting vista—an ecosystem that can, in all of its parts, be experimented with and understood within its evolutionary framework.

We know precious little about what actually goes on, socially, within the school envelope where spinner dolphins live, and it is here that the most interesting future advances are likely to be made. Their sounds, especially the burst-pulsed signals, are highly diverse and almost unknown in function. The dynamics of growing up at sea are only dimly understood. Likewise, we more suspect than know that spinner society is partitioned into a series of roles that may even engage an individual for life. For example, a male may become a professional alloparent for life, as Finger Dorsal and Four-Nip may have become. Do they mate, or are they culturally pressed into roles that partition their society socially and that preclude reproduction?

Mimicry of one dolphin by another begins to reveal itself as a means of identifying the respondents in a communication exchange for all schoolmates to hear. Is this likely the keystone of our sensory integration system, which describes in mere outline form the organization of an acoustical–visual society living in three-dimensional space? Is this likely to be what other mimicking group animals are up to? Can we expect a sensory integration system in parrot flocks, for instance?

We have seen that when dolphins are threatened they descend an organizational level downward to become a society much like that of a fish school. What sort of organization, of playing upon the cognitive capacities already in place, occurs when the dolphins are not so threatened? Do they, like the chimpanzees, enter into the world of politics, of second- and third-order complexity?

The perception that a spinner dolphin's daily round is mediated by a series of repetitive or rhythmically organized events is, to me, a fascinating one. Ever day of a spinner dolphin's life, from birth to death and on every day in between seems to be mediated by a complex of oscillations played in a kind of life-long counterpoint. Is this easier to see in dolphins than in terrestrial mammals because the dolphins are largely unconstrained by environmental complexity? Is this same set of patterns also seen, but less clearly, ashore in the howling of wolves, the roost finding of jackdaws, and the migratory decision making of pinyon jays? Does this also occur in humans, in our primal joy of the "jam" session, in which a group of us come together for the moment in the deepest coordination of aims and intents?

These were some of the thoughts that crossed our minds as we watched from the windows of our various ungainly viewing craft into the very separate world and lives of the spinner dolphin.

APPENDIX A

Those Who Helped

In retrospect, it is quite remarkable how many people's good will and assistance were needed to produce this volume about spinner dolphins. Our curiosity about these dolphins attracted and involved a large number of people in our various attempts to produce a broad-ranging view of their lives. Somewhere along the way we realized that there would be no conclusion to our work, only a story in the process of unfolding and of which we describe only a portion. Our project, Hana Nai'a, set out to outline the life patterns of a wild cetacean, but the assistance of many people was essential so that we could spend our precious days of quiet time with these wild animals and get as far as we did.

An appalling amount of money was needed by Hana Nai'a to fuel the chance to try at all. We are grateful to the National Science Foundation (Grant No. BNS-78-2463), the National Marine Fisheries Service (Contract 79-ABC-00090), and the National Institutes of Health (fellowship No. 5F32 HD 05710) and to the staffs of these organizations who shepparded our requests through proper channels. Dr. Jackie Jennings of the National Marine Fisheries Service made it possible for two of our team to return to Kealake'akua after the main work was over to extend our data set for another year.

The Cooke Foundation of Honolulu, Hawaii, and the Drager family of Ben Lomond, California, provided the funds to purchase our essential offshore vessel, the *Nai'a*.

We are most grateful that the Waikaloa Marine Life Fund has continued their support of our Hawaiian spinner dolphin studies beyond the time involved in this book, for support of Ph.D. candidate Jan Östman Lind, and Master's candidate Ania Driscoll Lind. Their work on male roles in spinner society and on chorusing should materially extend what we know of those aspects of spinner life.

There were officials who helped us establish camp and moor our vessels in the bay and those who found extra money when the rent went up (Dr. William Perrin's group at the Pacific Southwest Fisheries Center, especially Dr. Warren Stuntz.) We had endless help with making equipment such as trucks and cameras work once we were on the site, and many professional colleagues and students came through camp and gave us help and insights on what we were attempting to do. Others helped us smooth ruffled feathers and showed us how to behave in rural Hawaii, such as Roger Coryell. Leon Stirling who gave a benediction to our work, admonished Hana Nai'a to give back to Hawaii as good as we took away. This book is our contribution toward understanding Hawaii's spinner dolphins and hence our part in the conservation of this species.

We feel much gratitude to Dr. Ed and Ingrid Shallenberger, to Marley Breese, and to the entire gracious training staff at Sea Life Park Oceanarium, Makapuu Point, Hawaii, for their many courtesies to us at that institution from which so many of our insights about spinner dolphins sprang. Steve Kaiser helped us obtain the akule for the prey debilitation experiment.

There is a special group of volunteers who stayed for a while at the Kealake'akua Camp or at Long Marine Laboratory and who spent long hours with us on data collection, collation, and reduction. Although their names do not appear as authors in this book, they were true Hana Nai'aites, and we could hardly have succeeded without them. We especially thank Paula Wolfe, who was practically a member of the team working on nearly every aspect of the program; Karen Miller, who did just about the same; Rachel Smolker, who worked side by side with our underwater behavior team for weeks, as did Sallie Beavers. John Fitch and Dr. Richard Young assisted us with the identification of stomach remains.

One can hardly list all of the friends and colleagues who helped shape the ideas in this book. that list includes a fair proportion of those who came through our doors during the years of work involved because we were constantly evolving the ideas expressed in this book with anyone who would listen. We especially thank Carol Howard, who talked about communication and culture with us; Carl Schilt, who talked about schools, especially anchovies; and Alan and Carrie Strout, who were wonderful sounding boards during a long bouncy trip across East Africa. We also thank Jan Östman Lind, Ania Driscoll Lind, and Michael Poole, who are even now evolving their own understandings of how spinner dolphin society works, the Linds at our old stomping grounds on the

Kona coast and Michael in Moorea, French Polynesia. Jan played an integral part in building, designing, and naming the *Smyg Tittar'n* (see chap. 3), and we thank him a great deal for the many hours he spent on the vessel.

We are especially grateful to those who have spent time reviewing parts of the book. Drs. William Perrin, Peter Tyack and John Heyning reviewed the entire book. This has been a most valuable exercise in colleagueship for us, allowing us to compare and discuss ideas with three scientists for whom we have so much respect. Their thoughtful comments have helped us at many points, and knowing what a considerable job this was in the midst of busy careers, we are especially grateful to them.

The editorial collaboration we have had with the staff at University of California Press was splendid. Sponsoring editor Elizabeth Knoll and copyeditor Kathy Walker helped shape, clarify, and bring the manuscript into proper form.

Dr. Nicholas Littlestone, who knows about systems theory and the mathematics of machine learning; Nancy Norris Littlestone, who knows about the mathematics of aggregations; Dr. Richard Norris, who is deeply immersed in the paleontological view of evolutionary theory; and Dr. Adrienne Zihlman, who knows an immense amount about the evolution of higher mammals, especially primates, all helped immeasurably by providing a sounding board for the ideas of chapter 16, in which I (Norris) attempted to fit dolphins into the evolutionary scheme. They all read the chapter, and I debated with each of them, and the resultant ideas are my own views, but without this colleagueship the result would have been much less useful and much less subtly fitted to modern ideas.

Dr. Robert Trivers who has pioneered much of what we know about the evolution of social systems has been a valued sounding board, although he might not have known it. Sharing his classes has helped shape the views of chapter 16, and although we see evolution from somewhat different perspectives, I emerge with deep respect for the penetrating insights he has brought to the evolutionary debate.

In terms of ideas and information, no scientific colleagues have meant more than Drs. Bruce Bridgeman, Evan C. Evans, III, and Bill McFarland, who helped us shape various parts of this work.

And, of course, many of our dolphin friends such as Finger Dorsal, Four-Nip, E-fin, Temple Baby, and Mom became real parts of our lives. It seems that, with them, we brought our two species a little closer together, and that's bound to be good.

No operation like Hana Nai'a runs without a quiet support group back at the home institution who helps to keep the expenses, purchases, grant requests, and reports straight. We are very grateful to Maxine

Lane and her staff in the Contracts and Grants office and to Dennis Artman and his people in the Natural Sciences Business office. Among the many people in these offices we especially thank Mary Castro, Glenna Heller, Edie Donovan, Ken Kawazoe, and Jeri Ruby, although their names by no means complete the list. The staff at the Long Marine Laboratory helped in many ways. We especially thank Steve Davenport, Keith Skaug, and Maria Choy Vasquez.

Jim Simmons kept our old tired truck, Winifred, running at Kealake'akua, and he and Charlotte Simmons were friends at all times. Al Simmons provided us with extra boat support when we needed it most, and Mike Atwood and Terry Powers flew our aerial surveys. Drs. Louis Herman and Scott Baker of the University of Hawaii provided funds and expertise for the aerial survey effort.

The special cooperation of the scientists and staff at the Naval Ocean Systems Center laboratories allowed us to carry out our acoustic studies. Drs. Whitlow Au, Don Carder, Rick Johnson, Patrick H. T. U. Moore, Earl Murchison, Paul Nachtigall, Sam Ridgway, and Dick Soulé provided expertise, equipment, and use of their data reduction equipment, as well as a sounding board for ideas. They were essential, and they were constantly gracious about our many requests for help. Dr. Vicky Kirby, then of the San Diego Zoo, generously analyzed our serum samples for progesterone concentrations. We are also grateful to Drs. John Tamkun and Mary Zavanelli for introducing us to the important world of homeoboxes.

Help with laboratory analysis is not always as immediately rewarding as watching dolphins at sea, but it is just as essential to making sense of all the data that are gathered. Those who assisted our central team in cataloguing identifications, transcribing data tapes, and reducing these data to statistics were Mary Helen Garcia, Jenny Holder, Michael Halpern, Dr. Ken Marten, Dr. Sus Shane, Beth Mathews, Dr. David Bain, Keyt Fischer, Holly Semple, Dr. James Boran, and Don Croll. Marshall Sylvan helped steer us toward good statistical techniques. Jenny Wardrip contributed much to the book through her elegant illustrations. Helen Cole and Chris Carothers prepared many of our maps, tables, and graphs. Dr. Bob Given graciously provided the photograph of the "SSSM."

Nancy Griffith helped us get to sea. We used her vessels at sea to watch dolphins, while at other times we slept aboard them or moored alongside them. Rebekah Levy helped with so many things, especially ministering to my sanity at times when bureaucratic entanglements were beyond my ability to cope. We thank Mr. Stanley Altman for his generous financial support when we tried to understand spinner dolphin escape behavior.

Others who contributed support when we needed it most, including ideas, data acquisition skills, and logistical help, were Kirin Arellano, Dr. George Barlow, The Bishop Estate, Dick Butler, Clay Clifton, Dr. Dave Coder, Oscar Cota, Dr. Jim Darling, Dale Fellbaum, Dr. Larry Ford, Sherwin Greenwell, Madison Jones, David Kelly, Bob Leslie, Bob and Joan Lajala, Cissy Lucy, Richard Lyman, Marilyn Manildi, Phyl Norris, Dr. Roger Payne, Bill Schevill, Dr. Michael Scott, Tom Shay, Teka Tehanui, Francoise Francoise, Dr. Al Takiyama, Dr. Charles Walcott, Dr. Bill Walsh, Lorraine Medina, Dr. Bill Watkins, Peter Wussow, and Andy Yuen.

Our thanks to you all. We hope sharing a small piece of our odyssey felt good to you. We had one of those wonderful, comradely, challenging, sometimes desperately close to insolvent, sometimes maddening, occasionally frightening, relaxed, beautiful, exciting, magnificently rewarding times, during which unfolded before our eyes so much that we had not known about those lovely animals, the spinner dolphins.

APPENDIX B

Materials and Methods

BASE CAMP AND RESEARCH PARTY

A base camp was established in a rented waterfront house in the village of Napoopoo, at the back of Kealake'akua Bay. This house allowed us easy access to our two boats moored offshore and provided a panoramic view of the bay for monitoring school movements, arrival and departure times, and behavior. A VHF radio station was established in the house that allowed us contact with our theodolite tracking station part way up a nearby cliff and the boats working nearby at sea.

The scientific party varied in number from time to time but was usually about six people. It was housed in the base camp and its outbuildings. A fair proportion of the routine part of the study was carried out by volunteers, to whom we owe much. The full-time members were chosen for the variety of skills needed in such a broad program. While all contributed to every part of the program, there were also specific assignments, as follows:

Kenneth Norris	Principal investigator
Bernd Würsig	Field camp supervisor, behavior, scars and marks analysis, theodolite and radiotracking, electronics, and statistical analysis
Randall Wells	Assistant field program supervisor, boat operations, flights, reproductive studies, scars and marks analysis, oceanarium observations, tagging, and radiotracking
Melany Würsig	Theodolite tracking, photographic analysis, daily scans, statistics, and data reduction

Christine Johnson	*Maka Ala* behavioral observations and under-
	water film interpretation
Shannon Brownlee	Acoustic recording and analysis; graphics
Jody Solow	Free swimming studies

A separate camp was established at Makalawena Beach, about 45 km (28 mi) north of Kealake'akua Bay, because the free swimming effort required isolation from other activities common in Kealake'akua Bay. This additional camp allowed us to sample two distant points within the Kona dolphin population simultaneously. The Makalawena site is adjacent to an area of shallow water sand flats that sometimes accommodate the largest school of spinner dolphins on the Kona coast.

FLIGHTS

Twenty-three complete circuits of the coast of the "Big Island" (the island of Hawaii) were flown using a small plane, usually on a biweekly basis from August 22, 1979, to August 31, 1980. Aircraft used were either a low-winged Piper Cherokee or a Cessna 172. The crew on these flights consisted of a pilot and a primary observer/photographer in the front seats plus one or two secondary observers in the rear seats. Surveys were flown at 150-m (~500-ft) altitude at a velocity of 170 km/hr (~100 mph) and at a distance of about 0.5–1.0 km (~0.5 mi) from shore. Flight paths were flown alternately clockwise and counterclockwise around the island to standardize for the tendency of spinner dolphins to move offshore in the late afternoon. This prevented us from always surveying a given area either in the morning or the afternoon. Plane height and speed were chosen to maximize visibility of dolphin schools while still leaving them undisturbed by the aircraft.

Three additional flights were dedicated to the examination of dolphin school structure and to the movements of specific groups of dolphins along the Kona shore. These flights occurred on August 2, 1980, August 16, 1980, and September 7, 1980. They involved repeated observation of single dolphin schools during the 10 hr of usable daylight.

All sightings of dolphins, whales, sharks, and rays were recorded. Some schools were circled for as long as 45 min. During these circles, the dolphin species was identified, school size was estimated, school structure and activity were described, associated organisms and boat activity were noted, and environmental factors such as surface condition, distance from shore, characteristics of the coastline, and presence or absence of visible seafloor beneath the animals was recorded. Photographs were taken for school size estimates and for studies of school composition and geometry. In Kealake'akua Bay, simultaneous group size estimates were made on occasion from aircraft, a shore station, and boats for later comparison. This comparison provided a way of understanding the biases of each method of observation.

The regularly recorded sighting factors of greatest influence on the data set included sea height and prevalence of whitecapping, underwater light penetration as influenced by glare, cloud reflections, rain, turbidity, and turbulence. The most common conditions during each three-month season are shown on the sighting maps (see figs. 13–17 in chap. 2) and are categorized as follows:

1. Excellent: 10% whitecapped waves or less, little or no glare inshore of the aircraft (where most spinners are seen), good light penetration, and the absence of chop that might obscure dorsal fins at the surface
2. Good: Up to 20% whitecapped waves; some glare and little chop
3. Fair: 15–30% whitecapped waves, glare, chop or heavy seas, but still a reasonable chance of sighting dolphin schools
4. Poor: Over 30% whitecapped waves, glare, poor light penetration and/or rain squalls, heavy turbulence, with little chance of seeing dolphin schools

The aerial surveys were designed both to provide census data and to study the group structure and behavior of dolphin schools during morning and early afternoon when spinners were typically nearshore and when sighting conditions were usually best. Unavoidable nonuniform sighting conditions around the island certainly introduced bias into our counts, especially because the trade winds are frequent on some coasts and rare on others. Nonetheless, we feel the data do provide a reasonable indication of total population size and should clearly be regarded as minimum estimates.

DAILY SCANS

From May 1979 through May 1980, a twice-hourly log of dolphins activity within Kealake'akua Bay was kept. During the first two months of observation, these scans were recorded from Manini Point at the southern edge of the bay, and later they were taken from a cement platform 5 m (16 ft) above mean low water at our base camp. Some scan data were obtained from the higher vantage point of the theodolite station. Concurrent data from both sites showed that the information from each was comparable in spite of the difference in viewing angle.

The observation protocol was as follows. Beginning at the first half hour of light and for every half hour until and including 1800 (6:00 P.M.) (earlier during winter), an observer recorded weather information (cloud cover, wind, and sea state), the number and activity of boats in the bay, and miscellaneous information that the observer felt was important. The bay was scanned for dolphins with 7× binoculars. If no dolphins were seen, the scan lasted 10 min to make certain that diving animals were not missed. If dolphins were sighted, then the group size and general activity state (traveling, resting, socializing, etc.) were assessed, and all aerial activity was counted for a 5-min period. We also estimated the position of schools. These data are only rough indicators of position. Precise records of location depended upon theodolite tracks. Good consistency in estimates of group size, (fig. 102), counts of aerial activity, and assignment of activity state was, we believe, achieved among observers. This was tested by comparing the data derived from each observer on the same flight and finding that variance, while present, was small.

Data on group size and aerial activity were subjected to computer analysis deriving the amount of aerial activity per animal per day (figs. 25, 27, 36, 42, and 43). Because our scans were so frequent and performed relatively close to the resting area, we consistently knew when dolphins entered or left Kealake'akua Bay. Therefore, we were able to obtain information on the frequency of aerial activity, the amount of time spent in the bay by season, and the group size. Any errors incurred would have been small.

The scans tended to pick up schools after they had entered the mouth of the bay and were some distance inside. Hence, the descent into rest had probably begun before or during our first sightings, and these early morning records should not be taken as indicative of the level of activity of schools at sea.

Although daily scans restricted at least one member of our crew to camp for 2 to 3 hr before they were relieved, they allowed us to gain the information just described, but perhaps more important, they ultimately gave us an intuitive feel for how spinners spend their days near shore. For example, we became aware that our own observational activity on boats disturbed the dolphins early in their entry into the bay, while later on they were less easy to dislodge. The dolphins seemed to "settle in," and we were able to coordinate our activities as a result.

THEODOLITE TRACKING

To achieve precise tracking of dolphin schools in Kealake'akua Bay, we established an observation platform 72 m (236 ft) above mean low water, part way up a very steep slope at the back of the bay. One had to climb to it hand over hand, holding onto small trees. A model NT-2A theodolite (or high accuracy surveyor's transit) allowed us to obtain azimuths and dip angles from the station to the dolphin school swimming below.

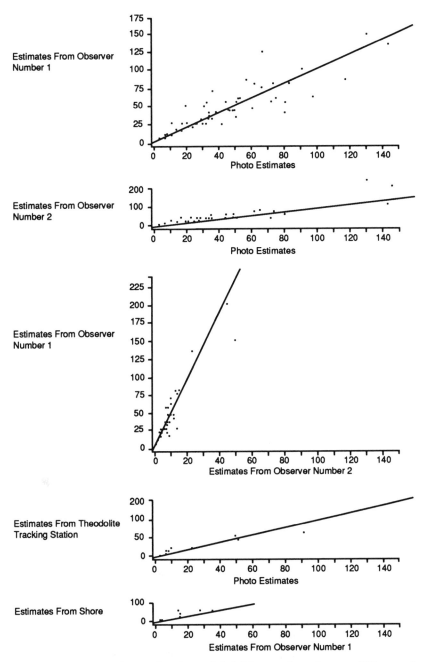

Figure 102. Comparisons of estimated dolphin numbers for two different observers and from different vantage points. Compared are estimates by eye and from counts taken from photographs.

These observations along with behavioral data were spoken into a cassette tape recorder and later transcribed into data logs. The readings were converted to *x* and *y* coordinates on a map of the area by use of a computer program utilizing an iterative correction for the curvature of the earth (developed by J. Wolitzky and described in B. Würsig 1978) (figs. 37–41).

The theodolite tracking technique, developed by Roger Payne, allows accurate tracking of marine mammals from onshore observation sites. In our case, the accuracy of the theodolite (= 10 sec of arc) and its placement 72 m above mean low water allowed determination of a dolphin's position within approximately ±10 m (±33 ft) at 3 km (~2 mi) distance, which we verified in the field. The readings accommodated for tidal state but not wave chop. Our records were plotted on a bathymetric chart that we constructed of Kealake'akua Bay. Dolphin speed and direction was calculated along a straight-line path between readings because not every shift in direction underwater made by the dolphins was recorded. Speed was determined only if readings were within 10 min of one another in order to minimize errors in such directional change.

The bottom type notations plotted on our bathymetric chart were determined by use of the fathometer on the *Nai'a* and by visual observation from the *Maka Ala* while these ship positions were being fixed by theodolite. This record became more and more imperfect at depths greater than about 10 m (33 ft) because decreasing visibility reduced the accuracy of classifying the bottom type. Nonetheless, it was clear that resting spinner schools stayed over clear sand bottom.

SCARS AND MARKS CATALOGUE

Although to the casual observer individual dolphins appear much alike, close examination usually uncovers differences in pigmentation pattern or scarring that allow individual identification. The trailing edge of the dorsal fin is especially useful because it is thin and easily tattered (Würsig and Würsig 1977, Shane 1980, Wells et al. 1980, all for the bottlenose dolphin). The smaller, more pelagic dolphins generally have fewer scars than coastal bottlenose dolphins, but Norris and Dohl (1980*a*) showed that similar analysis would also be fruitful for spinner dolphins.

A suite of photographs was obtained for every school encountered, taken of as many individuals as possible and of the various subgroups within the school. We used mainly Pentax MX 35-mm cameras with power winders and Kodachrome ASA 64 film. Lens sizes ranged from 50 to 300 mm, with the 200-mm lens being the most commonly used for our small boat work. Frame numbers were recorded for each suite of photographs, and the location and other relevant observations were noted in the film log.

Since the features of the trailing edge of the fin were especially useful, an attempt was made to photograph from both sides of a dolphin. Deformed or folded fins were also good identifiers. Scars or special pattern marks on the body were also recorded but were less useful in this analysis because they were infrequently resighted. (However, if spinners are studied underwater, such body scars might be the most important marks since they are very numerous.) Slides of scar-bearing dolphins were projected and distinctive fins were traced, resulting in a running catalog of individuals, dates, and places of sighting and of the animal's associates within the school.

Scarring was also recorded underwater by use of an Eumig Nautica Super 8-mm underwater cine camera and by 35-mm cameras from inside the viewing vehicle. Because our underwater work was confined to two inner bays, such records often did not tell us much about dolphin movements. Nevertheless, such marks did prove useful within single observation sessions and in a longer underwater effort would certainly prove valuable.

From these photographic data, a matrix of sightings, resightings, and associations was developed. These data allowed interpretation of individual movement, association patterns, and the temporal aspects of an individual dolphin's life (figs. 55–61). Only about 20% or fewer of the dolphins in a given school were identified. Our interpretations of school dynamics rest wholly upon this modest fraction of the total population. The absence of any particular individual in the photo record is not proof of its absence from any group because photo opportunities were limited and it seems likely that not all animals were photographed on every occasion.

Data from dolphins that were sighted five or more times were analyzed for frequency of location. A matrix of occurrence in different areas along the Kona coast was developed by Donald Croll and run on a Hewlett-Packard 9845 computer plotter at Moss Landing Marine Laboratories, Moss Landing, California.

TAGGING AND RADIOTRACKING

Four spinner dolphins were tagged during this study (see figs. 43–46). Bow-riding dolphins were captured with the tail-grab device designed by Roger Payne and Bernd Würsig and used successfully for dusky dolphins (*Lagenorhynchus obscurus*) (Würsig 1976). As the grab made contact with the dolphin, a trigger was released that allowed a padded grab to clamp around the animal's peduncle, and the grab then separated from its long handle. The dolphin was retrieved with an elastic shock cord and nylon line attached to the grab. The dolphin was then carefully lifted into a 2.7-m (~9 ft) inflatable boat alongside the capture boat. While one person on the capture boat monitored respiration and general condition of the dolphin, two others in the raft measured and determined its sex, drew blood samples from the flukes (two dolphins), marked the dolphin, and released it. One animal was freeze branded. A 7.6-cm (3-in.) numeral brand supercooled with liquid nitrogen was applied to both sides of the dorsal fin for 15 sec. Colored roto tags (Nasco, Inc.) were attached to the dorsal fins of these animals.

Radiotransmitters (Telonics, Mesa, Arizona, Models MK-11 and MK-V-4a) transmitting at 148+ mHz were affixed on the dorsal fins of the other three animals. A 47-cm (18.5-in.) steel wire antenna was attached to the back of the transmitter. The transmitter was fastened to a padded backing plate that was through-bolted on spring-loaded bolts with dissolving nuts to two padded washers on the other side of the fin. The first transmitter was designed to release within 48 hr, while the next two were supposed to remain in place for approximately three months. The transmitters operated continuously. Signals were transmitted at the rate of once per 0.9 sec. After tagging, each animal was photographed and released. The entire process from capture to release took from 10 to 30 min, depending upon the number of procedures performed.

The tracking effort used a Modified Automatic Direction Finder (ADF, Ocean Applied Research, Inc., San Diego, California) connected to an Adcock antenna mounted on the vessel *Nai'a*. An identical apparatus was used from a vehicle on shore. Signals received by the ADFs were presented visually as a bearing relative to the heading of the vessel. An audible signal accompanied each flash. The intensity of this signal was shown on a meter as a rough indication of distance from the transmitter (signal strength). Attempts were made to stay within approximately 1 km (0.6 mi) of the radiotagged dolphin. Transmitter range varied from less than 1 km to several kilometers depending upon the individual transmitter's performance and the sea state. The mobile shore station operated in an automobile on the island's peripheral road and for this reason was located at variable heights up to 450 m (1480 ft) and had significantly better reception than the shipboard receiver. The shore station helped in triangulating positions at night. Positions, depth, and environmental conditions were logged every hour.

Surfacing times were recorded from visual observation, such as during the night when we approached the animal every 4 hours to check on its condition and position in the school, as well as continuously on a Rustrak chart recorder that was actuated by the radio signal. During daylight, the school containing the radiotagged dolphin was approached hourly to observe behavior and school size, to describe the shape and dimensions of the school, to record interactions between the tagged animal and others in the school, and to photograph identifiable dolphins in the school.

OFFSHORE STUDIES

Our 7.6-m (25 ft) Skipjack inboard-outdrive vessel *Nai'a,* equipped with twin OMC 140-HP gasoline engines, supported offshore work outside the confines of Kealake'akua Bay (fig. 24). The vessel was equipped with a single sideband radio and a VHF radio system, including handheld portable stations that could be taken ashore or onto small boats. A Furuno Model FE500 Recording Depth Finder and an acoustic recording cabinet were mounted in the open cabin. A small inflatable life raft was aboard. The vessel had facilities for four workers, allowing us to cook and sleep on board for extended stays in sheltered waters, and it could operate to about 10-km (6 mi) offshore in the lee of the island of Hawaii. In our view, she was not sufficiently seaworthy to work in the windward areas of the island.

ACOUSTIC STUDIES

Acoustic recordings were made on 34 days for a total of 7 hr of taping during the following eight recognizable subdivisions of behavior:

1. *entering:* animals coming into bays to rest
2. *descent into rest:* schools in the process of changing from active patterns to quiescence
3. *rest:* schools with slow locomotion, synchronous diving, and suppression of active social patterns such as caressing
4. *zig-zag swimming:* the period following rest in which schools abruptly and repeatedly change course and swimming speed
5. *travel:* a very active pattern in which schools move rapidly, often out of bays
6. *spread formation:* after traveling to sea, schools abruptly disperse
7. *feeding:* near dusk and at night schools move and dive subsynchronously
8. *socializing:* some schools before and after rest may engage in intense caressing and reproductive patterns
9. *meeting:* when two schools meet and either merge or separate again

Recordings were made on a Lockheed 417 ¼-in. tape recorder at 15 in. per second. An H56 Naval USRD hydrophone attached to a 10-m (3.3-ft) coaxial cable was deployed with an elongate float attached to the cable 4 m (13 ft) above the hydrophone to reduce water noise. This equipment was calibrated once a year by the Navy Sound Reference Laboratory. This system was reasonably flat from DC to 65 kHz, limited by the hydrophone.

All recordings were made from the *Nai'a*. When a school was located, the boat was positioned ahead of the school, stopped, the engine turned off, and the gear deployed off the stern. The hydrophone hung 4 m below the surface. Schools that were recorded varied in size from 7 to more than 100 animals. Most recordings were made in the vicinity of Kealake-'akua Bay, although we recorded as far north as Hoona Bay and as far south as Milolii.

Recordings and surface observations were all made when the dolphins were within an estimated 300 m (980 ft) or less of the drifting boat. The recordings used in analysis were culled down to only those having loud, clear signals and were usually taken from a fraction of the maximum recording distance, typically less than about 70 m (230 ft).

Behavioral observations were of two kinds. First, the frequency and kind of aerial behavior (which is a general indicator of school activity state) and the number of animals, their distance from the boat, their orientation relative to the recording gear, and recording conditions were spoken into a voice channel while recordings were being made. These observations allowed us to typify behavioral state and to calculate the relationship of signal abundance to animal numbers. Because of the highly directional nature of some dolphin sounds, we selected portions of the tape when dolphins faced the recording gear and were within good recording range.

Second, for purposes of standardization, every half hour an additional 10-min scan was made for specific categories of dolphin behavior, weather and sea conditions, swimming behavior (such as synchrony of swimming movements and speed), and the dimensions of the school. We also quantified aerial behavior and attempted to make an assessment of social patterns in progress through the surface during this scan.

The original recordings were dubbed onto 1-in. tape and processed on an Ampex instrumentation recorder. By selection of the best recordings, 4 of the original 7 hr were dubbed. Recordings were excluded if (1) the dolphins were all 100 m (330 ft) or more away from the hydrophone and moving away from it, (2) the dolphins were generally facing away from the hydrophone or moving away from it, (3) the sound level was especially faint, or (4) background noise, especially that from snapping shrimp, was high enough to mask dolphin sounds.

These recordings were then analyzed using the waterfall display of a Spectral Dynamics spectrum analyzer. Tapes were run at one-fourth speed (3.75 ips). The analyzer was set for 25-Hz resolution, 20-kHz range, linear display, and transform size at 2048. This meant that the output range was from 0 to 80 kHz and that the real resolution was 100 Hz. From 18 to 32 samples were taken for each listed behavioral state except zig-zag swimming. The zig-zag period was excluded from statistical analysis because of its great variability and the small sample we had to typify it (figs. 72–78).

A total of 198 5-sec-long samples were printed on a Tektronix printer using photosensitive paper, which produced a time–frequency plot. Each sample was examined for the number of whistles, clicks, and broadband burst-pulsed signals. The instrument was not useful for a precise record of click rate due to its low scan rate. As a result, our figures indicate only relative abundance. The number of each signal type was divided by the estimated number of animals present to obtain the estimated number of sounds per animal for each sample. Unavoidable uncertainties in this method relate to the accuracy in estimating animal numbers and to the number of animals phonating toward the hydrophone. Nonetheless, we are confident that the animal numbers estimate is generally accurate within about 25% based upon other comparisons of our ability to estimate the number of dolphins in a school.

It is probably true that some of the dolphins did not swim in the general direction of our hydrophone even though we judged the school as a whole to be moving toward it, and this affected our total assessment of directional sounds such as clicks. It is also true that because dolphins are schooling animals, most tend to move in a single general direction, the tendency toward coherence increasing with school speed. So such an error is apt to be small, reducing the apparent correlation between number of dolphins and sounds per unit time slightly. This means that where our figures show a high correlation between behavioral class and click emission frequency, the actual correlation is probably even better than we note by an unknown factor.

UNDERWATER BEHAVIOR OBSERVATIONS

Three means of underwater behavioral study were used: direct observations were made from the viewing vessel *Maka Ala*, underwater photography, and direct observations made by a swimmer (ch. 15).

The viewing vessel *Maka Ala* (Hawaiian for "watchful one"), a modified Boston Whaler skiff, had a rectangular plexiglas and plywood chamber inset into the hull (fig. 22). The observer lay on a mattress with the head and forearms in the plexiglas bow and spoke his or her observations into a microphone recorder, or recorded events on a videotape camera or on film. At the same time, the observer was able to listen to the dolphins using headphones and a bow-mounted hydrophone.

Since the plexiglas bow chamber sloped with the vessel contour, it allowed useful visibility downward and to the sides, but except through the small side windows, not up toward the surface. This proved to be a serious detriment because determination of sex depended upon viewing dolphin genitalia from below, which we could seldom do.

Nonetheless, determining the sex of dolphins became easier as the project wore on and we became attuned to the subtle differences in body and fin conformation of spinners. Fins of older adult males tend to be erect, tall, black, and triangular, and the tail stocks of these same animals are much more obvious clues to sex since they are marked by a postanal hump and a generally thickened peduncle, as compared to those of adult females. The demeanor of adult males swimming in coalitions is also diagnostic and, with experience, immediately recognizable. Males often swim in such small, discrete groups, moving resolutely as a unit and frequently interposing themselves between the observer and the remainder of the school (Pryor and Kang 1978).

The listening system allowed us to hear animals in spite of both outboard motor noise and the water noise of the moving vessel. The system consisted of a Gould CH-17UT hydrophone mounted inside a water-filled plastic tube oriented forward into the water stream. The hydrophone was partially shielded from motor noise with a foam neoprene barrier. The system was mounted below the bow chamber. This arrangement allowed the observer to observe and record simultaneously, and these records could be keyed together on a two-track recorder.

Most observations from the *Maka Ala* were made well inside Kealake'akua Bay because of concerns about the vessel's seaworthiness, a suspicion that proved correct when the vessel sank at its mooring in a torrential rainstorm and had to be refloated by divers.

Underwater films obtained with the Eumig Nautica Super 8-mm underwater cine camera were made both by a swimmer and from the *Maka Ala*. The most successful mode consisted of insinuating a swimmer into a school from an outboard-propelled inflatable boat. The photographer was situated under the vessel's bow, holding onto the bow painter with feet on an athwartships line, and rode slowly into schools of spinners. Once in the midst of the school, the swimmer moved quietly away from the boat into the midst of the school, filming in the process. Sometimes as many as four rolls of film, each 3 min long could be obtained from a single sortie of this sort in which the diver contacted the boat only to reload the camera. In this situation, the dolphins were sometimes remarkably tolerant of a swimmer. Some sequences were obtained with the animals only 1 m (3 ft) or so away from the swimmer.

Usable films were later analyzed on an Elmo Film Editor, Model 912, frame by frame. Knowing the film frame rate allowed us to time behavior patterns and events such as breathing sequences within fractions of a second. Since these rates were subject to imprecision due to differing states of battery drain in the camera, we checked projected frame rates with a stopwatch to control for this possibility and found no inaccuracies.

Other attempts to insinuate a swimmer into wild schools were carried out first in Kealake'akua Bay and later off Makalawena Beach, where a camp was maintained from November 1979 to July 1980. Each morning, weather and the presence of dolphins permitting, the swimmer and companion went to sea in an inflatable boat powered by a 15-hp engine. The school was approached, and the swimmer entered the water equipped with a safety buoy, wet suit, and underwater notepad. When dolphins were sighted, the swimmer emitted a mimic of a dolphin's squeal at frequent intervals as an announcement to the animals and then attempted to swim among them.

OBSERVATIONS OF CAPTIVE SPINNER DOLPHINS

Through the courtesy of Sea Life Park Oceanarium at Makapuu Point, Oahu, Hawaii, we were able to observe a captive school of spinner dolphins held in Bateson's Bay, a roughly circular 17-m (88 ft) diameter pool, 5 m (16 ft) deep in the center, gently sloping on one side and vertical on the other where the ports of an underwater viewing room were located. These animals were scarred and not considered fit for public display and therefore were undisturbed except for daily feeding times and periodic medical check-ups.

These animals included (1)Kahe, a female from the Waianae Coast of Oahu, estimated by the Sea Life Park staff to be 11 years old judging from time of capture and age class at that time; (2) Lioele, a male from Maile Point, Oahu, estimated at 16 years old, and (3) Kehaulani, a female from Five Needles, Lanai Island, estimated at 12 years old (ch. 14).

The viewing room had four viewing ports; two were plastic hemispheric bubbles 1.5m (5 ft) in diameter curved into the tank (which causes them to act as reducing lenses, but also to be panoramic), and two flat glass ports 1 m (3.3 ft) square (see fig. 87). Cables for listening gear could be run through the roof to hydrophones suspended in the tank. Outside light stanchions set on the periphery of the tank could be used to illuminate the tank at night. Other more distant park lights remained on all night which meant that darkness was never complete. The tank was supplied with fresh seawater at a high volume through a peripheral entry box and taken off at a center weir box in the tank bottom.

A regular series of behavioral observations of the captive school was made between September 1979 and October 1980 and in June 1981, primarily to define reproductive events. These are discussed in chapter 9. A

24-hr viewing session was made every two weeks, with observations recorded for 10 min every half hour. Unscheduled observations of various lengths were made to study particular behavioral questions, and these sometimes lasted for a 24-hr day or more (chap. 14). These studies concerned bout behavior, aerial patterns and their precursors, the dynamics of school maneuvering, echolocation "manners," acoustic intimidation, affiliative patterns, play, and the use of sound in prey debilitation.

REPRODUCTIVE STUDIES

Reproductive seasonality, behavior, and physiology were examined in various ways. The percentage of calves in schools photographed from the air was determined against the time of year. A *calf* was defined as a dolphin equal to or less than three-fourths the size of the nearest large animal. Calves were required to maintain this size relationship through a series of at least three photographs to be recorded as a calf, as a control on various swimming attitudes a dolphin might take in a single photograph relative to the course of the aircraft.

The dates of captive births (courtesy of Ingrid Shallenberger, head trainer at Sea Life Park) were combined with age data on captives determined by tooth growth layer analysis, with assistance from A. Myrick of the U.S. National Marine Fisheries Service, La Jolla, California (Wells 1984), to provide information on reproductive seasonality for ten captive spinner dolphins.

Serum steroid hormone concentrations were measured every other week for a group of three to five captive spinners at Sea Life Park from September 1979 through October 1980 and again in June 1981. Serum was obtained from 24–36 cc blood samples drawn from fluke vessels. Serum samples were frozen for delivery to analyzing laboratories.

Male serum was analyzed for concentrations of testosterone by radioimmunoassay technique (Smith-Kline Laboratory, Honolulu, Hawaii; sensitivity = 0.05 ng/mL). Similar methods were used to analyze female serum for estradiol (Smith-Kline Laboratory; sensitivity = 33 pg/mL), through the courtesy of Vicky Kirby, San Diego Zoo Research Laboratory.

Behavior related to or indicative of seasonal reproductive state was recorded from the side of Bateson's Bay at Sea Life Park. The following categories of behavior were recorded: (1) associations among individuals including duration, frequency, relative position, and interactions; (2) individual patterns such as aerial behavior; (3) activity level; (4) respiration rates; and (5) location and headings. One animal was considered to be associated with another if they swam or surfaced more or less in synchrony within about 2 m (6.5 ft) or less of each other.

TERMINOLOGY

Certain terms are used throughout this study with very specific intent. We define them as follows:

School—A group of dolphins swimming together as a discrete recognizable unit for more than about 5 min time. Spinner schools (as defined here) usually existed for longer than we observed them. Definition sometimes became difficult when two schools met or during behavior periods when animals spread widely. But spinners are more fully group animals than many dolphin species, especially nearshore taxa. So schools were typically easily recognizable entities. In this study, schools ranged in number from groups of 4 animals to about 225. We never saw spinners alone or in a group of less than 4, and we saw a four-dolphin school only once. If a large school was approached, it might fragment into two or more smaller groups that we then called schools.

Subgroup—Any recognizable subdivision within the overall envelope of a school. Subgroups seem to form for a variety of reasons. We recognized ones related to age, sexual activity, nurture, feeding, and defense.

Bout—Any episode in which dolphins engage for a significant period of time (typically for more than 1 min to about 1 hr for active patterns such as caressing and spinning) in a single, repetitive behavior pattern, alone or with another animal or animals. Usually such bouts were marked by recognizable onset and termination behavior, that is, the participants in a bout underwent a regular increase and decrease in the expression of a given behavior during its course.

School envelope—The outside three-dimensional shape of a school at any instant.

Newborn—A dolphin up to about 105 cm (41 in.) total length (Perrin and Henderson 1984), which we roughly estimated, sometimes showing faint fetal folds (not very evident in spinner dolphins), folded fins, and flukes.

Calf—A dolphin estimated to be older than a newborn but less than 1 year of age. This meant that if the dolphin was less than three-fourths the adult total length ($\sim$ 105–128 cm, or 41–50 in.) it was considered to be a calf. The heads of calves are relatively larger than those of adults, and they have relatively smaller fins, show juvenile locomotion (see chap. 14), and are less boldly marked with color pattern than adults.

Subadult—A dolphin of length approximately 128–170 cm (50–67 in.). These animals approached adult proportions and adult locomotion patterns. Their fins were noticeably smaller and their bodies less robust than adults.

Adult—A dolphin of length greater than 170 cm (67 in.) (Perrin 1975a and b). They have proportionately smaller heads and taller fins, while males have larger postanal humps and thickened tail stocks.

Literature Cited

Akin, P. A. 1988. Geographic variations in tooth morphology and dentinal patterns in the spinner dolphin, *Stenella longirostris*. *Marine Mamm. Sci.*, 4(2): 132–140.

Alberts, B., D. Bray, J. Lewis, M. Raff, K. Roberts, and J. D. Watson. 1989. *Molecular Biology of the Cell*, 2d ed. Garland Publishing, Inc., New York and London.

Altmann, S. A. 1967. The structure of primate social communication. *In* Altmann, S. A., ed., *Social Communication among Primates*, Univ. of Chicago Press, Chicago.

American Institute of Biological Science. 1967. Conference on the shark–porpoise relationship. Amer. Inst. Biol. Sci., Symposium Proceedings, Washington, D.C.

Arnold, P. W. 1972. Predation on harbor porpoise, *Phocoena phocoena*, by a white shark, *Carcharodon carcharias*. *Jour. Fish. Res. Bd. Canada*, 29: 1213–1214.

Au, D. W. K. 1991. Polyspecific nature of tuna schools: Shark dolphin and seabird associates. *Fish. Bull.*, *U.S.*, 89: 343–354.

Au, D. W. K., and R. L. Pitman. 1986. Seabird interactions with dolphins and tuna in the eastern tropical Pacific. *Condor*, 88: 304–317.

Au, D. W. K., and D. Weihs. 1980. At high speeds dolphins save energy by leaping. *Nature*, 284(5756): 548–550.

Au, D. W. K., W. L. Perryman, and W. F. Perrin. 1979. Dolphin distribution and the relationship to environmental features in the eastern tropical Pacific. Natl. Mar. Fish. Serv., Southwest Fish. Center, La Jolla, CA, Admin. Rept. LJ-79-43, pp. 1–59.

Au, W. W. L. 1980. Echolocation signals of the Atlantic bottlenose dolphin (*Tursiops truncatus*) in open waters. *In* Busnel, R. -G., and J. F. Fish, eds., *Animal Sonar Systems*. N.A.T.O. Adv. Studies Inst. Ser. A, Plenum Press, New York, pp. 251–282.

Au, W. W. L., and P. W. B. Moore. 1984. Receiving beam patterns and directivity indices of the Atlantic bottlenose dolphin, *Tursiops truncatus. Jour. Acoust. Soc. Amer.*, 75(1): 255–262.

Au, W. W. L., R. W. Floyd, R. H. Penner, and A. E. Murchison. 1978. Propagation of Atlantic bottlenose dolphin echolocation signals. *Jour. Acoust. Soc. Amer.*, 64(2): 411–422.

Au, W. W. L., D. A. Carder, R. H. Penner, and B. L. Scronce. 1985. Demonstration of adaptation in beluga whale echolocation signals. *Jour. Acoust. Soc. Amer.*, 77(2): 726–730.

Axelrod, R., and W. D. Hamilton. 1981. The evolution of cooperation. *Science*, 211: 1390–1396.

Bain, D. 1986. Acoustic behavior of *Orcinus:* Sequences, periodicity, behavioral correlates, and an automated technique for call classification. *In* Kirkevold, B. C., and J. S. Lockhard, eds., *Behavioral Biology of Killer Whales.* Alan Liss, New York.

Bain, D. 1989. An evaluation of evolutionary processes: Studies of natural selection, dispersal, and cultural evolution in killer whales (*Orcinus orca*). Ph.D. Dissertation, Univ. of California, Santa Cruz.

Barlow, J. 1984. Reproductive seasonality in pelagic dolphins (*Stenella* spp.): Implications for measuring rates. *In* Perrin, W. F., R. L. Brownell, Jr., and D. P. DeMaster, eds., *Reproduction in Whales, Dolphins, and Porpoises.* Rept. Int. Whaling Comm., Spec. Issue 6, Cambridge, U.K., pp. 191–198.

Barnes, L. G. 1984. Search for the first whale. *Oceans*, 17(2): 20–23.

Barnes, L. G., and E. D. Mitchell. 1978. Cetacea. *In* Maglio, V. J., and H. B. S. Cooke, eds., *Evolution of African Mammals.* Harvard Univ. Press, Cambridge, MA, pp. 582–602.

Bartholomew, G. A. 1970. A model for the evolution of pinniped polygyny. *Evolution*, 24: 546–559.

Bateson, G. 1963. The role of somatic change in evolution. *Evolution*, 17: 529–539.

Bateson, G. 1966. Problems in cetacean and other mammalian communication. *In* Norris, K. S., ed., *Whales, Dolphins, and Porpoises.* Univ. of California Press, Berkeley and Los Angeles, pp. 569–579.

Bateson, G. 1974. Observations of a cetacean community. *In* McIntyre, J., ed., *Mind in the Waters.* C. Scribner's Sons, New York, pp. 146–165.

Bateson, G. 1979. *Mind and Nature: A Necessary Unity.* E. P. Dutton, New York.

Bateson, G., and M. C. Bateson. 1988. *Angels Fear: Towards an Epistemology of the Sacred.* Bantam Books, New York.

Batteau, D. W. 1968. Role of pinna in localization: Theoretical and physiological consequences. *In* De Reuck, A. V. S., and J. Knight, eds., *Hearing Mechanisms in Vertebrates.* Churchill, London.

Belanger, L.-F. 1940. A study of the histological structure of the respiratory portion of the lungs of aquatic mammals. *Contrib. L'Institut Zool. (Univ. Montreal)*, 8: 437–461.

Bel'kovich, V. R., and N. A. Dubrovskiy. 1977. Sensory basis of cetacean orientation. U.S. Joint Publ. Res. Serv. L-7157

Bel'kovich, V. R., and A. V. Yablokov. 1963. Marine mammals. "share" experience with designers. *Nauka i Zhzn (Science and Life), Moscow,* 30(5): 61–64.

Bel'kovich, V. R., and A. V. Yablokov. 1969. School structure of toothed cetaceans (*Odontoceti*). *In Marine Mammals.* Moscow, pp. 65–69.

Bell, R. H. V. 1971. A grazing ecosystem in the Serengeti. *Sci. Amer.,* 225(1): 86–93.

Benirschke, K., M. L. Johnson, and R. J. Benirschke. 1980. Is ovulation in dolphins, *Stenella longirostris* and *Stenella attenuata,* always copulation induced? *Fish. Bull.,* 78: 507–528.

Bezrukov, Y. F., and V. V. Natarov. 1976. Formation of abiotic conditions over submarine eminences in some regions of the Pacific Ocean. Izvestiya Tikhookeanskiy Nauchno-Issledovatel'skiy Institut Rybnogo Khozyaystva i Okeanografii (*TNIRO*) 100: 93–99.

Boehlert, G. W., and A. Genin. 1987. A review of the effects of seamounts on biological processes. *In* Keating, B. H., P. Fryer, R. Batiza, and G. H. Boehlert, eds. *Seamounts, Islands, and Atolls.* Geophys. Mon. 43, Amer. Geophys. Union, Washington, D.C., pp. 319–334.

Bonner, J. T. 1980. *The Evolution of Culture in Animals.* Princeton Univ. Press, Princeton, NJ.

Borchert, C. M., and A. L. Zihlman. 1990. The ontogeny and phylogeny of symbolizing. *In* Foster, M. L., and L. J. Botscharow, *The Life of Symbols.* Westview Press, Bouler, CO, pp. 15–44.

Boulding, K. E. 1968. General systems theory: The skeleton of science. *In* Buckley, W., ed., *Modern Systems Research for the Behavioral Scientist.* Aldine, Chicago, pp. 3–10.

Boyd, R., and P. J. Richerson. 1985. *Culture and the Evolutionary Process.* Univ. of Chicago Press, Chicago.

Brett, J. R. 1956. Some principles in the thermal requirements of fishes. *Quart. Rev. Biol.,* 31: 75–87.

Brody, S. 1945. *Bioenergetics and Growth: With Special Reference to the Efficiency Complex in Domestic Animals.* Reinhold Publ. Corp., New York.

Brower, K., and W. R. Curtsinger. 1979. *Wake of the Whale.* Friends of the Earth, New York.

Brown, D. H., and K. S. Norris. 1956. Observations of captive and wild Cetacea. *Jour. Mammal.,* 37(3): 120–145.

Brown, D. H., D. K. Caldwell, and M. C. Caldwell. 1966. Observations on the behavior of wild and captive false killer whales, with notes on associated behavior of other genera of captive delphinids. *Los Angeles County Mus. Nat. Hist., Contrib. Sci.,* 95: 1–32.

Brown, J. L. 1975. Helpers among Arabian babblers, *Turdoides squamiceps. Ibis,* 117: 243–244.

Bullock, T. H. 1955. Compensation for temperature in the metabolism and activity of poikilotherms. *Biol. Rev.,* 30: 311–342.

Bullock, T. H., and S. H. Ridgway. 1972. Evoked potentials in the central auditory system of alert porpoises to their own and artificial sounds. *Jour. Neurobiol.,* 3(1): 79–99.

Busnel, R.-G., and A. Dzeidzic. 1966. Acoustic signals of the pilot whale *Globicephala melaena* and of the porpoises *Delphinus delphis* and *Phocoena phocoena*. *In* Norris, K. S., ed., *Whales, Dolphins, and Porpoises*. Univ. of California Press, Berkeley and Los Angeles, pp. 606–646.

Busnel, R.-G., and A. Dzeidzic. 1968. Étude des signaux acoustiques associes a des situations de detresse chez certains cetaces odontocetes. *Ann. L'Institut Oceanogr., NS*, 46: 109–144.

Cadenat, J., and M. Doutre. 1959. Notes sur les Delphinides Ouest-Africains. 5, Sur un *Prodelphinus* à long bec capturé au large des cotes du Senegal, *Prodelphinus longirostris* (Gray 1828). *Bull. Inst. Franc. Afrique Noire, Ser. A.*, 21(2): 787–792.

Caldwell, D. K., and M. C. Caldwell. 1972*a*. *The World of the Bottlenose Dolphin*. J. B. Lippincott, New York.

Caldwell, D. K., and M. C. Caldwell. 1972*b*. Senses and communication. *In* Ridgway, S. H., ed., *Mammals of the Sea, Biology, and Medicine*. Charles Thomas, Publ., Springfield, IL, pp. 466–502.

Caldwell, D. K., M. C. Caldwell, and D. W. Rice. 1966. Behavior of the sperm whale, *Physeter catodon* L. *In* Norris, K. S., ed., *Whales, Dolphins, and Porpoises*. Univ. of California Press, Berkeley and Los Angeles, pp. 677–717.

Caldwell, D. K., M. C. Caldwell, W. F. Rathjen, and J. R. Sullivan. 1971. Cetaceans from the Lesser Antillean island of St. Vincent. *Fish. Bull.*, 69(2): 303–312.

Caldwell, M. C., and D. K. Caldwell. 1965. Individualized whistle contours in bottlenosed dolphins (*Tursiops truncatus*). *Nature*, 207: 434–435.

Caldwell, M. C., and D. K. Caldwell. 1966. Epimeletic (care-giving) behavior in Cetacea. *In* Norris, K. S., ed., *Whales, Dolphins, and Purpoises*. Univ. of California Press, Berkeley and Los Angeles, pp. 755–789.

Caldwell, M. C., and D. K. Caldwell. 1968. Vocalizations of naive captive dolphins in small groups. *Science*, 159: 1121–1123.

Caldwell, M. C., and D. K. Caldwell. 1979. The whistle of the Atlantic bottlenosed dolphin (*Tursiops truncatus*): Ontogeny. *In* Winn, H. E., and B. L. Olla, eds., *The Behavior of Marine Animals, Vol. 3, Cetaceans*. Plenum Press, New York, pp. 369–401.

Caldwell, M. C., D. K. Caldwell, and J. F. Miller. 1973. Statistical evidence for individual signature whistles in the spotted dolphin, *Stenella plagiodon*. Cetology, 16: 1–21.

Caldwell, M. C., D. K. Caldwell, and P. Tyack. 1990. Review of the signature-whistle hypothesis for the Atlantic bottlenose dolphin. *In* Leatherwood, S., and R. R. Reeves, eds., *The Bottlenose Dolphin*. Academic Press, San Diego, pp. 199–234.

Campbell, H. W. 1967. The effects of temperature on the auditory sensitivity of lizards. Ph.D. Dissertation, Univ. of California, Los Angeles.

Campbell, J. F., and D. Erlandson. 1979. OTEC-1 Anchor Site Survey. Final Rept. Hawaii Institute of Geophysics to U.S. Dept. of Energy, Hawaii Institute of Geophysics, 79-6, pp. 1–55.

Capitanio, J. P. 1991. Levels of integration and the "inheritance of dominance." *Animal Behav.*, 42: 495–496.

Carpenter, F. L. 1976. Plant-pollinator interactions in Hawaii: Pollination energetics of *Metrosideros collina* (Myrtaceae). *Ecology*, 57: 1125–1144.

Cheney, D. L., and R. M. Seyfarth. 1990. *How Monkeys See the World: Inside the Mind of Another Species*. Univ. Chicago Press, Chicago and London.

Cheney, D. L., and R. W. Wrangham. 1987. Predation. *In* Smuts, B. B., D. L. Cheney, R. M. Seyfarth, R. W. Wrangham, and T. T. Struhsaker, eds., *Primate Societies*. Univ. of Chicago Press, Chicago.

Clark, C. W., and M. Mangel. 1984. Foraging and flocking strategies: Information in an uncertain environment. *Amer. Nat.*, 123(5): 626–641.

Clay, C. S., and H. Medwin. 1977. *Acoustical Oceanography: Principles and Applications*. Wiley-Interscience, New York.

Cody, M. L. 1971. Finch flocks in the Mohave Desert. *Theor. Pop. Biol.*, 2: 142–158.

Coe, J. M. 1980. Passive behavior by the spotted dolphin. *Fish. Bull.*, 78(2): 535–537.

Connor, R. C. 1987. Aggressive herding of females by coalitions of male bottlenose dolphins (*Tursiops* sp.) (abst.). Seventh Biennial Conf. Biol. Marine Mamm., Miami, FL, Dec. 5–9, 1987.

Connor, R. C. 1990. Alliances among male bottlenose dolphins and a comparative analysis of mutualism. Ph.D. Dissertation, Univ. of Michigan, Ann Arbor.

Connor, R. C., and K. S. Norris. 1982. Are dolphins reciprocal altruists? *Amer. Nat.*, 119: 358–374.

Connor, R. C., R. A. Smolker, and A. F. Richards. 1992. Dolphin alliances and coalitions. *In* Harcourt, A. H., and F. M. de Waal, eds., *Coalitions and Alliances in Humans and Other Animals*. Oxford Univ. Press, Oxford, U.K.

Cosgrove, R., ed. 1991. Reducing dolphin mortality from tuna fishing. *In* Policansky, ed., National Research Council, Rept. Committee on Reducing Porpoise Mortality from Tuna Fishing, Board on Biology, Commission on Life Sciences, R. C. Francis, Chair. National Academy of Science Press, Washington D.C., pp. 1–193.

Cowan, D. F., and W. W. Walker. 1979. Disease factors in *Stenella attenuata* and *Stenella longirostris* taken in the eastern tropical Pacific yellowfin tuna purse seine fishery. Natl. Mar. Fish. Serv., Southwest Fish. Center, La Jolla, CA, Admin. Rept. LJ-79-32C, pp. 1–19.

Cranford, T. 1988. The anatomy of acoustic structures in the spinner dolphin forehead as shown by X-ray computed tomography and computer graphics. *In* Nachtigall, P. E., and P. W. B. Moore, eds., *Animal Sonar, Processes, and Performance*. Plenum Press, New York, pp. 66–77.

Cupps, P. T., L. L. Anderson, and H. H. Cole. 1969. The estrous cycle. *In* Cole, H. H., and P. T. Cupps, eds., *Reproduction in Domestic Animals*. Academic Press, New York, pp. 217–250.

Dailey, M. D., and R. L. Brownell, Jr. 1972. A checklist of marine mammal parasites. *In* Ridgway, S. H., ed., *Mammals of the Sea; Biology and Medicine*. C. Thomas, Publ., Springfield, IL, pp. 528–589.

Dailey, M. D., and W. F. Perrin. 1973. Helminth parasites of porpoises of the genus *Stenella* in the tropical eastern Pacific, with descriptions of two new species: *Mastigonema stenellae* gen. et spec. N. (Nematoda: Spiruroidea) and *Zalophotrema pacificum* sp. N. (Trematoda: Digenea). *Fish. Bull.,* 71(2): 455–471.

Davis, J. M. 1980. The coordinated aerobatics of dunlin flocks. *Animal Behav.,* 28: 668–673.

Dawkins, R. 1976. *The Selfish Gene.* Oxford Univ. Press, Oxford, U.K.

Dawkins, R. 1982. *The Extended Phenotype: The Long Reach of the Gene.* Oxford Univ. Press, Oxford, U.K.

Dawson, W. W. 1980. The cetacean eye. *In* Herman, L. H., ed., *Cetacean Behavior: Mechanisms and Processes.* John Wiley, New York, pp. 53–98.

Dawson, W. W., L. Birndorf and J. Perez. 1972. Gross anatomy and optics of the dolphin eye. *Cetology,* 10: 1–12.

Delius, J. D. 1970. Irrelevant behavior, information processing, and arousal homeostasis. *Psych. Forschung.,* 33: 165–188.

Denison, D. M., and G. Kooyman. 1973. The structure and function of the small airways in pinniped and sea otter lungs. *Resp. Physiol.,* 17: 1–10.

de Waal, F. B. M. 1982. *Chimpanzee Politics: Power and Sex among the Apes.* Harper and Row, New York.

Diercks, J., R. Trochta, C. Greenlaw, and W. E. Evans. 1971. Recording and analysis of dolphin echolocation signals. *Jour. Acoust. Soc. Amer.,* 49: 1749–1732.

Dizon, A. E., and W. F. Perrin. 1987. Genetic distances among and within species, populations, and schools of spinner and spotted dolphins. (abst.) Seventh Biennial Conf. Biol. Marine Mammals, Miami, FL.

Dizon, A. E., S. O. Southern, and W. F. Perrin. 1990. Molecular analysis of mtDNA types in exploited populations of spinner dolphins (*Stenella longirostris*). *Rept. Int. Whaling Comm.,* 13: 183–202.

Dobzhansky, T. 1941. *Genetics and the Origin of Species,* 2d ed., Columbia Univ. Press, New York.

Dormer, K. J. 1979. Mechanism of sound production and air recycling in delphinids: Cineradiographic evidence. *Jour. Acoust. Soc. Amer.,* 65(1): 229–239.

Doty, M. S. 1968. Biological and physical features of Kealake'akua Bay, Hawaii. Univ. of Hawaii Botan. Sci. Pap. 8, pp. 1–34.

Dral, A. D. G. 1972. Aquatic and aerial vision in the bottle-nosed dolphin. *Netherlands Jour. Sea Res.,* 5: 510–513.

Dreher, J. J. 1966. Cetacean communication: Small group experiment. *In* Norris, K. S., ed., *Whales, Dolphins, and Porpoises.* Univ. of California Press, Berkeley and Los Angeles, pp. 529–543.

Eisenberg, J. F. 1981. *The Mammalian Radiations: An Analysis of Trends in Evolution, Adaptation, and Behavior.* Univ. of Chicago Press, Chicago.

Erdman, D. S., J. Harms, and M. M. Flores. 1973. Cetacean records from the northwestern Caribbean region. *Cetology,* 17:1–14.

Essapian, F. S. 1962. Courtship in saddle-back porpoises, *Delphinus delphis* L 1758. *Zeit. für Saugetierkunde,* 27: 211–217.

Estes, R. D., and J. Goddard. 1967. Prey selection and hunting behavior of the African wild dog. *Jour. Wildlf. Mgmt.,* 31: 52–70.

Evans, E. C. III, and K. S. Norris. 1988. On the evolution of acoustic communication systems in vertebrates, Pt. II, Cognitive aspects. *In* Nachtigall, P. E., and P. W. B. Moore, eds., *Animal Sonar: Processes and Performance.* Plenum Press, New York, pp. 671–681.

Evans, W. E. 1974. Radio-telemetric studies of two species of small odontocete cetaceans. *In* Schevill, W. E., ed., *The Whale Problem: A Status Report.* Harvard Univ. Press, Cambridge, MA.

Evans, W. E., and J. Bastian. 1969. Marine mammal communication: social and ecological factors. *In* Andersen, H. T., ed., *The Biology of Marine Mammals.* Academic Press, New York, pp. 425–475.

Feibleman, J. K. 1954. Theory of integrative levels. *Brit. Jour. Phil. Sci.,* 5: 59–66.

Fentress, J. C. 1967. Observations on the behavioral development of a hand-reared male timber wolf. *Amer. Zool.,* 7: 339–351.

Fish, F. E., and C. Hui. 1991. Dolphin swimming—a review. *Mammal. Rev.,* 21: 181–195.

Fitch, J. E., and R. L. Brownell, Jr. 1968. Fish otoliths and their importance in interpreting feeding habits. *Jour. Fish. Res. Bd. Canada,* 25:2561–2574.

Fleischer, G. 1976. Hearing in extinct cetaceans as determined by cochlear structure. *Jour. Paleont.,* 50(1): 133–152.

Ford, J. K. B. 1984. Call traditions and dialects of killer whales (*Orcinus orca*) in British Columbia. Ph.D. Dissertation, Univ. of British Columbia, Vancouver, B.C.

Ford, J. K. B., and H. D. Fisher. 1983. Group-specific dialects of killer whales (*Orcinus orca*) in British Columbia. *In* Payne, R., ed., *Communication and Behavior of Whales.* Westview Press, Boulder, CO, pp. 129–161.

Fordyce, R. E. 1980. Whale evolution and oligocene southern ocean environments, *Paleogeog., Paleoclim., Paleoecol.,* 31:319–336.

Fowler, C. W. 1984. Density dependence in cetacean populations. *In* Perrin, W. F., R. L. Brownell, Jr, and D. P. DeMaster, eds., *Reproduction in Whales, Dolphins, and Porpoises.* Rept. Int. Whaling Comm., Spec. Issue 6, Cambridge, U.K., pp. 373–387.

Fowler, C. W., and T.D. Smith, eds. 1981. *Dynamics of Large Mammal Populations.* Wiley-Interscience, New York.

Fowler, C. W., W. T. Bunderson, M. B. Cherry, R. J. Ryel, and B. B. Steel. 1980. Comparative population dynamics of large mammals: A search for management criteria. U.S. Marine Mamm. Comm. Rept., Natl. Info. Serv., Springfield, VA, PB80-1786327.

Frame, L. H., J. G. Malcolm, G. W. Frame, and H. van Lawick. 1979. Social organization of African wild dogs (*Lycaon pictus*) on the Serengeti Plains, Tanzania, 1967–1970. *Z. Tierpsychol,* 50: 225–249.

Fraser, F. C. 1966. Comments on the Delphinoidea. *In* Norris, K. S., ed., *Whales, Dolphins, and Porpoises,* Univ. of Calif. Press, Berkeley and Los Angeles, pp. 7–31.

Fraser, F. C., and P. E. Purves. 1960. Hearing in cetaceans: Evolution of the accessory air sacs and the structure and function of the outer and middle ear in recent cetaceans. *Bull. British Mus. (Nat. Hist.), Zool.,* 7(1): 1–140.

Fry, F. E. J. 1947. Effects of the environment on animal activity. Univ. of Toronto Studies, Biol. Ser. No. 55, Publ. Ontario Fish. Res. Lab, pp. 1–91.

Futuyama, D. J., and M. Slatkin, eds. 1983. *Coevolution.* Sinauer Assoc., Sunderland, MA.

Galliano, R. E., P. J. Morgane, W. L. McFarland, E. L. Nagle, and R. L. Catherman. 1966. The anatomy of the cervicothoracic arterial system in the bottlenosed dolphin (*Tursiops truncatus*), with a surgical approach suitable for guided angiography. *Anat. Rec.*, 155: 325–337.

Gaskin, D. E. 1982. *The Ecology of Whales and Dolphins.* Heinemann, London.

Gates, D. 1962. *Energy Exchange in the Biosphere.* Bio. Monograph Ser., Harper and Row, New York.

Gautier-Hion, A. 1973. Social and ecological features of talapoin monkeys: Comparisons with sympatric Cercopithecines. *In* Michael, R. P., and J. H. Crook, eds., *Comparative Ecology and Behavior of Primates.* Academic Press, New York.

Gavrilov, L. R. 1969. On the size distribution of gas bubbles in water. *Sov. Phys.-Acoust.*, 15: 22–24.

Gawain, E. 1981. *The Dolphin's Gift.* Whatever Publishing, Mill Valley, CA.

Geist, V. 1971. *Mountain Sheep.* Univ. of Chicago Press, Chicago.

Gentry, R. L. 1974. The development of social behavior through play in the Steller Sea Lion. *Amer. Zool.*, 14: 391–403.

Geraci, J. R., M. D. Dailey, and D. J. St. Aubin. 1978. Parasitic mastitis in the Atlantic whitesided dolphin, *Lagenorhynchus acutus*, with implications as a probable factor in herd productivity. *Jour. Fish. Res. Bd. Canada*, 35: 1350–1355.

Gingerich, P. D. 1977. A small collection of fossil vertebrates from the middle Eocene Kuldana and Kohat formations of Punjab (Pakistan). *Contrib. Mus. Paleont., Univ. Michigan*, 24: 190–203.

Gingerich, P. D., and D. E. Russell. 1981. *Pakicetus inachus,* a new archaeocete (Mammalia, Cetacea) from the early middle Eocene Kuldana Formation of Kohat (Pakistan). *Contrib. Mus. Paleont., Univ. Michigan*, 25(11): 235–246.

Gingerich, P. D., N. A. Wells, D. E. Russell, and S. M. Ibrahim Shah. 1983. Origin of whales in epicontinental remnant seas: New evidence from the early Eocene of Pakistan. *Science*, 220: 403–406.

Glotov, V. P. 1962. Investigation of the scattering of sound of bubbles generated by an artificial wind in seawater and the statistical distribution of bubble sizes. *Sov. Phys.-Acoust.*, 7: 341–345.

Goldizen, A. W. 1987. Tamarins and marmosets: Communal care of offspring. *In* Smuts, B. B., D. L. Cheney, R. M. Seyfarth, R. W. Wrangham, and T. H. Struhsaker, eds., Primate Societies. Univ. of Chicago Press, Chicago.

Goldizen, A. W., and J. Terborgh. 1987. Cooperative polyandry and helping behavior in saddle-backed tamarins (*Saguinus fuscicollis*). *In* Else, J. G., and P. C. Lee, eds., *Primate Ecology and Conservation.* Cambridge Univ. Press, Cambridge, U.K., pp. 191–198.

Goodenough, W. H. 1957. Cultural anthropology and linguistics. *In* Garim, P., ed., *Report of the Seventh Annual Round Table Meeting on Linguistics and Language Study.* Georgetown Univ. Monogr. Ser. Lang. and Ling. No. 9, pp. 167–173.

Gooding, R. M., and J. J. Magnuson. 1967. Ecological significance of drifting objects to pelagic fishes. *Pacific Sci.*, 21: 486–497.

Gordon, D. 1989. Caste and change in social insects. *In* Harvey, P. H., and L. Partridge, eds., *Oxford Surveys in Evolutionary Biology.* Oxford Univ. Press, Oxford, U.K., pp. 55–72.

Gosline, W. A., and V. E. Brock. 1960. *Handbook of Hawaiian Fishes.* Univ. of Hawaii Press, Honolulu.

Gould, S. J. 1989. *Wonderful Life: The Burgess Shale and the Nature of History.* W. W. Norton, New York.

Gray, J. E. 1828. *Spicilegia Zoologica.* Pt. 1: 1–8, Treüttel, Wurtz, London.

Green, S. 1975. Variation in vocal pattern with social situation in the Japanese monkey (*Macaca fuscata*). *Prim. Behav.,* 4: 1–102.

Greenblatt, P. R. 1976. Associations of tuna with flotsam in the eastern tropical Pacific. *Fish. Bull.,* 77: 147–155.

Hamilton, W. D. 1964a. The genetical evolution of social behaviour, I. *Jour. Theor. Biol.,* 7: 1–16.

Hamilton, W. D. 1964b The genetical evolution of social behaviour, II. *Jour. Theor. Biol.,* 7: 17–32.

Hamner, W. M. 1984. Aspects of schooling in *Euphausia superba. Jour. Crustacean Biol.,* 4: 67–74.

Hampton, I. F. G., and C. Whittow. 1976. Body temperature and heat exchange in the Hawaiian spinner dolphin, *Stenella longirostris. Comp. Biochem. Physiol.,* 55A: 195–197.

Harrison, R. 1969. Reproduction and reproductive organs. *In* Andersen, H. T., ed., *The Biology of Marine Mammals.* Academic Press, New York, 253–348.

Harrison, R. J., R. L. Brownell Jr., and R. C. Boyce. 1972. Reproduction and gonadal appearances in some odontocetes. *In* Harrison, R. J., ed., *Functional Anatomy of Marine Mammals,* vol. 1. Academic Press, New York.

Hecht, S., S. Shlaer, and C. D. Verrijp. 1934. Intermittant stimulation by light, II. The measurement of critical fusion frequency for the human eye. *Jour. Gen. Physiol.,* 17: 237–249.

Hempleman, H. V., and A. P. M. Lockwood. 1978. The physiology of diving in man and other animals. The Institute of Biology, Studies in Biology, no. 99, E. Arnold, London.

Herman, L. M. 1980a. Cognitive characteristics of dolphins. *In* Herman, L. M., ed., *Cetacean Behavior: Mechanisms and Functions.* Wiley-Interscience, New York.

Herman, L. M., ed. 1980b. *Cetacean Behavior: Mechanisms and Functions.* Wiley-Interscience, New York.

Herman, L. M., M. Peacock, M. Yunker, and C. Madsen. 1975. Bottlenosed dolphin: Double-slit pupil yields equivalent aerial and underwater diurnal acuity. *Science,* 189: 650–652.

Hertel, H. 1969. Hydrodynamics of swimming and wave riding dolphins. *In* Andersen, H., ed., *The Biology of Marine Mammals.* Academic Press, New York, pp. 31–63.

Hester, F. J., J. R. Hunter, and R. R. Whitney. 1963. Jumping and spinning behavior in the spinner porpoise. *Jour. Mammal.,* 44(4): 586–588.

Heyning, J. E. 1980. Functional morphology involved in intraspecific fighting of the beaked whale. *Mesoplodon carlhubbsi* (abst.). Paper presented at So. Calif. Academy of Sciences, Long Beach CA.

Heyning, J. E. 1984. Functional morphology involved in intraspecific fighting of the beaked whale *Mesoplodon carlhubbsi. Canadian Jour. Zool.*, 62(8): 1645–1654.

Heyning, J. E. 1989. Comparative facial anatomy of beaked whales (Ziphiidae) and a systematic revision among the families of extant Odontoceti. *Contrib. Sci. Los Angeles County Natural History Museum*, 405: 1–64.

Hobson, E. S. 1978. Aggregating as defense against predators in aquatic and terrestrial environments. *In* Reese, E. S., and F. J. Lighter, eds., *Contrasts in Behavior.* Wiley, New York.

Hoffman, A. 1979. Community paleoecology as an epiphenomenal science. *Paleobiol.*, 5(4): 357–379.

Hoffman, A. 1983. The status of ecological communities. *Zbl. Geol. Palaont.*, 2(3/4): 81–87.

Hofstadter, D. R. 1979. *Gödel, Escher, Bach: An Eternal Golden Braid.* Vintage Books, Random House, New York.

Hohn, A. A., S. J. Chivers, and J. Barlow. 1985. Reproductive maturity and seasonality in male spotted dolphins, *Stenella attenuata*, in the eastern tropical Pacific. *Marine Mamm. Sci.*, 1(4): 273–293.

Holbrook, J. R. 1980. Dolphin mortality related to the yellowfin tuna purse seine fishery in the eastern tropical Pacific: An annotated bibliography. Porpoise Rescue Foundation, San Diego, CA, *Tech. Bull.*, 2: 1–131.

Howell, A. B. 1930. *Aquatic Mammals: Their Adaptations to Life in the Water.* C. Thomas, Publ., Springfield, Ill.

Hubbs, C. L. 1965. Data on speed and underwater exhalation of a humpback whale accompanying ship. *Hvalradets-Skrifter*, 48: 42–44.

Hui, C. A. 1979. Undersea topography and distribution of dolphins of the genus *Delphinus* in the southern California bight. *Jour. Mammal.*, 60(3): 521–527.

Hunter, J. R. 1968a. Affects of light on schooling and feeding of jack mackerel (*Trachurus symmetricus*). *Jour. Fish. Res. Bd. Canada*, 25: 393–407.

Hunter, J. R. 1968b. Fishes beneath flotsam. *Sea Frontiers*, 14(5): 280–288.

Hunter, J. R., and C. T. Mitchell. 1966. Association of fishes with flotsam in the offshore waters of Central America. *Fish. Bull.*, 66(1): 13–29.

Hunter, J. R., and C. T. Mitchell. 1968. Field experiments on the attraction of pelagic fish to floating objects. *Int. Council Explor. Sea, Jour. du Conseil*, 31: 427–434.

Huntley, A. C., D. P. Costa, G. A. J. Worthy, and M. A. Castellini, eds. 1987. Approaches to Marine Mammal Energetics, Soc. Mar. Mamm., Spec. Pub. 1, Lawrence, Ks., pp. 1–253.

Irvine, B., R. S. Wells, and P. Gilbert. 1973. Conditioning an Atlantic bottle-nose dolphin, *Tursiops truncatus*, to repel various species of sharks. *Jour. Mammal.*, 54(2): 503–505.

Irvine, A. B., M. D. Scott, R. S. Wells, and J. H. Kaufmann. 1981. Movements and activities of the Atlantic bottlenose dolphin, *Tursiops truncatus*, near Sarasota, FL. *Fish. Bull.*, 80: 671–688.

Irvine, A. B., R. S. Wells, and M. D. Scott. 1982. An evaluation of techniques for tagging small odontocete cetaceans. *Fish. Bull.*, 80: 135–143.

Itani, J. 1958. On the acquisition and propagation of new food habits in a troop of Japanese monkeys at Takasakiyama. *Primates,* 1: 84–98.

Jacobs, M., P. Morgane, and W. McFarland. 1975. Degeneration of visual pathways in the bottlenose dolphin. *Brain Res.,* 88: 346–352.

Jakobson, R. 1960. Linguistics and poetics. *In* Sebeok, T. A., ed., *Style in Language.* MIT Press, Cambridge, MA, pp. 350–377.

Jameson, D., and L. M. Hurvich. 1961. Complexities of perceived brightness. *Science,* 133: 174–179.

Jarman, M. V. 1979. Impala social behavior, territory, hierarchy, mating and use of space. *In Adv. in Ethology, Suppl. 21. Jour. Comp. Ethology,* Berlin, pp. 1–93.

Jarman, P. J. 1974. The social organization of antelope in relation to their ecology. *Behaviour,* 68(3–4): 215–267.

Jerison, H. J. 1976. Paleoneurology and the evolution of the mind. *Sci. Amer.,* 234(1): 90–101.

Johnson, C. M., and D. L. Herzing. 1991. The social use of tacticle effects of dolphin vocalizations (abst.). *Ninth Biennial Conf. on Biology of Marine Mammals,* December 1991, Chicago, IL.

Johnson, C. M., and K. S. Norris. 1986. Delphinid social organization and social behavior. *In* Schusterman, R. J., J. A. Thomas, and F. G. Wood, eds., *Dolphin Cognition and Behavior: A Comparative Approach.* L. Erlbaum, Assoc., Hillsdale, NJ, pp. 335–346.

Johnson, C. S. 1967. Sound detection thresholds in marine mammals. *In* Tavolga, W. N., ed., *Marine Bio-acoustics,* vol. 2, Proc. 2nd Symp. Marine Bio-acoustics (Amer. Mus. Nat. Hist), Pergamon Press, New York, pp. 247–260.

Johnson, R. H., and D. R. Nelson. 1973. Agonistic display in the gray reef shark, *Carcharhinus menisorah,* and its relation to attacks on man. *Copeia,* 1: 70–83.

Jones, E. C. 1971. *Isistius brasiliensis,* a squaloid shark, the probable cause of crater wounds on fishes and cetaceans. *Fish. Bull.,* 69(4): 791–798.

Kandel, E. R., and J. H. Schwartz. 1985. *Principles of Neural Science,* 2d ed. Elsevier, New York.

Kasuya, T., and H. Marsh. 1984. Life history and reproductive biology of the short-finned pilot whale, *Globicephala macrorhynchus,* off the coast of Japan. *In* Perrin, W. F., R. L. Brownell, Jr., and D. P. DeMaster, eds., *Reproduction in Whales, Dolphins, and Porpoises.* Rept. Int. Whaling Comm., Spec. Issue 6. Cambridge, U.K., pp. 259–310.

Kasuya, T., M. Miyazaki, and W. H. Dawbin. 1974. Growth and reproduction of *Stenella attenuata* on the Pacific coast of Japan. *Sci. Rept. Whales Res. Inst.,* 26: 157–226.

Kavanau, J. L. 1987. *Love birds, Cockatiels, and Budgerigars: Behavior and Evolution.* Science Software Systems, Los Angeles.

Kelley, H. R. 1959. A two-body problem in the echelon-formation swimming of porpoises. U.S. Naval Ordinance Test Sta., China Lake, Tech. Notes, 40606-1, pp. 1–7.

Kenagy, G. J., and S. C. Trombulak. 1986. Size and function of mammalian testes in relation to body size, *Jour. Mammal.,* 67(1): 1–22.

Kennett, J. P. 1982. *Marine Geology.* Prentice-Hall, Englewood Cliffs, NJ.

Kielhorn, W. V., K. S. Norris, and W. E. Evans. 1963. Bathing behavior of frigate birds. *The Condor,* 65: 240–241.

Kirby, V. L., and S. H. Ridgway. 1984. Hormonal evidence for spontaneous ovulation in captive dolphins, *Tursiops truncatus* and *Delphinus delphis. In* Perrin, W. F., R. L. Brownell Jr., and D. P. DeMaster, eds., *Reproduction in Whales, Dolphins, and Porpoises.* Rept. Int. Whaling Comm. Spec., Issue 6, Cambridge, U.K., pp. 459–464.

Kooyman, G. L. 1973. Respiratory adaptations in marine mammals. *Amer. Zool.,* 13: 457–468.

Kooyman, G. L., and H. T. Andersen. 1969. Deep diving. *In* Andersen, H. T., ed., *The Biology of Marine Mammals,* Academic Press, New York.

Kooyman, G. L., and L. H. Cornell. 1981. Flow properties of expiration and inspiration in a trained bottlenose porpoise. *Physiol. Zool.,* 54(1): 55–61.

Kooyman, G. L., K. S. Norris, and R. L. Gentry. 1975. Spout of the gray whale: Its physical characteristics. *Science,* 190: 908–910.

Kooyman, G. L., E. A. Wahrenbrock, M. A. Castellini, R. W. Davis, and E. E. Sinnett. 1980. Aerobic and anaerobic metabolism during voluntary diving in Weddell Seals: Evidence of preferred pathways from blood chemistry and behavior. *Jour. Comp. Physiol.,* 138: 335–346.

Kooyman, G. L., M. A. Castellini, and R. W. Davis. 1981. Physiology of diving marine mammals. *Ann. Rev. Physiol.,* 43: 343–356.

Kummer, H. 1968. *Social Organization of Hamadryas Baboons: A Field Study.* Univ. of Chicago Press, Chicago.

Landeau, L, and J. Terborgh. 1986. Oddity and the "confusion effect" in predation. *Animal Behav.,* 34: 1372–1380.

Lang, T. G. 1966. Hydrodynamic analysis of cetacean performance, *In* Norris, K. S., ed., *Whales, Dolphins, and Porpoises.* Univ. of California Press, Berkeley and Los Angeles, pp. 410–432.

Lang, T. G., and K. S. Norris. 1966. Swimming speed of a Pacific bottlenose porpoise, *Science,* 151: 588–590.

Lang, T. G., and K. Pryor. 1966. Hydrodynamic performance of porpoises *(Stenella attenuata). Science,* 152: 531–533.

Lawrence, B., and W. E. Schevill. 1956. The functional anatomy of the delphinid nose. *Bull. Mus. Comp. Zool.,* 114: 103–152.

Leatherwood, J. S., W. F. Perrin, R. Garvie, and J. LaGrange. 1973. Observations of sharks attacking porpoises *(Stenella* spp. and *Delphinus* cf. *D. delphis). Naval Undersea Tech. Note,* 908: 1–7.

Leatherwood, S., R. R. Reeves, and L. Foster. 1983. *The Sierra Club Handbook of Whales and Dolphins.* Sierra Club Books, San Francisco.

LeBoeuf, B. J. 1974. Male–male competition and reproductive success in elephant seals. *Amer. Zool.,* 14: 163–176.

Lehner, P. N. 1978. Coyote vocalizations: A lexicon and comparisons with other canids. *Animal Behav.,* 26: 712–722.

Lissaman, P. B. S., and C. A. Schollenberger. 1970. Formation flight of birds. *Science,* 168: 1003–1005.

Loew, E. R., and W. N. McFarland. 1990. The underwater visual environment. *In* Douglas, R. H., and M. B. A. Djamgoz, eds., *The Visual System of Fish.* Chapman and Hall, Ltd., London, pp. 1–41.

Longhurst, A. R., and D. Pauly. 1987. *Ecology of Tropical Oceans.* Academic Press, New York.

Longuet-Higgins, M. S. 1990. Bubble noise spectra. *Jour. Acoust. Soc. Amer.,* 87(2): 652–661.

Lorenz, K. 1952. *King Solomon's Ring: New Light on Animal Ways.* T. Crowell Co., New York.

Lott, D. F. 1984. Intraspecific variation in the social systems of wild vertebrates. *Behaviour,* 88(3–4): 266–325.

Lythgoe, J. N. 1966. Visual pigments and underwater vision. *In* Bainbridge, R., G. C. Evans, and O. Rackham, eds., *Light as an Ecological Factor.* Blackwell Press, Oxford, pp. 375–391.

Lythgoe, J. N. 1979. *The Ecology of Vision.* Clarendon Press, Oxford.

Macdonald, G. A., A. T. Abbott, and F. L. Peterson. 1983. *Volcanoes in the Sea: The Geology of Hawaii.* Univ. of Hawaii Press, Honolulu.

Mackay, R. S. 1982. Dolphins and the bends. *Science,* 216: 650.

Mackay, R. S. 1988. Whale heads, magnetic resonance images, ray diagrams and tiny bubbles. *In* Nactigall, P. E., and P. W. B. Moore, eds., *Sonar: Processes and Performance.* Plenum Press, New York, pp. 79–86.

Mackay, R. S., and C. Liaw. 1981. Dolphin vocalization mechanisms. *Science,* 212(8): 676.

Mackintosh, N. 1965. *The Stocks of Whales.* Fishing News (Books) Ltd., London.

Madsen, C. J., and L. M. Herman. 1980. Social and ecological correlates of cetacean vision and visual appearance. *In* Herman, L., ed., *Cetacean Behavior: Mechanisms and Functions.* Wiley-Interscience, New York.

Major, P. F. 1978. Predator–prey interactions in two schooling fishes, *Caranx ignobilis* and *Stolephorus purpureus. Animal Behav.* 26: 760–777.

Marler, P., and L. Hobbett. 1975. Individuality in long range vocalization of wild chimpanzees. *Z. Tierpsych.* 38: 97–109.

Marsh, H., and T. Kasuya. 1984. Changes in the ovaries of the short-finned pilot whale, *Globicephala macrorhynhus,* with age and reproductive activity. *In* Perrin, W. F., R. L. Brownell, Jr., and D. P. DeMaster, eds., *Reproduction in Whales, Dolphins, and Porpoises.* Rept. Int. Whaling Comm., Spec. Issue 6. Cambridge, U.K., pp. 311–341.

Marsh, H., and T. Kasuya. 1991. An overview of the changes in the role of a female pilot whale with age. *In* Pryor, K., and K. S. Norris, eds., *Dolphin Societies: Discoveries and Puzzles.* Univ. of California Press, Berkeley and Los Angeles, pp. 281–285.

Marshall, N. B. 1954. *Aspects of Deep Sea Biology.* Hutchinson and Co., London.

Marten, K., K. S. Norris, P. W. B. Moore, and K. Englund. 1988. Loud impulse sounds in odontocete predation and social behavior. *In* Nachtigall, P. E., and P. W. B. Moore, eds., *Animal Sonar: Processes and Performance,* vol. 156. N.A.T.O. Adv. Sci. Inst. Ser., Life Sciences, Plenum Press, New York.

McBride, A. F. 1940. Meet Mr. Porpoise. *Nat. Hist.,* 45(1): 16–29.

McBride, A. F., and D. O. Hebb. 1948. Behavior of the captive bottlenose dolphin (*Tursiops truncatus*). *Jour. Comp. Physiol. and Psychol.,* 41: 111–123.

McBride, A. F., and H. Kritzler. 1951. Observations on pregnancy, parturition, and postnatal behavior in the bottlenose dolphin. *Jour. Mammal.,* 32: 251–266.

McCann C. 1974. Body scarring on Cetacea-Odontoceti. *Sci. Repts. Whales Res. Inst. (Tokyo)*, 26: 145–155.

McCarley, H. 1975. Long distance communication in coyotes. *Jour. Mammal.*, 56: 847–856.

McCormick, J. G. 1969. Relationships of sleep, respiration and anaesthesia in the porpoise: A prelimary report. *Proc. Nat. Acad. Sci.*, 62: 697–703.

McGuire, M. T., ed. 1988. *Ethology and Sociobiology*, vol. 9. Elsevier Science Publ., New York.

McFarland, W. N. 1971. Cetacean visual pigments *Vision Res.*, 11: 1065–1076.

McFarland, W. N. 1986. Light in the sea: Correlations with behaviors of fishes and invertebrates. *Amer. Zool.*, 26: 389–401.

McFarland, W. N. 1991. The visual world of coral reef fishes. *In* Sale, P. F., ed., *The Ecology of Fishes on Coral Reefs*. Academic Press, New York.

McFarland, W N., and E. R. Loew. 1983. Wave produced changes in underwater light and their relations to vision. *Env. Biol. Fish.*, 8(3–4): 173–184.

McFarland, W. N., and F. W. Munz. 1975. Part II: The photic environment of clear tropical seas during the day. *Vision Res.*, 15: 1063–1070.

McGinnis, S. M., G. C. Whittow, C. A. Ohata, and H. Huber. 1972. Body heat dissipation and conservation in two species of dolphins. *Comp. Biochem. Physiol.*, 43A: 417–423.

McGowan, J. A. 1972. The nature of oceanic ecosystems. *In The Biology of the Oceanic Pacific*. Proc. 33rd Ann. Biol. Colloquium, Oregon State Univ. Press, Corvallis, OR, pp. 9–28.

Mead, J. G. 1975. Anatomy of the external nasal passages and facial complex in the Delphinidae (Mammalia: Cetacea). *Smithsonian Contrib. Zool.*, 207: 1–72.

Mead, J. G., D. K. Odell, R. S. Wells, and M. D. Scott. 1980. Observations on a mass stranding of spinner dolphin, *Stenella longirostris*, from the west coast of Florida. *Fish. Bull.*, 78: 353–360.

Mech, L. D. 1970. *The Wolf: The Ecology and Behavior of an Endgangered Species*. Univ. of Minnesota Press, Minneapolis.

Medwin, H., and N. D. Breitz. 1989. Ambient and transient bubble spectral densities in quiescent seas under spilling breakers. *Jour. Geophys. Res. (Oceans)*, 94(C9): 12751–12759.

Melville, H. 1851. *Moby-Dick; or, the Whale*. Harper and Brothers, New York.

Meyer, E. 1957. Air bubbles in water. *In* Richardson, E. G., ed., *Technical Aspects of Sound*. Elsevier Publ., London, pp. 222–239.

Miller, G. A. 1956. The magical number seven, plus or minus two: Some limits on our capacity for processing information, *Psych. Rev.*, 63: 81–97.

Miller, R. C. 1977. The significance of the gregarious habit. *Ecology*, 3: 122–129.

Mitchell, E. D. 1970. Pigmentation pattern evolution in delphinid cetaceans: An essay in adaptive coloration. *Canadian Jour. Zool.*, 48(4): 717–740.

Miyazaki, N. 1977. School structure of *Stenella coeruleoalba*. *Rept. Int. Whal. Comm.* 27. pap. L, 18: 498–499.

Miyazaki, N., and M. Nishiwaki. 1978. School structure of the striped dolphin of the Pacific coast of Japan. *Sci. Rept. Whales Res. Inst. (Toykyo)*, 30: 65–115.

Miyazaki, N., T. Kasuya, and M. Nishiwaki. 1974 Distribution and migration of two species of *Stenella* in the Pacific coast of Japan. *Sci. Rept. Whales Res. Inst. (Tokyo)*, 26: 227–243.

Mukhametov, L. M., A. Y. Supin, and I. G. Polyakova. 1977. Interhemispheric asymmetry of the electroencephalographic sleep patterns in dolphins. *Brain Res.*, 134: 581–584.

Murchison, A. E. 1980. Detection range and range resolution in echolocating bottlenose porpoise, *Tursiops truncatus. In* Busnel, R.-G., and J. F. Fish, eds., *Animal Sonar Systems.* Plenum Press, New York, pp. 43–70.

Murphy, G. I., and I. I. Ikehara. 1955. A summary of sightings of fish schools and bird flocks and of trolling in the central Pacific. U.S. Fish. and Wildlf. Serv., *Spec. Sci. Rept., Fish.,* 154: 1–19.

Nachtigall, P. E., and P. W. B. Moore, eds. 1988. *Animal Sonar: Processes and Performance,* vol. 156. N.A.T.O. Adv. Sci. Ser. Life Sciences, Plenum Press, New York.

National Research Council. 1992. *Reducing Dolphin Mortality from Tuna Fishing.* Committee on Reducing Porpoise Mortality from Tuna Seining, Board on Biology, Bd. of Environ. Stud. and Toxicol. Comm. on Life Sci., Natl. Res. Council, Rept. National Academy Press, Washington D.C., pp. 1–179.

Nordmark, J. 1960. Perception of distance in animal echo-location. *Nature,* 188(4755): 1009–1010.

Nordmark, W. R., P. W. Lipman, J. P. Lockwood, and J. G. Moore. 1978. Bathymetry and geology, Kealakekua Bay, Hawaii. *U.S. Geological Survey Misc. Field Studies Map,* MF-986.

Norris, K. S. 1963. The functions of temperature in the ecology of the percoid fish, *Girella nigricans* (Ayres). *Ecol. Mon.,* 33: 23–62.

Norris, K. S. 1967. Aggressive behavior in Cetacea. *In* Clemente, C. D., and D. B. Lindsley, eds., *Aggression and Defense: Neural Mechanisms and Social Patterns,* vol. 5. Proc. Fifth Conf. Brain Function, Brain Res. Institute (UCLA), Univ. of California Press, Berkeley and Los Angeles, pp. 225–241.

Norris, K. S. 1968. The evolution of acoustic mechanisms in odontocete cetaceans. *In* Drake, E. T., ed., *Evolution and Environment.* Yale Univ. Press, New Haven, CT, pp. 297–324.

Norris, K. S. 1974. *The Porpoise Watcher.* W. W. Norton, New York.

Norris, K. S. 1991. *Dolphin Days: The Life and Times of the Spinner Dolphin.* W. W. Norton, New York.

Norris, K. S., and T. P. Dohl. 1980a. Behavior of the Hawaiian spinner dolphin, *Stenella longirostris. Fish. Bull.,* 77(4): 821–849.

Norris, K. S., and T. P. Dohl. 1980b. The structure and functions of cetacean schools. *In* Herman, L. M. ed., *Cetacean Behavior: Mechanisms and Functions.* Wiley-InterScience, New York, pp. 211–261.

Norris, K. S., and W. E. Evans. 1967. Directionality of echolocation clicks in the rough-tooth porpoise, *Steno bredanensis* (Lesson). *In* Tavolga, W. N., ed., *Marine Bio-Acoustics,* 2d Symp. Pergamon Press, New York, pp. 305–316.

Norris, K. S., and E. C. Evans III. 1988. On the evolution of acoustic communication systems in vertebrates, Part 1: Historical Aspects. *In* Nactigall, P. E., and P. W. B. Moore, eds., *Animal Sonar: Processes and Performance,* vol. 156. N.A.T.O. Adv. Sci. Inst. Ser. Life Sciences, Plenum Press, New York, pp. 655–669.

Norris, K. S., and B. Møhl. 1983. Can odontocetes debilitate prey with sound? *Amer. Nat.,* 122(1): 85–104.

Norris, K. S., and J. H. Prescott. 1961. Observations of Pacific cetaceans in Californian and Mexican waters. *Univ. Calif. Publ. Zool,* 63(4): 291–402.

Norris, K. S., and C. R. Schilt. 1988. Cooperative societies in three-dimensional space: On the origins of aggregations, flocks, and schools, with special reference to dolphins and fish. *Ethol. Sociobiol.,* 9: 149–179.

Norris, K. S., W. E. Evans, and R. Turner. 1967. Echolocation in an Atlantic bottlenose porpoise during discrimination. *In* Busnel, R.-G. ed., *Les Systemes Sonars Animaux, Biologie et Bionique.* N.A.T.O. Adv. Study Inst., Lab Physiol. Acoust., Juoy-en-Josas, France, pp. 409–436.

Norris, K. S., K. J. Dormer, J. Pegg, and G. J. Liese. 1971. The mechanism of sound production and air recycling in porpoises: A preliminary report. Proc. Eight Ann. Conf. on Biol. Sonar and Diving Mammals, Stanford Res. Inst. Menlo Park, Calif., pp. 113–129.

Norris, K. S., W. E. Stuntz, and W. Rogers. 1978. The behavior of porpoises and tuna in the Eastern Tropical Pacific yellowfin tuna fishery: Preliminary studies. *U.S. Mar. Mamm. Comm. Rept.,* Natl. Info. Serv., Springfield, VA, PB-283970.

Oelofsen, B., and D. C. Araujo. 1983. Paleoecological implications of the distribution of mesosaurid reptiles in the Permian Irati Sea (Parana Basin), South America. *Rev. Brasileira de Geosci.,* 13(1): 1–6.

Orr, R. T. 1976. *Vertebrate Biology,* W. B. Saunders Co., Philadelphia.

Oshumi, S. 1972. Catch of marine mammals, mainly small cetaceans, by local fisheries along the coast of Japan. *Far Seas Fish. Res. Lab. Bull.,* 7: 137–166.

Oster, G., and E. O. Wilson. 1978. *Caste and ecology in the Social Insects.* Princeton Univ. Press, Princeton, NJ.

Overstrom, N. A. 1982. Association between burst-pulse sounds and aggressive behavior in captive Atlantic bottlenosed dolphins *(Tursiops truncatus). Zoo. Biol.,* 2: 93–103.

Owings, D. H., D. F. Hennessy, D. W. Leger, and A. B. Gladney. 1986. Different functions of "alarm" calling for different time scales: A preliminary report on ground squirrels. *Behaviour* 99(1–2): 101–116.

Pabst, D. A. 1990. Axial muscles and connective tissues of the bottlenose dolphin. *In* Leatherwood, S., and R. R. Reeves, eds., *The Bottlenose Dolphin.* Academic Press, New York, pp. 51–67.

Parker, G. A. 1984. Evolutionarily stable strategies. *In* Krebs, J. R., and N. B. Davies, eds., *Behavioral Ecology: An Evolutionary Approach.* Sinauer Assoc., Sunderland, MA.

Partridge, B. L. 1981. Lateral line function and the internal dynamics of fish schools. *In* Tavolga, W. N., A. N. Popper, and R. R. Fay, eds., *Hearing and Sound Communication in Fishes.* Springer-Verlag, New York, pp. 515–522.

Partridge, B. L. 1982. The structure and function of fish schools. *Sci. Amer.,* 245: 114–123.

Pearse, J. S., and S. W. Arch. 1969. The aggregation behavior of *Diadema* (Echinodermata, Echinoidea). *Micronesica,* 5(1): 165–171.

Pepperberg, I. 1987. Acquisition of anomalous communicatory systems: Implications for studies of interspecies communication. *In* Schusterman, R. J., J. A.

Thomas, and F. G. Woods, eds., *Dolphin Cognition and Behavior.* L. Erlbaum, Assoc., Hillsdale, NJ, pp. 289–302.

Perrin, W. F. 1969a. Using porpoise to catch tuna. *World Fishing,* 18(6): 42–45.

Perrin, W. F. 1969b. Color pattern of the Eastern Pacific spotted porpoise *Stenella graffmani* Lonnberg (Cetacea, Delphinidae). *Zoologica,* 54(4): 135–142.

Perrin, W. F. 1972. Color patterns of spinner porpoises *(Stenella* cf. *S. longirostris)* of the eastern Pacific and Hawaii, with comments on dephinid pigmentation. *Fish. Bull:* 70(3): 983–1003.

Perrin, W. F. 1975a. Variation and taxonomy of spotted and spinner porpoises (genus *Stenella*) of the Eastern Tropical Pacific and Hawaii. *Bull. Scripps Inst. Oceanog.,* 21: 1–206.

Perrin, W. F. 1975b. Distribution and differentiation of populations of dolphins of the genus *Stenella* in the Eastern Tropical Pacific. *Jour. Fish. Res. Bd. Canada,* 32(7): 1059–1067.

Perrin, W. F. 1990. Subspecies of *Stenella longirostris* (Mammalia: Cetacea, Delphinidae). *Proc. Biol. Soc. Washington,* 103(2): 453–463.

Perrin, W. F., and J. W. Gilpatrick, Jr. in press. Spinner dolphin, *Stenella longirostris* (Gray, 1846). *In* Ridgway, S. H., and R. J. Harrison, eds., *Handbook of Marine Mammals,* vol. 5 Academic Press, New York.

Perrin, W. F., and J. R. Henderson. 1984. Growth and reproductive rates in two populations of spinner dolphins, *Stenella longirostris,* with different histories of exploitation. *In* Perrin, W. F., R. L. Brownell, Jr., and D. P. DeMaster, eds., *Reproduction in Whales, Dolphins, and Porpoises.* Rept. Int. Whaling Comm., Spec. Issue 6, Cambridge, U.K., pp. 417–430.

Perrin, W. F., and J. R. Hunter. 1972. Escape behavior of the Hawaiian spinner porpoise *(Stenella* cf. *S. longirostris). Fish. Bull.,* 70: 49–60.

Perrin, W. F., and C. W. Oliver. 1982. Time/area distribution and composition of the incidental kill of dolphins and small whales in the U.S. purse-seine fishery for tuna in the eastern tropical Pacific, 1979–1980. *Rept. Int. Whaling Comm.,* 32: 429–444.

Perrin, W. F., and J. Powers. 1980. Role of a nematode in natural mortality of spotted dolphins. *Jour. Wildlife Mgt.,* 44(4): 960–963.

Perrin, W. F., and S. B. Reilly. 1984. Reproductive parameters of dolphins and small whales of the family Delphinidae. *In* Perrin, W. F., R. L. Brownell, Jr., and D. P. DeMaster, eds., *Reproduction in Whales, Dolphins, and Porpoises.* Rept. Int. Whaling Comm., Spec. Issue 6, Cambridge U.K., pp. 97–133.

Perrin, W. F., and E. L. Roberts. 1972. Organ weights of non-captive porpoise *(Stenella* spp). *Bull. So. Calif. Acad. Sci.,* 71: 19–32.

Perrin, W. F., R. R. Warner, C. H. Fiscus, and D. B. Holts. 1973. Stomach contents of porpoise, *Stenella* spp., and yellowfin tuna, *Thunnus albacares,* in mixed species aggregations. *Fish. Bull.,* 71(4): 1077–1092.

Perrin, W. F., J. M. Coe, and J. R. Zweifel. 1976. Growth and reproduction of the spotted porpoise, *Stenella attenuata,* in the offshore Eastern Tropical Pacific. *Fish. Bull.,* 74(2): 229–269.

Perrin, W. F., D. B. Holts, and R. B. Miller. 1977a. Growth and reproduction of the eastern spinner, a geographic form of *Stenella longirostris* in the eastern tropical Pacific. *Fish. Bull.,* 75(4): 725–750.

Perrin, W. F., R. B. Miller, and P. A. Sloan. 1977*b*. Reproductive parameters of the offshore spotted dolphin, a geographical form of *Stenella attenuata,* in the eastern tropical Pacific, 1973–1975. *Fish. Bull.,* 75: 629–633.

Perrin, W. F., P. A. Sloan, and J. R. Henderson. 1979. Taxonomic status of the "southwestern" stocks of spinner dolphin *Stenella longirostris* and spotted dolphin *S. attenuata. Rept. Int. Whaling Comm.,* 29: 175–184.

Perrin, W. F., E. D. Mitchell, J. G. Mead, D. K. Caldwell, and P. J. H. van Bree. 1981. *Stenella clymene,* a rediscovered tropical dolphin of the Atlantic. *Jour. Mammal.,* 62(3): 583–598.

Perrin, W. F., R. L. Brownell, Jr., and D. P. DeMaster, eds. 1984. *Reproduction in Whales, Dolphins, and Porpoises.* Proc. Conf. Cetacean Reprod.: Estimating Parameters for Stock Assessment and Management, La Jolla, CA, Repts. Int. Whaling Comm., Spec. Issue 6, Cambridge, U.K.

Perrin, W. F., N. Miyazaki, and T. Kasuya. 1987. A dwarf form of the spinner dolphin (*Stenella longirostris*) from Thailand. *Marine Mamm. Sci.,* 5(3): 213–227.

Perrin, W. F., P. A. Akin, and J. V. Kashiwada. 1991. Geographic variation in external morphology of the spinner dolphin, *Stenella longirostris,* in the eastern Pacific and implications for conservation. *Fish. Bull.,* 89(3): 411–428.

Perryman, W. L., and T. C. Foster. 1980. Preliminary report on predation by small whales, mainly the false killer whale *Pseudorca crassidens* on dolphin (*Stenella* spp. and *Delphinus delphis*) in the Eastern Tropical Pacific. Natl. Mar. Fish. Serv., Southwest Fish. Center, La Jolla, CA, Admin. Rept. LJ-80-05, pp. 1–9.

Peters, R. H. 1983. *The Ecological Implications of Body Size.* Cambridge Univ. Press, Cambridge, U.K.

Petter, J. F., and P. Charles-Dominique. 1979. Vocal communication in prosimians. *In* Doyle, G., and R. D. Martin, eds., *The Study of Prosimian Behavior.* Academic Press, New York, pp. 247–304.

Pilleri, G., ed. 1976. Comparative study of the skin and general myology of *Platanista indi* and *Delphinus delphis* in relation to hydrodynamics and behaviour. *In* Pilleri, G., ed., *Investigations on Cetacea,* 6: 90–127. Institute of Brain Anatomy, Univ. of Berne, Switzerland.

Pilleri, G., and M. Gihr. 1971. The central nervous system of the mysticete and odontocete whales. *In* Pilleri, G., ed., *Invest. on Cetacea,* 2. Institute of Brain Anatomy, Univ. of Berne, Switzerland, pp. 89–128.

Pitcher, T. J., ed. 1986. *The Behavior of Teleost Fishes.* The Johns Hopkins Univ. Press, Baltimore, MD.

Popper, A. N. 1980. Sound emission and detection by delphinids. *In* Herman, L. M., ed., *Cetacean Behavior: Mechanisms and Functions.* Wiley-Interscience, New York.

Potts, W. K. 1984. The chorus line hypothesis of manoeuvre coordination in avian flocks. *Nature,* 309: 344–345.

Powell, B. A. 1966. Periodicity of vocal activity of captive Atlantic bottlenosed dolphins, *Tursiops truncatus. Bull. So. Calif. Acad. Sci.,* 65: 237–244.

Premack, D., and G. Woodruff. 1978. Does the chimpanzee have a theory of mind? *Behav. and Brain Sci.,* 4: 515–526.

Pryor, K., and I. Kang. 1978. Social behavior and school structure in pelagic porpoises (Stenella attenuata and S. longirostris) during purse seining for tuna. Natl. Mar. Fish. Serv., Southwest Fish. Center, La Jolla, CA, Admin. Rept. LJ-80-11C, pp. 1–86.

Pryor, K., and K. S. Norris, eds. 1991. Dolphin Societies: Discoveries and Puzzles. Univ. of California Press, Berkeley and Los Angeles.

Pryor, K., and I. Shallenberger. 1991. Social structure in spotted dolphins (Stenella attenuata) in the tuna purse seine fishery in the eastern tropical Pacific. In Pryor, K., and K. S. Norris, eds. Dolphin Societies: Discoveries and Puzzles. Univ. of California Press, Berkeley and Los Angeles, pp. 161–196.

Pryor, T. A., K. Pryor, and K. S. Norris. 1965. Observations on a pygmy killer whale (Feresa attenuata Gray) from Hawaii. Jour. Mammal., 46: 450–461.

Pryor, K., J. Lindbergh, S. Lindbergh, and R. Milano. 1990. A dolphin–human fishing cooperative in Brazil. Mar. Mammal Sci., 6(1): 77–82.

Puente, A. E., and D. A. Dewsbury. 1976. Courtship and copulation behavior of bottlenosed dolphins, Tursiops truncatus. Cetology, 21: 1–9.

Purbrick, E. I. 1977. Daily activity patterns of captive Hawaiian spinner dolphins, Stenella longirostris (Gray). Bachelor's Thesis, Wilamette Univ., McMinnville, OR.

Purves, P. E., and G. Pilleri. 1973–1974. Observations on the ear, nose, throat and eye of Platanista indi. In Pilleri, G., ed., Investigations on Cetacea, vol. 5. Institute of Brain Anatomy, Univ. of Berne, Switzerland, pp. 13–57.

Purves, P. E., and G. Pilleri. 1978. The functional anatomy and general biology of Pseudorca crassidens (Owen) with a review of the hydrodynamics and acoustics in Cetacea. In Pilleri, G., ed., Investigations on Cetacea, 9: 67–227. Institute of Brain Anatomy, Univ. of Berne, Switzerland.

Purves, P. E., and G. Pilleri. 1983. Echolocation in Whales and Dolphins. Academic Press, London.

Radakov, D. V. 1973. Schooling and the Ecology of Fish. Israeli Sci. Translation Ser. John Wiley, New York.

Ramirez, M. F., C. H. Freese, and J. Revilla. 1978. Feeding ecology of the pygmy marmoset, Cebuella pygmaeâ, in northeastern Peru. In Kleinman, D. G., ed., The Biology and Conservation of the Callitrichidae. Smithsonian Institution Press, Washington, D.C.

Reilly, S. 1990. Seasonal changes in distribution and habitat differences among dolphins in the eastern tropical Pacific. Mar. Ecol. Prog. Ser., 66: 1–11.

Reilly, S., and P. C. Fiedler. 1991. Interannual variability in dolphin habitats in the eastern tropical Pacific, 1986–89. Natl. Mar. Fish. Serv., Southwest Fish. Center, La Jolla, CA, Admin. Rept. LJ-90-29, pp. 1–40.

Richards, D. G., J. P. Wolz, and L. M. Herman. 1984. Vocal mimicry of computer-generated sounds and vocal labeling of objects by a bottlenosed dolphin, Tursiops truncatus. Jour. Comp. Psych, 98(1): 10–28.

Richman, B. 1980. Did human speech originate in coordinated vocal music? Semiotica, 32: 3–4.

Ridgway, S. H. 1990. The central nervous system of the bottlenose dolphin. In Leatherwood, S., and R. R. Reeves, eds., The Bottlenose Dolphin. Academic Press, San Diego, pp. 69–97.

Ridgway, S. H., and R. H. Brownson. 1984. Relative brain sizes and cortical surface areas in odontocetes. *Acta Zool., Fennica,* 172: 149–152.

Ridgway, S. H., and M. D. Dailey. 1972. Cerebral and cerebellar involvement of trematode parasites in dolphins and their possible role in stranding. *Jour. Wildlf. Diseases,* 8: 33–43.

Ridgway, S. H., and R. Howard. 1979. Dolphin lung collapse and intramuscular circulation during free diving: Evidence from nitrogen washout, *Science,* 206: 1182–1183.

Ridgway, S. H., D. A. Carder, R. F. Green, A. S. Gaunt, S. L. L. Gaunt, and W. E. Evans. 1980. Electromyographic and pressure events of dolphins during sound production. *In* Busnel, R.-G., and J. F. Fish, eds., *Animal Sonar Systems.* N.A.T.O. Adv. Studies Inst., Ser. A., Plenum Press, New York, pp. 239–249.

Ritzmann, R. E. 1974. Mechanisms for the snapping behavior of two alpheid shrimp. *Alpheus californiensis* and *Alpheus heterochelis. Jour. Comp. Physiol.,* 95: 217–236.

Robineau, D., and J.-M. Rose. 1983. Note sur le *Stenella longirostris* du Golfe d' Aden. *Mammalia,* 47(2): 237–245.

Robineau, D., and J.-M. Rose. 1984. Les Cétacés de Djibouti. Bilan des connaissances actuelles sur la faune cétologique de la mer Rouge et du golfe d'Aden. *Bull. Mus. Natn. Hist. Nat., Paris (4th Ser.),* 6(A) 1: 219–249.

Robinson, J. G., and C. Janson. 1987. Capuchins, squirrel monkeys and atelines: Socioecological convergence with old world primates. *In* Smuts, B. B., D. L. Cheney, R. M. Seyfarth, R. W. Wrangham, and T. T. Struhsaker, eds., *Primate Societies.* Univ. of Chicago Press, Chicago, pp. 69–82.

Roden, G. I. 1987. Effect of seamounts and seamount chains on ocean circulation and thermohaline structure. *In* Keating, B. H., P. Fryer, R. Batiza, and G. W. Boehlert, eds., Seamounts, islands, and atolls. *American Geophysical Union,* Geophysical Mon. 43, pp. 335–354.

Rowell, T., and R. Hinde. 1962. Vocal communication by the rhesus monkey, *Macaca mulatta. Proc. Zool. Soc. London,* 138(II): 279–294.

Saayman, G. S., and C. K. Tayler. 1977. Observations on the sexual behavior of Indian Ocean bottlenosed dolphins *(Tursiops aduncus). In* Ridgway, S. H., and K. Benirschke, eds., *Breeding Dolphins: Present Status, Suggestions for the Future.* U.S. Mar. Mamm. Comm. Rept., Springfield, VA, MMC-76/07, pp. 113–129.

Saayman, G. S., C. K. Tayler, and D. Bower. 1973. Diurnal activity cycles in captive and free-ranging Indian Ocean bottlenose dolphins (*Tursiops aduncus* Ehrenburg). *Behavior,* 44: 212–233.

Savage-Rumbaugh, E. S., D. M. Rumbaugh, and S. Boysen. 1980. Do apes use language? *Amer. Sci.,* 68: 49–61.

Sawyer-Steffan, J. E., and V. L. Kirby. 1980. A study of serum steroid hormone levels in captive female bottlenose dolphins, their correlation with reproductive status, and their application to ovulation induction in captivity. Natl. Tech. Info. Serv., PB80-177199.

Schenck, H., Jr. 1957. On the focusing of sunlight by ocean waves. *Jour. Optical Soc. Amer.,* 47(7): 653–657.

Schilt, C. R. 1991. Fish schools, impulse sounds and sensory integration; or the Ballad of J-Dock. Master's Thesis, Univ. of California, Santa Cruz.

Schmidt-Nielsen, K. 1984. *Scaling: Why is Animal Size So Important?* Cambridge Univ. Press, Cambridge, U.K.

Schnell, G. D., M. E. Douglas, and D. J. Hough. 1982. Geographic variation in morphology of spotted and spinner dolphins (*Stenella attenuata* and *S. longirostris*) from the eastern tropical Pacific. Natl. Mar. Fish. Serv. Southwest Fish. Center, La Jolla, CA, Admin. Rept. LJ-82-15C, pp. 1–213.

Schnitzler, H-U, D. Menne, R. Kober, and K. Heblich. 1983. The acoustical image of fluttering insects in echolocating bats. *In* Huber, F., and H. Markl, eds., *Neuroethol. and Behav. Physiol.* Springer-Verlag, Berlin, pp. 235–250.

Schusterman, R. J., and K. Kreiger. 1984. California sea lions are capable of semantic comprehension. *Psych. Record,* 34: 3–23.

Scott, M. D. 1991. Diurnal patterns in aggregations of pelagic dolphins and tunas in the eastern Pacific, Pap. 2. Ph.D. Dissertation, Univ. of California, Los Angeles.

Scott, M. D., and S. J. Chivers. 1990. Distribution and herd structure of bottlenose dolphins in the eastern tropical Pacific Ocean. *In* Leatherwood, S., and R. R. Reeves, eds, *The Bottlenose Dolphin.* Academic Press, New York, pp. 387–402.

Scott, M. D., and W. L. Perryman. 1991. Using aerial photogrammetry to study dolphin school structure. *In* Pryor, K., and K. S. Norris, eds., *Dolphin Societies: Discoveries and Puzzles.* Univ. of California Press, Berkeley and Los Angeles, pp. 227–241.

Scott, M. D., and P. C. Wussow. 1983. Movements of a Hawaiian spotted dolphin. *In* Pincock, D. G., ed., *Proc. Fourth Int. Conf. Wildlife Biotelemetry.* Halifax, Nova Scotia, pp. 353–364.

Seyfarth, R. M., D. L. Cheney, and P. Marler. 1980. Monkey responses to three different alarm calls: evidence of predator classification and semantic communication. *Science,* 201: 801–803.

Shallenberger, E. W. 1991. The status of Hawaiian cetaceans. U.S. Marine Mammal Commission Rept., Washington, D.C., MMC-77/23.

Shane, S. H. 1980. Occurrence, movements and distribution of the bottlenose dolphin, *Tursiops truncatus,* in southern Texas. *Fish. Bull.,* 78(3): 593–601.

Silk, J. B. 1987. Social behavior in evolutionary perspective. *In* Smuts, B. B., D. L. Cheney, R. M. Seyfarth, R. W. Wrangham, and T. H. Struhsaker, eds., *Primate Societies.* Univ. of Chicago Press, Chicago, pp. 318–329.

Simpson, J. G., and M. B. Gardner. 1972. Comparative microscopic anatomy of selected marine mammals. *In* S. H. Ridgway, ed., *Mammals of the Sea: Biology and Medicine,* Chas. Thomas, Springfield, IL, pp. 298–418.

Simpson, J. H., P. Tett, M. L. Argote-Espinoze, A. Edwards, K. L. Jones, and G. Sividge. 1983. Mixing and phytoplankton growth around an island in a stratified (shelf) sea. *Continental Shelf Res.,* 1: 15–31.

Slijper, E. 1936. Die Cetacean, Vergleichend-Anatomisch und Systematisch. *Capita Zoologica,* 7: 1–590.

Slijper, E. 1966. Functional morphology of the reproductive system in Cetacea. *In* Norris, K. S., ed., *Whales, Dolphins, and Porpoises.* Univ. of California Press, Berkeley and Los Angeles, pp. 278–319.

Slobochikoff, C. N., and W. C. Schultz. 1988. Cooperation, aggression, and the evolution of social behavior. *In* Slobodchikoff, C. N., ed., *The Ecology of Social Behavior.* Academic Press, New York, pp. 13–32.

Slobodchikoff, C. N., and W. M. Shields. 1988. Ecological trade-offs and social behavior. *In* Slobodchikoff, C. N., ed., *The Ecology of Social Behavior.* Academic Press, New York, pp. 3–10.

Smith, T. D. 1983. Changes in the size of three dolphin (*Stenella* spp.) populations in the eastern tropical Pacific. *Fish. Bull.* 18(1): 1–13.

Smuts, B. B., D. L. Cheney, R. M. Seyfarth, R. W. Wrangham, and T. T. Struhsaker, eds. 1987. *Primate Societies.* Univ. of Chicago Press, Chicago.

Stearns, H. T. 1985. *Geology of the State of Hawaii,* 2d ed. Pacific Books, Palo Alto, CA.

Stearns, S. C. 1989. Trade-offs in life-history evolution. *Function. Ecol.,* 3:259–268.

Stein, J., and L. Urdang. 1967. *The Random House Dictionary of the English Language.* Random House, New York.

Steiner, W. W. 1981. Species-specific differences in pure tonal whistle vocalizations of five Western North Atlantic dolphin species. *Behav. Ecol. Sociobiol.,* 9: 241–246.

Stevens, J. D., ed. 1987. *Sharks.* Facts on File Publications, New York.

Taruski, A. G. 1979. The whistle repertoire of the North Atlantic pilot whale (*Globicephala melaena*) and its relationship to behavior and environment. *In* Winn, H. E., and B. L. Olla, eds., *Behavior of Marine Animals: Current Perspectives in Research.* Plenum Press, New York, pp. 345–368.

Tavolga, M. C. 1966. Behavior of the bottlenose dolphin (*Tursiops truncatus*): Social interactions in a captive colony. *In* Norris, K. S., ed., *Whales, Dolphins, and Porpoises.* Univ. of California Press, Berkeley and Los Angeles, pp. 718–730.

Tavolga, M. C., and F. S. Essapian. 1957. The behavior of the bottlenosed dolphin (*Tursiops truncatus*): Mating, pregnancy, parturition and mother–infant behavior. *Zoologica,* 42: 11–31.

Tayler, C. K., and G. S. Saayman. 1972. The social organization and behaviour of dolphins (*Tursiops aduncus*) and baboons (*Papio ursinus*): Some comparisons and assessments. *Ann. Cape Prov. Mus. Nat. Hist.,* 9: 11–49.

Tayler, C. K., and G. S. Saayman. 1973. Imitative behavior by Indian Ocean bottlenose dolphins (*Tursiops aduncus*) in captivity. *Behaviour,* 44(3–4): 277–298.

Terborgh, J. W., and A. Goldizen. 1985. On the mating system of the cooperatively breeding saddle-backed tamarin (*Saguinus fuscicollis*). *Behav. Ecol. Sociobiol.,* 16: 293–299.

Theberge, J. B., and B. J. Fall. 1967. Howling as a means of communication in timber wolves. *Amer. Zool.,* 7: 331–338.

Thomas, J. A., S. R. Fisher, and L. M. Ferm. 1983. Acoustic surveys of marine mammals using a towed hydrophone array (abst.). Fifth Biennial Conf. on the Biology of Marine Mammals, Boston, MA.

Tolstoy, I., and C. S. Clay. 1966. *Ocean Acoustics.* McGraw-Hill, New York.

Tomich, P. Q. 1986. *Mammals in Hawaii: A Synopsis and Notational Bibliography,* 2d ed. Bishop Museum Spec. Publ. 76, Bishop Mus. Press, Honolulu, HI.

Trivers, R. L. 1971. The evolution of reciprocal altruism. *Quart. Rev. Biol.*, 46: 35–57.

Trivers, R. L. 1985. *Social Evolution*. Benjamin/Cummings Publ., Menlo Park, CA.

Trivers, R. L., and H. Hare. 1976. Haplodiploidy and the evolution of social insects. *Science*, 191: 249–263.

Turl, C., and R. H. Penner. 1987. Differences in patterns of click trains between the beluga and bottlenose dolphin (abst.). Seventh Biennial Conf. Biol. of Marine Mammals, Miami, FL.

Twain, M. 1851. *Roughing It*. American Publishing Co., Hartford, CT. (Reprinted 1962, Signet Classics, The New Library of World Literature, New York.)

Tyack, P. 1976. Patterns of vocalization in wild *Tursiops truncatus*. Senior Thesis, Harvard Univ., Cambridge, MA.

Tyack, P. 1986. Whistle repertoires of two bottlenosed dolphins, *Tursiops truncatus:* Mimicry of signature whistles? *Behav. Ecol. Sociobiol.*, 18: 251–257.

Tyack, P. 1991. Use of a telemetry device to identify which dolphin produces a sound. *In* Pryor, K., and K. S. Norris, eds., *Dolphin Societies: Discoveries and Puzzles*. Univ. of California Press, Berkeley and Los Angeles, pp. 319–344.

Urick, R. J. 1975. *Principles of Underwater Sound for Engineers*, 2d ed. McGraw-Hill, New York.

van Bergeijk, W. A. 1966. Evolution of the sense of hearing in vertebrates. *Amer. Zool.*, 6(3): 371–377.

Van Dorn, W. G. 1974. *Oceanography and Seamanship*. Dodd, Mead and Co., New York.

Waddington, C. H. 1960. Evolutionary adaptation; Evolution after Darwin. *In The Evolution of Life*, vol. 1. Univ. of Chicago Press, Chicago, pp. 381–402.

Waddington, C. H. 1968. The theory of evolution today. *In* Koestler, A., and J. R. Smythies, eds., *Beyond Reductionism*. Columbia Univ. Press, New York.

Waddington, C. H. 1975. *The Evolution of an Evolutionist*. Cornell Univ. Press, Ithaca, NY.

Walters, J. R. 1987. Transition to adulthood. *In* Smuts, B. B., D. L. Cheney, R. M. Seyfarth, R. W. Wrangham, and T. T. Struhsaker, eds., *Primate Societies*. Univ. of Chicago Press, Chicago.

Walther, F. R. 1984. *Communication and Expression in Hoofed Mammals*. Indiana Univ. Press, Bloomington, IN.

Watkins, W.A., and K.E. Moore. 1982. An underwater acoustic survey for sperm whales (*Physeter catodon*) and other cetaceans in the southeast Caribbean. *Cetology*, 46: 1–7.

Watkins, W. A., and W. E. Schevill. 1974. Listening to Hawaiian spinner porpoises, *Stenella* cf. *longirostris* with a 3-dimensional hydrophone array. *Jour. Mamm.*, 55: 319–328.

Webb, T. III, 1988. Eastern North America. *In* Huntley, B., and T. Webb III, eds., *Vegetation History*. Kluwer Academy Publ., Dordrecht, pp. 385–414.

Webb, T. III, P. J. Bartlein, and J. E. Kutzbach. 1987. Climatic change in eastern North America during the past 18,000 years: Comparisons of pollen data with model results. *In* Ruddiman, W. F., and H. E. Wright, Jr., eds., *North*

America and Adjacent Oceans During the Last Deglaciation. DNAG V. K-3, Geol. Soc. Amer., Boulder, CO.

Wells, R. S. 1978. Home range characteristics and group composition of Atlantic bottlenosed dolphins, *Tursiops truncatus,* on the west coast of Florida. Master's Thesis, Univ. of Florida, Gainesville, FL.

Wells, R. S. 1984. Reproductive behavior and hormonal correlates in Hawaiian spinner dolphins, *Stenella longirostris. In* Perrin, W. F., R. L. Brownell, Jr., and D. P. DeMaster, eds., *Reproduction of Whales, Dolphins, and Porpoises.* Rept. Int. Whaling Comm., Spec. Issue 6, Cambridge, U.K., pp. 465–472.

Wells, R. S. 1991. The role of long-term study in understanding the social structure of a bottlenose dolphin community. *In* Pryor, K., and K. S. Norris, eds., *Dolphin Societies: Discoveries and Puzzles.* Univ. of California Press, Berkeley and Los Angeles, pp. 199–225.

Wells, R. S., A. B. Irvine, and M. D. Scott. 1980. The social ecology of inshore odontocetes. *In* Herman, L. M., ed., *Cetacean Behavior, Mechanisms, and Functions.* Wiley-Interscience, New York, pp. 263–317.

Wells, R. S., M. D. Scott, and A. B. Irvine. 1983. Reproductive and social patterns of free-ranging female bottlenose dolphins, *Tursiops truncatus* (abst.). Fifth Biennial Conf. Biol. of Marine Mammals, Boston, MA.

West-Eberhardt, M. J. 1975. The evolution of social behavior by kin selection. *Quart. Rev. Biol.,* 50: 1–33.

Whitehead, A. N., and B. Russell. 1910–1913. *Principia Mathematica* (3 vols.). Cambridge Univ. Press, Cambridge, U.K.

Wiener, N. 1948. *Cybernetics, or Control and Communication in the Animal and the Machine.* MIT Press, Cambridge, MA.

Williams, G. C. 1964. Measurement of consociation among fishes and comments on the evolution of schooling. *Michigan State Univ. Mus. Publ. Biol. Ser.,* 2: 351–383.

Williams, G. C. 1966. *Adaptation and Natural Selection: A Critique of Some Current Evolutionary Thought.* Princeton Univ. Press, Princeton, NJ.

Williams, G. C. 1975. *Sex and Evolution.* Princeton Univ. Press, Princeton, NJ.

Wilson, E. O. 1975. *Sociobiology: The New Synthesis.* Belknap Press of Harvard Univ. Press, Cambridge, MA.

Wilson, E. O. 1984. The relation between caste ratios and division of labor in *Pheidole. Behav. Ecol. Sociobiol.,* 16(1): 89–98.

Wilson, M. E., T. P. Gordon, and D. C. Collins. 1982. Variation in ovarian steroids associated with annual mating period in female rhesus monkeys (*Macaca mulatta*). *Biol. Reprod.,* 27: 530–539.

Wood, F. G. III, D. K. Caldwell, and M. C. Caldwell. 1970. Behavioral interactions between porpoises and sharks. *In* Pilleri, G., ed., *Investigations on Cetacea,* vol. 2. Institute of Brain Anatomy, Univ. of Berne, Switzerland, pp. 264–277.

Würsig, B. 1976. Radio tracking of dusky porpoises (*Lagenorhynchus obscurus*) in the South Atlantic. FAO-ACMR Scientific Consultation on Marine Mammals, Bergen, Norway, pp. 1–21.

Würsig, B. 1978. Occurrence and group organization of Atlantic bottlenose porpoises (*Tursiops truncatus*) in an Argentine bay. *Biol. Bull.,* 154: 348–359.

Würsig, B., and M. Würsig. 1977. The photographic determination of group size, composition, and stability of coastal porpoises *(Tursiops truncatus)*. *Science*, 198: 755–756.

Würsig, B., and M. Würsig. 1979. Behavior and ecology of the bottlenose dolphin, *Tursiops truncatus*, in the South Atlantic. *Fish. Bull.*, 77: 399–412.

Würsig, B., and M. Würsig. 1980. Behavior and ecology of the dusky dolphin, *Lagenorhynchus obscurus*, in the South Atlantic. *Fish. Bull.*, 77: 871–890.

Würsig, B., F. Cipriano, and M. Würsig. 1991. Dolphin movement patterns. Information from radio and theodolite tracking studies. *In* Pryor, K. W., and K. S. Norris, eds., *Dolphin Societies: Discoveries and Puzzles*. Univ. of California Press, Berkeley and Los Angeles, pp. 79–111.

Zar, J. H. 1974. *Biostatistical Analysis*. Prentice-Hall, Englewood Cliffs, NJ.

INDEX

Tuna
    bond with dolphins, 223–225
    in multispecies aggregations, 291–296
Tuna seine fishery
    behavior of trapped dolphins, 274
    and dolphin mortality, 15–16, 22
    dolphin studies from, 135–139
    and shark attacks, 289
Tursiops, 149–150
Tursiops truncatus, 192
Twain, Mark, 103–104
Twin Peaks, 132

Underwater observation
    of dolphins, 54–64, 365–367
    vehicles for, 55–64
Ungulate herds, 295, 302, 326
Upstroke, 202–205
Upwelling
    around Hawaiian Islands, 36–37
    obstruction, 35
    at seamounts, 36
Urick sonar equation, 162
U.S. Marine Mammal Protection Act of
    1972, 16

Variation, among dolphins, 20–25, 27,
    303–304
Vehicles, for underwater observation,
    55–64
Veiling brightness, definition of, 143
Viewing vault, 244
Visibility, in water, 142–144
Visual domain, 141–160, 238–239
Visual fields, 152–154
Visual pigments, 151
Visual signals, 152–157
Visual system, of dolphin, 149–150
Vocalization. *See* Click trains; Sounds
Vocative signals, 181

Wahoo, 294
Warning system, 237, 239
Water depth, of spinners, 83, 219–222
Water noise, 57
Wave-produced flicker, 144
Waves
    capillary, 144–145
    gravity, 144–145
Whales
    breaching by, 103
    false killer, 18–19, 37, 296
    killer, 18–19, 23, 296
    pilot, 269, 295
    pygmy killer, 296
    short-finned pilot, 296
    sperm, 19, 207, 282
Whistles, 165–166, 170, 171, 198
    as communication, 180–181
    harmonic, 179
    mimicry of, 176, 181–182
    nonharmonic, 178
    number of, 176
    signature, 166, 176, 180–181, 181–182,
        182–183
    socialization through, 183–184
White-belly spinner dolphin, 20
Whitetip shark, 294–296
Whole-body signals, 154–156
Wild dolphin studies, 1–5, 54
Worldwide distribution, of spinner dol-
    phin, 14–17
Wuzzles, 138, 250, 263, 315

Yellowfin tuna, bond with dolphins,
    223–225

Zig-zag swimming
    behavior during, 87–90
    definition of, 28, 71, 87
    vocalization during, 174
Ziphiidae family, 19